Sustainable Coastal Design and Planning

Sustainable Coastal Design and Planning

edited by Elizabeth Mossop

CRC Press
Taylor & Francis Group
Boca Raton London New York

CRC Press is an imprint of the
Taylor & Francis Group, an **informa** business

CRC Press
Taylor & Francis Group
6000 Broken Sound Parkway NW, Suite 300
Boca Raton, FL 33487-2742

First issued in paperback 2020

ISBN 13: 978-0-367-57075-0 (pbk)
ISBN 13: 978-1-4987-7454-3 (hbk)

Library of Congress Cataloging-in-Publication Data

Names: Mossop, Elizabeth, editor.
Title: Sustainable coastal design and planning / Elizabeth Mossop.
Description: Boca Raton, FL : Taylor & Francis, 2018. | Includes bibliographical references.
Identifiers: LCCN 2018014032 | ISBN 9781498774543 (hardback : alk. paper)
Subjects: LCSH: Coastal zone management--Case studies. | Coastal engineering--Case studies. | Shore protection--Case studies.
Classification: LCC HT391 .M66 2018 | DDC 333.91/7--dc23
LC record available at https://lccn.loc.gov/2018014032

Visit the Taylor & Francis Web site at
http://www.taylorandfrancis.com

and the CRC Press Web site at
http://www.crcpress.com

Contents

PART 1   NEW WAYS OF THINKING AND WORKING

Section I   Shifting Realities

Section II   Methods and Practices

PART 2   NEW STRATEGIES

Section III   Deltas, Bays, and Estuaries

Section IV   Sites and Structures

Preface

This book originated from the experience of living in Louisiana during Hurricane Katrina (and subsequently Rita, Gustav, Ike, Lee, Isaac, and so on) and its aftermath in New Orleans and the Gulf Coast. Trite though it may be, it profoundly changed my understanding of the potential impacts of climate change and gave me an education on the cultural dimensions of disaster resilience. While I was personally little affected, people I knew lost close family and friends, their houses and all their possessions, jobs and livelihoods, and the entire fabric of their lives. The experience of spending time in familiar, yet totally devastated, landscapes post-disaster and seeing the intimate debris of people's lives, is one that forever changes your understanding of the fragility of existence.

In the 12 years since the storm, I have been struck by the unevenness of the region's recovery. For some it has brought new life and opportunity and for others the exacerbation of ongoing poverty and dysfunction. The post-Katrina environment has been rich with exploration of what it means to live with uncertainty, and, for those interested in landscape interventions of all kinds, fruitful speculation about future strategies.

My involvement in coastal resilience was initiated by conversations with coastal ecologist Robert Twilley at Louisiana State University (LSU), after his collaboration with planners and designers in the first coastal masterplan process (in 2007), he opened his ideas to how these disciplines approach problem solving. Like many in recent years, Twilley was struck by the studio methods and the creative processes used in the development of planning and design strategies. We then worked together in the establishment of the Coastal Sustainability Studio (CSS) at LSU. The CSS was conceived as a trans-disciplinary design laboratory to develop new strategies for coastal settlements that reduce risk to social, economic, and natural resources. Twilley has continued to lead the studio. Its other board members from 2008 until my departure at the end of 2015 were Clint Willson (civil engineering), Jori Erdman (architecture), John White (coastal ecology), studio director Jeff Carney, and assistant directors Lynne Carter and Mary Bergeron.

The CSS and its many and varied collaborations provided a fertile testing ground for ideas of resilience, for ways of communicating complex coastal issues, and for design speculation about future strategies. Community resilience projects included "River to Bayou: Restoration and Resilience—The Central Wetland Unit and the Lower 9th Ward" and the "Louisiana Resiliency Assistance Program." The issues of visual communication were extensively explored in CSS projects for the Coastal Protection and Restoration Authority and for the Center for River Studies Exhibition Design. Design speculation was a theme for the CSS from the beginning when an early river diversions project, "In the Mississippi Delta: Constructing with Water" became an exhibit for the 2010 Venice Biennale. My colleagues in the Coastal Sustainability Studio and the many faculty and students involved provided a productive and sustained collaboration over many years and gave me the opportunity to initially develop the ideas behind this publication.

From 2012 to 2015, I also served as an expert in regional landscape and planning issues on the technical advisory panel for the Changing Course Design Competition run by the Environmental Defense Fund and the Van Alen Institute. Competing teams were asked to develop 100-year visions for restoring and sustaining Louisiana's Lower Mississippi Delta for the people, communities, and industries that call the region home. The experts of the technical advisory team and the participants in the three finalist design teams provided a wealth of insights, as well as opportunities for the testing of ideas and strategies.

Through practice with Spackman Mossop Michaels, I have also had the opportunity to be involved in the planning and design of numerous projects in New Orleans, Baton Rouge, and across the Gulf Coast region. Many of these site investigations have involved the development of new strategies for increasing urban and coastal resiliency. My ongoing collaboration and conversation with business partner Wes Michaels has been particularly important in developing the ideas behind this work.

Book contributions are from colleagues involved in all of these endeavors and from the conversations that have come from them. I would also like to thank my colleagues at the University of Technology in Sydney, research assistants Christopher Cunningham-Reid and Judy Robinson, and especially research project manager Jane Reynaud, without whose help I could not have completed this project.

Elizabeth Mossop
Sydney, Australia

Editor

Elizabeth Mossop is Professor and Dean of the School of Design, Architecture and Building at the University of Technology Sydney (UTS). A landscape architect and urbanist with wide-ranging experience in both landscape design and urban planning, Mossop is a founding principal of Spackman Mossop Michaels landscape architects based in Sydney and New Orleans. Her professional practice concentrates on urban infrastructure and open space projects, such as the multiple award-winning Bowen Place Crossing in Canberra, Press Street Gardens in New Orleans, and Sydney's Cook and Phillip Park. She has been involved in many aspects of the post-hurricane reconstruction of New Orleans and the Gulf Coast and the ongoing revitalization of Detroit.

With an academic career spanning 25 years, Mossop has held key roles at universities in both the United States and Australia. Before joining UTS, she was Professor of Landscape Architecture and Director of the Robert Reich School of Landscape Architecture at Louisiana State University, one of the highest-ranked landscape architecture programs in the United States. Previously, she was the Director of the Masters of Landscape Architecture program at the Harvard Graduate School of Design.

Her research and teaching focuses on landscape and urbanism, through investigation of contemporary landscape design both at the urban scale and at the site scale. Previous publications include *Contemporary Landscape Design in Australia* (BT Latitude, Sydney, 2006); *Hong Kong: Defining the Edge* (Harvard Graduate School of Design, Cambridge, MA, 2001); and *City Spaces: Art and Design* (Craftsman House, Sydney, 2001).

Contributors

Claire Agre
West 8 Urban Design & Landscape
    Architecture
New York, New York

Penny Allan
School of Architecture
University of Technology Sydney
Sydney, Australia

Edzo Bindels
West 8 Urban Design & Landscape
    Architecture
Rotterdam, The Netherlands

Riette Bosch
Rijksvastgoedbedrijf
The Hague, The Netherlands

Samuel Bowstead
James Davidson Architect
Brisbane, Australia

Martin Bryant
School of Architecture
University of Technology Sydney
Sydney, Australia

Jeffrey A. Carney
Coastal Sustainability Studio
School of Architecture
Louisiana State University
Baton Rouge, Louisiana

Craig E. Colten
Department of Geography and
    Anthropology
Louisiana State University
Baton Rouge, Louisiana

Dilip da Cunha
Mathur/DaCunha
Philadelphia, Pennsylvania

James Davidson
James Davidson Architect
Brisbane, Queensland, Australia

John W. Day
Department of Oceanography and Coastal
    Sciences
Louisiana State University
Baton Rouge, Louisiana

Jill Allen Dixon
Sasaki
Watertown, Massachusetts

Abbas El-Zein
School of Civil Engineering
University of Sydney
Sydney, Australia

Jori Erdman
School of Architecture
Louisiana State University
Baton Rouge, Louisiana

Heather Fenyk
Lower Raritan Watershed Partnership
New Brunswick, New Jersey

Gina Ford
Agency Landscape
Cambridge, Massachusetts

Ioannis Georgiou
Department of Earth and Environmental
    Sciences
University of New Orleans
New Orleans, Louisiana

Adriaan Geuze
West 8 Urban Design & Landscape
    Architecture
Rotterdam, The Netherlands

Steven N. Handel
Center for Urban Restoration Ecology
Rutgers University
New Brunswick, New Jersey

Briana Hensold
Agency Landscape
Cambridge, Massachusetts

Kristina Hill
Department of Landscape Architecture
    and Environmental Planning
University of California
Berkeley, California

Richard L. Hindle
Department of Landscape Architecture
    and Environmental Planning
University of California
Berkeley, California

Jonathan Hird
Moffatt & Nichol
Long Beach, California

Kathleen John-Alder
Department of Landscape Architecture
Rutgers University
New Brunswick, New Jersey

Robbert de Koning
Robbert de Koning Landschapsarchitect
Arnhem, The Netherlands

Forbes Lipschitz
Knowlton School of Architecture
The Ohio State University
Columbus, Ohio

Nina-Marie Lister
School of Urban & Regional Planning
Ryerson University
Toronto, Ontario, Canada

Stanley Lung Wai Cham
Turenscape
Beijing, China

Anuradha Mathur
School of Design
University of Pennsylvania
Philadelphia, Pennsylvania

Bruno De Meulder
Department of Architecture, Urban Design
    and Regional Planning
University of Leuven
Leuven, Belgium

Elizabeth Mossop
Faculty of Design Architecture and
    Building
University of Technology Sydney
Sydney, Australia

Karen M. O'Neill
Department of Human Ecology
Rutgers University
New Brunswick, New Jersey

Mary G. Padua
School of Architecture
Clemson University
Clemson, South Carolina

Rob Roggema
School of Architecture
University of Technology Sydney
Sydney, Australia

Catherine Seavitt Nordenson
Spitzer School of Architecture
City College of New York
New York, New York

Kelly Shannon
Department of Architecture, Urban Design
    and Regional Planning
University of Leuven
Leuven, Belgium

Jeff Shelden
Moffatt & Nichol
Raleigh, North Carolina

Dirk Sijmons
H+N+S Landscape Architects
Amersfoort, The Netherlands

Robert R. Twilley
Louisiana Sea Grant College Program
Louisiana State University
Baton Rouge, Louisiana

Jane Wolff
Daniels Faculty of Architecture
Landscape & Design
University of Toronto
Toronto, Ontario, Canada

Introduction

New Coastal Paradigms

Coasts are a locus of attraction for many cultures: rich in opportunity, naturally diverse, and endlessly fascinating. Australians are particularly tied to the coast, concentrating their inhabitation in the coastal cities and towns of the continent's edge, blessed with spectacular beaches and coastal landscapes. For many it's a big part of cultural identity, bound up in childhood holidays, beach culture, surfing, swimming, sailing, and so on. Waterfronts are our most prized urban landscapes, and the water view is the ultimate symbol of success, whether beachfront or harbor. But here, as across the globe, these relationships may all change as the impacts of climate change profoundly alter the coast.

This book gathers essays on various aspects of coastal resilience in the face of climate change. Through exploring the questions of coastal change at different scales, many strategies are illustrated in different geographies and cultures, some speculative and some more concrete. When taken as a whole they canvas a broad palette of approaches and techniques for engaging with complex problems. I believe that it is through an exploration of how these issues play out in solutions for real places that we as a society come to understand the issues and are better equipped to make more intelligent decisions on how to move forward.

Rather than examining aspects of coastal resilience scientifically, or addressing the issues comprehensively, these essays are engaged with design speculation on coastal futures. While the details of sea level rise projections are not discussed here, all of the essays assume significant but unpredictable rise in sea level and an increase in the frequency and intensity of storms, tracking the Intergovernmental Panel of Climate Change predictions as a conservative indicator.[1]

This book is divided into two parts. The first focuses on how we can think about the coast as it changes, as well as on specific methods and practices that will be needed to work effectively in this dynamic space. The second part explores case studies: more broadly for deltas, bays, and estuaries, and then at the site scale of specific forms and structures. All of the essays are grounded in real places and make proposals specific to both geography and culture.

The Coast Is Not a Line

If we think about how the coast is represented, as a line providing demarcation between land and sea, it conveys the idea of something located physically and clearly defined. But as soon as we look in more detail, even at maps, we begin to see low and high tide marks and the hints that this coastal line is something infinitely more complex. Beaches, bays, cliffs, reefs, estuaries, deltas, wetlands, lagoons—all of these coastal landscapes are changed by the natural forces of the sea, weather, and climate. Daily, monthly, seasonally, and randomly, tides, waves, currents, and storms all influence the form of the coast from one moment to the next. The coast can be a different place depending on the time in which it is experienced. This changeability is essential to the coast's magnetism and suggests that we need to find more complex ways of understanding it.

Coasts are made up of connected ecosystems and in their undisturbed form are often protected by broad and shifting landscapes of dunes or wetlands that shelter the inland from the forces of waves and storms. These natural systems of coastal protection—sandbars, barrier islands, dunes, wetlands, marshes—that often create a wide zone of dynamism and indeterminacy between land and water, are not particularly conducive to human settlement. As development pressures have increased in coastal areas, with housing and infrastructure encroaching into these dynamic zones, we have seen myriad strategies for trying to fix the landscape in a more stable form to suit the conveniences of real estate. The natural landscapes have historically been squeezed or reduced by human settlement and coastal development into constructed barriers (seawalls, levees, promenades) that have proven to be much more fragile than the soft systems they replaced. This development has been driven by the desire to inhabit all of the land area close to the waterfront, resulting in persistent hardening of its landscape. This has led to a greatly increased vulnerability that we are only beginning to understand, as storms and floods increasingly deluge our coastal cities and towns.

As the distribution of coastal sand through long shore drift is interrupted by the coastal structure of ports and piers, beach nourishment is prevented, and erosion threatens the coast and foredune. As dune systems are destroyed by development, their protective function is lost, and wetland loss removes their attenuation of waves and storm surges. As deltas have their sediment removed through channelization and new land is not built, coastal erosion speeds up.

The engineering structures that typically replace these soft landscapes, such as seawalls, flood walls, and levees, are singular elements that are very effective until there is a failure. They do not embody the multiple lines of defense characteristic of the natural systems they supplant. In plan, an expansive natural buffer is converted to a narrow construction, so that the protection of distance is also lost.

Anuradha Mathur and Dilip da Cunha have a body of work stretching back to *Mississippi Floods*[2] that explores the dynamic landscapes of deltas and of coastal estuaries in their complexity and flux. In Chapter 1, they look at both the Bengal Delta and the Chesapeake Bay as exemplars of these dynamic places, and question the way that cartography fixes our understanding of watery landscapes in the quest for an unrealistic certainty of landscape form.

Shifting Realities

The complexity of issues around coastal resilience is enough to make anyone's head spin. At the heart of the dilemma, the uncertainty around climatic conditions, the rate and scale of sea level rise, and the frequency and intensity of storms suggest that we need a mode of operating that allows us to move forward without the comfort of predetermined outcomes. We must be able to simultaneously deal with the impacts of what is happening today at the same time as we are planning for a changing future.

In recent decades, the impacts of storms and floods, as well as other natural disasters, have been the drivers of change, precipitating the greatest progress in changing attitudes to the urgency of action to adapt to climate change. Disasters have also been influential in shifting people's thinking from notions of sustainability to an emphasis on resilience. The implications of this concept are explored in Nina-Marie Lister's essay (see Chapter 3). Are we now able to advance our thinking on these questions more broadly rather than having to wait for the next disaster?

The impact of these questions has been intensified in the last two decades by ever-increasing coastal risk due to a greater concentration of both people and investment at the water's edge. In some instances, this risk is ratcheted up by climate change denial in government policy and actions, such as in Florida, where despite its vulnerability to sea level rise and extreme weather events, state employees are discouraged from using the terms "climate change" and "global warming."[3]

Uncertainty infects every aspect of these issues as we struggle to understand the future through different types of modeling and predictions. The level of trust that we have in these models may drive the types of solutions that we can support. Do we want grand engineering designs, or do we want the adaptive approaches of ecological restoration? To what extent can we trust the scenarios that are generated through mathematical modeling? How much do we understand of their reliability in describing the future, and how can we approach that uncertainty? Different disciplines have fundamentally different conceptions of how to address the unknowable future. At the risk of gross generalization, scientists proceed with caution based on the gathering and interpretation of data over time, engineers propose solutions applying engineering techniques of building structures, while designers speculate promiscuously about the future and experiment with scenarios in real time. These disciplinary attitudes are also culturally inflected, as we see in the Netherlands with its long history of broad-scale land and sea modification, which over time has bred a level of comfort with grand-scale landscape engineering. Jane Wolff discusses the cultural specificity of resilience design and how this needs to be understood in parallel with the technical questions of landscape change in Chapter 4. Cultural context drives our approach to resilience and our fundamental conceptions of the coastal landscape and its processes. Profound differences emerge in agency around questions of centralized planning, or whether there is broad understanding of the issues in society. How is it possible to reconcile these different approaches in the context of the urgency of climate adaptation?

What are the appropriate time frames in which to think and act? When we contemplate landscape interventions on a large scale, they are approached with long periods of investigation and development as befits their massive investments. But do we have that luxury any longer? We are faced with the conundrum of, on the one hand, needing real long-term strategic thinking, far beyond the scope of political terms, and on the other hand, needing effective procedures for swift decision making and implementation, in the face of infrastructure projects at a huge scale. Many would argue that we must proceed immediately to address mitigation and adaptation prior to exhaustive research and conclusive deliberation if we are to have any hope of a level of resilient survival as the climate changes. Abbas El-Zein discusses the myriad issues around the human inhabitation of the coast as it is affected by sea level rise in Chapter 2. Governance, land tenure, and insurance are unable to deal with the long-term view that is needed, and these outmoded frameworks drive the kinds of futures that we can imagine. On what values do we draw to rethink our institutions and what appetite do we collectively have for bold solutions versus adaptive change?

Methods and Practices

We are faced with ecologically complex sites that are highly variable, in contexts of enormous social and economic disruption. The threats are extreme, and there is pressure to find solutions whose outcomes can be guaranteed. While this may not be within the realm of possibility, it is clear that we do need new ways of working that address the complexity. They have to be socially inclusive and trans-disciplinary, as well as open to new ways of thinking. What kind of procedures will give us outcomes that balance competing interests, as well as addressing the public good? Rob Roggema discusses the switch from traditional Dutch land and water management to methods working with natural processes and letting water in (see Chapter 5), and Dirk Sijmons describes the role of a "Quality Team" in the Dutch "Room for the River" project, that applies these same principles (Chapter 6). Kathleen John-Alder's essay describes the dilemma of conflicting cultural and disciplinary attitudes played out in the "Blue Dunes" proposal by the West 8 team from the Rebuild by Design Competition (Chapter 7).

How can we represent the complexity of place and process on the coast? How do we even communicate with each other about these questions and bring disparate groups of interest into meaningful dialogue and decision making? Forbes Lipschitz discusses the benefits of interactive mapping and modeling in describing the landscape and as a vehicle to increase cross-disciplinary communication (Chapter 8), and Jill Dixon describes the issues specific to interdisciplinary collaboration (Chapter 9). In parallel, Karen O'Neill and Heather Fenyk discuss the imperative for, and the processes of, community engagement in the imagining of coastal futures (Chapter 10).

New Infrastructure for Deltas, Bays, and Estuaries

The questions of coastal resilience play out in different geographies, and the nature of the coastal landscapes is varied, as are density, scale, and culture. The essays in the book's third section look at these different landscapes and proposals across broad territories that imagine new strategies for infrastructure. Strategies for flood mitigation and storm attenuation through green or soft infrastructure along the coastline can involve the use of earth berms or dunes, the creation or restoration of wetlands and marshes, and the planting of mangroves. A number of the essays also describe strategies for controlled flooding that create landscapes that can tolerate periodic inundation. These landscapes can be in floodways or greenways, in parks, and in urban water gardens.

Catherine Seavitt Nordenson describes her work in Jamaica Bay, New York, that explores how a new U.S. Army Corps of Engineers paradigm might be developed to include the green infrastructure of berms and wetlands in managing floods and improving water quality (Chapter 11). Steven Handel, Gina Ford, and Briana Hensold write about a "Rebuild by Design" competition proposal for the Jersey Shore on the East Coast of the United States that rethinks how we conceptualize the coast (Chapter 12). From an understanding of ecological systems and the way communities inhabit coastal areas, they propose a much broader idea of coast and beach that builds resilience into future planning and design. Also thinking about coastal inhabitation, at the low density of agricultural and suburban communities, Penny Allan and Martin Bryant's essay looks at the West Coast of New Zealand's North Island, the importance of design-led speculation, and how to combine Western and indigenous knowledge in planning for a changing future (Chapter 17).

Craig Colten and John Day's essay gives us a picture of the Mississippi Delta today through examining its history of inhabitation and the modification of its natural processes

(Chapter 13). These authors explore the interrelationships of social, economic, and ecological systems that underpin resilience and the cultural landscape that so clearly needs new visions. One such vision for the future is explored by Jeff Carney et al. in Chapter 15, as they discuss a proposal to reengineer the Lower Mississippi and rebuild the delta articulated in the "Changing Course" competition. In the same region, Elizabeth Mossop discusses the relationship between the delta, the coast, and the city of New Orleans, and how the approach of "letting the water in" could reshape the city (Chapter 14).

Robbert de Koning describes part of the "Room for the River" program in the Rhine delta, which exemplifies the revision of the Dutch approach to flooding, finding ways to work with natural processes, and creating more resilient landscapes that allow water in (Chapter 16). Bruno De Meulder and Kelly Shannon's essay describes the development of the Mekong Delta over time and the relationship between its shifting landscape of mud and the very specific pattern of desakota urbanism. They explore how new planning frameworks can harness soft infrastructure to reshape future inhabitation of a changing landscape (Chapter 18).

Sites and Structures

At the smaller scale we can design individual sites and structures to be resilient to flooding, to protect coastal areas, and to adapt to coastal change. It is here that we really test what it means to live with water or to design for habitat and how these ideas play out in the places that we use and inhabit. In the final section of the book, essays look at site-specific projects, the scale at which we can talk about materials and form, and the processes of human interaction. These essays explore ways of restoring or adapting elements from the natural coast—dune systems, rocky shores, wetlands, swamps, and beaches. Protective structures can mimic nature or harness natural processes; they can create habitat, trap sediment, or nurture plant life; and we are yet to fully explore their formal possibilities.

If we take the approach of "letting water in," we can design landscapes to store, clean, and infiltrate water. These landscapes can be deployed in key locations to help cities and towns to function like sponges and become more resilient in the face of storms and flooding. Structures can be elevated or designed so that they can be modified as floods approach. James Davidson and Samuel Bowstead and Jori Erdman's essays look at architectural strategies for housing and other building types and for accommodating water and flooding in ways that can be resilient (Chapters 21 and 22). Soft infrastructure solutions are demonstrated in Riette Bosch's essay on West 8's Hondsbossche Dunes, and Mary Padua and Stanley Lung's exploration of Turenscape's waterfront parks (Chapters 19 and 20). Chapter 23 deals with new ways of designing coastal protection structures. Richard Hindle looks at seawalls in Sydney and New York and the way in which materials and design can contribute to the enrichment of habitat.

Kristina Hill describes estuary cities and their historical means of navigating the water's edge. She then explores what it means for cities to be resilient and uses Christchurch in New Zealand to understand what this might mean when translated into physical design strategies for the coastal city's reconstruction (Chapter 24).

Future Visions

This collection of essays describes ideas and proposals for the future of coastal resilience. The expertise is hard won and comes largely from the threat or experience of disasters. For many of us our way of thinking about the world has shifted to encompass a higher level of uncertainty and risk. We acknowledge the ongoing impacts of sea level rise and climate change and understand how they will influence our various geographies. Every part of the globe now has its watershed moment of coastal change, specific storms, or floods that have changed perceptions forever. There is a global sharing of expertise from these events and from experience developed over decades in vulnerable places. The Netherlands plays a particular role in the international conversation, with its highly developed landscape engineering and planning around coastal resilience dating back over a century.

Professional experts, government agencies, and universities continue to investigate ideas and strategies, as well as to provide better data about the state of the coast. Where this global conversation becomes more public, however, is in the aftermath of disasters, and this often provides a moment of traction for innovative ideas and strategies that are otherwise hard to be recognized in the conservatism of a business-as-usual mind-set. This time is also often the genesis of international design competitions, such as "Changing Course" or "Rebuild by Design," that bring collaborative groups of designers, scientists, engineers, and others together to propose solutions to coastal challenges. These events present an enormous intellectual production opportunity to start conversations around new ideas.

Through these various investigations it is possible to imagine a shared vision for a new resilient coast that looks ahead to the next centuries and comes from a broad coalition of interests. This shared vision would be shaped by, and responsive to, natural processes, its infrastructure defended against disruption. The approaches proposed in the following essays suggest this would bring opportunities across a broad range of new economic activities. Green enterprises would implement restoration ecology, remaking dunes, wetlands, and the other landscapes that can protect the coast. These landscapes can also support new conservation sites and related tourism and research. There is potential for new economies around sustainable and low-carbon living in forms that are viable for new coastal settlements. When will we imagine a whole new ecological, economic, and societal pattern of inhabitation for the coast?

Endnotes

1. IPCC, 2014: Climate Change 2014: Synthesis Report. Contribution of Working Groups I, II, and III to the Fifth Assessment Report of the Intergovernmental Panel on Climate Change [Core Writing Team, R.K. Pachauri and L.A. Meyer eds.]. IPCC, Geneva, Switzerland, 151 pp.
2. This work began in the Mississippi Delta (Mathur, A. and Da Cunha, D. 2001. *Mississippi Floods: Designing a Shifting Landscape*. New Haven, CT: Yale University Press) and has moved to the Mumbai Estuary (Mathur, A. and DaCunha, D. 2009. *Soak: Mumbai in an Estuary*. New Delhi: Rupa Publications). In 2013/2014, the authors of these publications led a PennDesign Team for the project "Structures of Coastal Resilience" (http://structuresofcoastalresilience.org) for Tidewater Virginia with a focus on Norfolk supported by the Rockefeller Foundation.
3. Korten, T. "In Florida Officials Ban Term 'Climate Change,'" *Miami Herald*, March 8, 2015, http://www.miamiherald.com/news/state/florida/article12983720.html.

PART 1

NEW WAYS OF THINKING AND WORKING

Section I

Shifting Realities

An introduction to the context of planning and design for coastal areas in the age of climate change, sea level rise, and increased storm intensity.

1

Designing the Coast in the Moment of Rain

Anuradha Mathur and Dilip da Cunha

CONTENTS

Introduction

Designers are familiar with a coast visualized in a particular moment of the hydrologic cycle when water appears separable from land on the earth surface. This moment is one of choice rather than necessity, a choice that turns down other moments of the cycle when water is precipitating, soaking, evaporating, clouding, misting, in short, blurring and erasing lines. The choice of this moment, though, is significant. In its time, maps are drawn, properties are demarcated, histories are written, places are described, and futures are envisioned. Also in its time, precipitations, such as the monsoons, hurricanes, snowfalls, and dew are made visitors to places where rivers, lakes, seas, and other containments are resident. What does the awareness of this choice do for design?

What if designers cast anchor in another moment of the water cycle, such as precipitation? Does it afford another approach to designing the coast, particularly in the face of a sea projected to rise not in millimeters but meters over the next century? We explore these questions in the Bengal Delta of South Asia and the Chesapeake Bay, Virginia.

Challenging the Moment of Flows

A delta and an estuary are two ways by which rivers meet the sea. In deltas, rivers dominate the sea, depositing sediment and, in the process, extending land. The conditions are such that upon meeting the sea, the river slows down, dropping its load of sediment, dividing and spreading, and in the process constructing ground in the

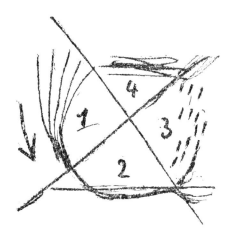

FIGURE 1.1
"The Water Cycle" in *Paul Klee: The Thinking Eye* (New York: George Wittenborn Inc. 1964, 402).

form of the Greek letter Δ. The layers of silt upon silt conceal a past, which makes it not surprising that some have attributed the term delta to Herodotus, the "father of history." This keen observer of the ancient world writes about seeing the Nile's sediment over "a day's journey" at sea, an observation that led him to conclude that Egypt was once a gulf and that the river, over thousands of years, made it land. Egypt, he famously writes, is the gift of the river.

An estuary, in contrast, does not layer as dramatic a history as the delta, perhaps because it allows the sea, with its affinity for erasure, to have more of a say. Indeed, some choose to define an estuary as a body of water through which rivers enter the sea, in comparison to a delta, a landmass of alluvium made by a roving, dividing, and extending river. Certainly, estuaries allow the sea to come inland, its tides operating a scale of salinity that varies diurnally, at times exceptionally, with storms and sea surges on one side and the river on the other. A number of beings reside in this dynamic transition, adapted to or perhaps naturally tuned to accommodating its variabilities, or at least a range of them.

This difference between a delta and an estuary, however, is articulated in a particular moment when a river, defined as "flowing water in a channel with defined banks," can be assumed to exist (*Encyclopaedia Britannica*). What if the chosen moment is in rain when water is everywhere rather than somewhere, when the banks of channels are blurred, even erased, when there is no clear directionality of water by which to call out a flow? Here, rain does more than challenge a difference in the way rivers meet the sea; it says that rivers are products of choice in the extended presence of water, popularly known as the hydrologic, or water cycle. This cycle, which some call an "infinite river" because its water keeps circulating, is described in distinct moments. Biologist William Amos writes that this cycle is an "infinite river," flowing from sea to sky to earth and back to sea, where it is "born again as it had been and would be, every moment of earthly time" (Amos, 1971). The artist Paul Klee called out four—precipitation from clouds to earth, flow formation from earth to sea, evaporation from sea to air, and cloud formation from sea to land (Figure 1.1). Hydrologists, of course, count many more. The water cycle, writes Robert Horton, can begin anywhere, anytime, in an "isolated tree, even a single leaf or twig of a growing plant, the roof of a building, the drainage basin of a river system or any of its tributaries, an undrained glacial depression, a swamp, a glacier, a polar ice cap, a group of sand dunes, a desert playa, a lake, an ocean, or the Earth as a whole" (Horton, 1931). It can also flow in many ways. "We

FIGURE 1.2

Rivers are products of anchoring in a particular moment of the water cycle. With this moment chosen to be the time of reality, other moments become moments of ephemerality. Thus rain, mist, clouds are visitors in a place where rivers are chosen to be residents. (Image adapted by the author, based on Paul Klee, in *Paul Klee: The Thinking Eye* (New York: George Wittenborn Inc. 1964, 402.)

speak of *the* water cycle as if there were only one," notes Robert Kandel, "but in fact there are many: some circuits are completed in a few days; but in others it can take months, years, millennia, even millions of years to go the course" (Kandel, 2003). The diagram of the water cycle that they draw is consequently more complex, with cycles within cycles.

Irrespective of its complexities, however, drawings of the water cycle show rivers as a presence in one moment of the cycle. In Klee's drawing, this moment is the second quadrant when water can be assumed to have a place on the earth's surface and *not* when it is precipitating, evaporating, soaking the earth with moisture and wetness, saturating the air with mist, clouds, and humidity, circulating in living beings, generally being in unseen places and taking forms that cannot be easily demarcated. Indeed rivers, together with other "water bodies," such as seas and lakes, not only come with a particular choice of moment; they also come with a line that is granted the ability to separate water from land and keep water in place. In the case of rivers, the line is granted the added ability to calibrate a flow of water from a point source to a destination, or at least from an earlier place to a later place. On the ground, this line takes the form of a riverbank; on maps, it is familiar as a geometric line.

The choice of the moment of flow formation is significant. In this moment, not only is water kept to a certain place and maps are drawn, but properties are demarcated, histories are written, places are described, and the future is envisioned. Arguably, this moment has inspired the very ideas of property, history, and place. In other words, the world has cast anchor in the moment of flow formation, making it the time of reality while turning other moments of the water cycle into moments of ephemerality. Thus, rain, mist, snow, and so on, are visitors in places where the river is resident (Figure 1.2). It is not surprising, then, that rivers are taken to be natural entities. Ecologists see them gathering a unique ecosystem. They even see their floods as natural events, floods that are nothing but water crossing a line that

FIGURE 1.3
G.W. Colton, "Mountains & Rivers," *Colton's Atlas of the World, Illustrating Physical and Political Geography.* (New York: J.H. Colton and Company, 1856 [David Rumsey Historical Map Collection].)

has been drawn in a moment of choice. Geographers, too, see rivers as natural features on the earth's surface, tasked to drain water off landmasses, while geologists see them shaping the earth, eroding ground, and building alluvial plains. Historians and archaeologists certainly do not hesitate to place the existence of rivers beyond culture. With rivers serving critical infrastructural needs for water supply, irrigation, drainage, transportation, and power, and with an edifice of powerful ideas built upon its banks, including the city and civilization, we would be hard-pressed to find anybody today who would want to challenge the choice of anchoring time in the moment of flow formation (Figure 1.3).

Yet the moment of flow formation cannot be assumed to be everyone's choice. We suspect that there are people who have chosen other moments of the hydrologic cycle in which to anchor their existence, their language, and the things they see constituting nature. As such, what archaeologist Brian Fagan calls the "clash of cultures" that followed the European discoveries in the fifteenth to eighteenth centuries, which he saw as "a progressive confrontation between […] societies living in totally incompatible worlds," could well be a clash of lives anchored in different moments of the water cycle. These are places where efforts were made by erstwhile colonial governments to impose, through education, maps, technologies, and a "development" agenda, the separation of land from water; and to cultivate people to appreciate the economic, ecological, aesthetic, and hydraulic possibilities that result from this separation. Their efforts continue today in the hands of independent governments. These are also places where infrastructures, ways of life, and imagination grounded in rain and other moments of water continue to be reduced to the informal, the uncertain, the ephemeral, the relative, the nomadic, the primitive, the exotic, and, not surprisingly, the unmappable.

We see here a more fundamental starting point of design; one that constructs possibilities that stem from asking, what if we drop anchor in another moment of the water cycle? What if India, a country that continues to be ridden with the "informal," is seen as a rain terrain rather than a river landscape, a place where the monsoon is resident rather than an annual visitor that stays for a season? What if the East Coast of the United States, where hurricanes—carriers of rain for a shorter and less predictable time than the monsoon—have been made terrifying and unwanted visitors to a coast, is firmly anchored in the moment of flow formation of the hydrologic cycle? Does this moment afford another approach to designing the coast?

The Case of the Bengal Delta

Mouth of the Ganges River

Despite the overwhelming presence of monsoons in India, rain is presented in school textbooks, government gazetteers, and scientific treatises as a visitor in a land where rivers have been granted residency. Indeed, rivers here are not only an intrinsic part of the subcontinent's infrastructure, they are also considered sacred, particularly the Ganges. Millions venerate this river as a goddess, Ganga. Her "real" abode, they say, is in "the highest heaven" (Eck, 2004). The story of her descent to earth is told in a number of ancient Sanskrit texts as being initiated by King Bhagiratha. Only the celestial waters of this goddess, he was told, could reconstitute the 60,000 sons of his ancestor, King Sagara. They had been reduced to ashes by Sage Kapila, whose meditation they had interrupted when they found him in possession of their father's sacrificial horse that had been sent out by the king to roam the world. If someone possessed this horse, it meant a call to war. After others before him failed to persuade Ganga to come down from her abode in the sky, it was left to Bhagiratha. His asceticism impressed the gods, and they convinced Ganga to descend; but they worried that the earth would shatter under her fall. Shiva offered to mediate by taking her fall on his head. Ganga, it is said, then flowed down his hair and, led by Bhagiratha, reached the netherworld, where she reconstituted Sagara's sons.

Millions of people read this story as portraying a river descending by the "locks" of Shiva's hair. They see this river beginning at the end of the Gangotri glacier in the Himalayas and extending 2,525 kilometers to meet the sea via the Bengal Delta. Here, writes Amitav Ghosh in *The Hungry Tide*, "the braid comes undone," dividing "into hundreds, maybe thousands, of tangled strands," channels which form "an immense archipelago of islands" (Ghosh, 2004) (Figure 1.4).

Mapmakers have long tried to capture these strands on paper, beginning with the geographer Ptolemy in the second century, who, with information from sailors, drew five channels fanning out from a point. The number varies in later maps, particularly those drawn following the arrival of the Portuguese in the early 1500s, and then again following the arrival in the 1600s and 1700s of a number of European trading companies who set up "factories" in the delta. When the English East India Company turned from merchant to ruler and acquired the delta in the second half of the 1700s, they sought to map it with precision. James Rennell was assigned the task. He pursued it with as much creativity as diligence, and in a place where he saw water rise and fall as much as four and a half meters,

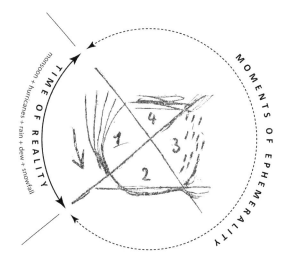

FIGURE 1.4
People could choose to anchor in the moment of precipitation. It will be the ground of another time and another place. (Image adapted by the author, based on Paul Klee, in *Paul Klee: The Thinking Eye* [New York: George Wittenborn Inc. 1964, 402].)

"tracts of land [...] swept away in the course of one season as would astonish those who have not been eye witnesses to the magnitude and force of the mighty streams occasioned by the periodical rains of the tropical regions," and where vast swaths of ground become "inland seas" and "swampy morasses" for a time, he separated land from water, forming a ground of clear and distinct islands and channels, at least on paper.

His language of islands and channels became the official language of government. W. W. Hunter, a pioneer of the government gazetteer that informs officials about their region and locality, describes the delta as a multiplicity of islands gradually falling in height from an apex toward the sea, where they become "a sort of drowned land, covered with jungle, smitten by malaria, and infested by wild beasts; broken up by swamps, intersected by a thousand river channels and maritime backwaters" (Hunter, 1875). Those to the north were relatively fixed, "locked into their channels," while the islands in the south were "in an unfinished state." But many of these unfinished islands were settled following Rennell's survey, a feat that required clearing the jungle and raising the surveyor's line into an embankment to keep water out, often after extending and filling earth in an act known as "reclamation." For much of the time these islands are below the level of the sea, their vulnerability exacerbated by heavy rains and their mud embankments in constant need of vigilance.

The few islands left unsettled, but still foraged by settlers on adjacent islands, are protected as a UNESCO World Heritage Site for the "ecological processes" of "island formation" that they display and for their jungles, which, far from being smitten by malaria and infested by wild beasts, are seen to be populated by a diversity of mangrove species that are believed to be exceptional and of great value. Besides that, they provide a barrier against a surging sea—a "coastal defense" that absorbs the force of storms (Figure 1.5).

Janapadas in Ganga

There is much to suggest that the mouth of the Ganges is an estuary rather than a delta. For one thing, the tide extends far inland. A number of maps from the fifteenth and sixteenth

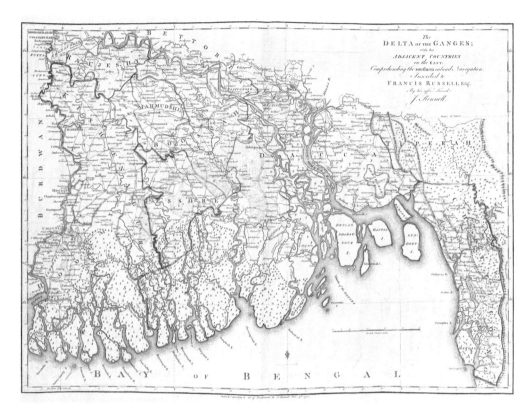

FIGURE 1.5
Rennell counted eight mouths to the Ganges. "In tracing the sea coast of the delta, we find no less than eight openings; each of which, without hesitation, one pronounces to have been in its time, the principal mouth of the Ganges." (From James Rennell, *The Delta of the Ganges with the Adjacent Countries on the East: Comprehending the Southern Inland Navigation, in A Bengal Atlas* [London, 1780].)

centuries show a meeting ground of river and sea that is more bay-like than delta-like. Charles Lyell, author of the influential text *Principles of Geology*, which ran into 12 editions between 1830 and 1875, describes the sea reaching to the "heads of the delta" over 320 kilometers inland on a regular basis when rivers ran low in the non-rainy season. He singled out two heads for attention, one on the Ganges and the other on the Brahmaputra, but the bay, which is still discernible, receives numerous other rivers that have long defied the single apex of a delta. They introduce a complexity and variation from across the diverse terrain of northern India, the Himalayas, Tibet, and the eastern ranges. Here, one can imagine a moment when runoff consumes the entirety of the bay with rain, and another time when a surging sea consumes it just as entirely.

However, even though an estuary accommodates more fluidity and complexity than a delta, it still anchors the moment of flow formation when water seems containable with lines. This is a moment that is increasingly defied not merely by exceptional events, such as hurricanes and tsunamis, but by tides and monsoon downpours on a regular basis. Is it possible to anchor Bengal in a moment of precipitation? In this moment the story of the descent of Ganga would be heard differently, with another view and imagination. Shiva here would let the goddess down not the locks of his hair, but down each of his infinite hairs. Her descent then would not picture a river as much as rain. Here, Ganga does not

FIGURE 1.6
(a) and (b) Mud embankments protect islands in the Bengal Delta from a sea that rises and falls as much as 4.5 meters. (Image courtesy of the author.)

flow as the Ganges does, in a course to the sea; she is rather held in soils, aquifers, glaciers, living things, snowfields, agricultural fields, tanks, terraces, wells, cisterns, even the air, all for a multiplicity of durations that range from minutes and days to centuries and eons. She soaks, saturates, and fills these places before overflowing in a multiplicity of ways only to be held again. Bhagiratha's task of leading her to a netherworld is much more challenging, much more mysterious, much more befitting of the infinite capacity of the gods. Unlike the Ganges, her source is not in a point or points, but in clouds. Also unlike the Ganges, her course cannot be drawn in a map, her "routes" being too complex, emergent, and changing, besides occurring across a vast cross section extending from the sky deep into the earth. The only anchor she offers people is the time of her descent. It is celebrated each year at the coming of the monsoon with the Ganga Dasahara festival.

Here, Ganges and Ganga, names that scholars use interchangeably in the belief that they refer to the same thing—namely, a river—in fact refer to two different things. Ganges refers to a river, a linear flow of water draining land to the sea or some lower ground. Ganga, on the other hand, refers to a rain terrain, a nonlinear field of holdings of rain driven to saturation and then overflow, only to be held again and again. Everything participates in holding rain when it arrives on the winds of the monsoon. As one author put it, "With the monsoon the tempo of life and death increases. Almost overnight grass begins to grow, and leafless trees turn green. Snakes, centipedes, and scorpions are born out of nothing. At night, myriad[…] moths flutter around the lamps. […] Inside rooms, the hum of mosquitoes is maddening" (Singh, 1987). A world such as this is not divided between land and water; instead, there is an appreciation for gradients of wetness, a depth of material across air, water, and earth.

Settlement in a rain terrain such as the Ganga does not occur on land across a line from water (or exotically, on water) (Figure 1.6a and b). It rather occurs on the slopes of *janapadas*, a word that translates as "footholds" (Figure 1.7). These raised grounds, many of them built up with earth, anchor in a rain terrain with infrastructure that facilitates practices on a slope, practices that respond to the rise and fall of wetness. Importantly, janapadas

FIGURE 1.7
The mangrove forests in the Bengal Delta are a UNESCO World Heritage Site protected for their diversity of species which include the Bengal Tiger, but also for the barrier that they afford against a surging sea. (Image courtesy of the author.)

are called out not by area but by trajectories that intersect upon them and extend toward open horizons. These trajectories, which are activated by movements, anchor janapadas in a network; in turn, janapadas anchor networks. Raised grounds like the janapadas were noted in other places, such as in Egypt and Mesopotamia by Herodotus, and in India by Alexander's men. Of the monsoon that arrived in Egypt from the Ethiopian highlands around the summer solstice, Herodotus writes: "The whole country is converted into a sea, and the towns, which alone remain above water, look like the islands in the Aegean. At these times water transport is used all over the country, instead of merely along the course of the river" (Herodotus, 2003). It is said that he believed that "the size of the Nile at its swelling is its natural one" (*Diodorus of Sicily*, 1933). If this is true, he saw the equivalent of janapadas *in* the Nile as a rain terrain, rather than *on* the banks of a river (Herodotus, *The Histories*).

Today, janapadas in rain terrains have lost out to settlements on riverbanks. Operations on their slopes that once held the rains of the monsoon and worked their overflow by means of tanks, fields, and jungles have given way to leveled grounds protected from the high waters in rivers by mud embankments while being fed by canals, pipes, drains, and other extensions of a river system. In a time when these embankments are being considered to be built in concrete amid disasters blamed on rain and a rising sea, reinstituting the janapada, its gradients, and its open network offers a more resilient alternative (Figure 1.8a and b).

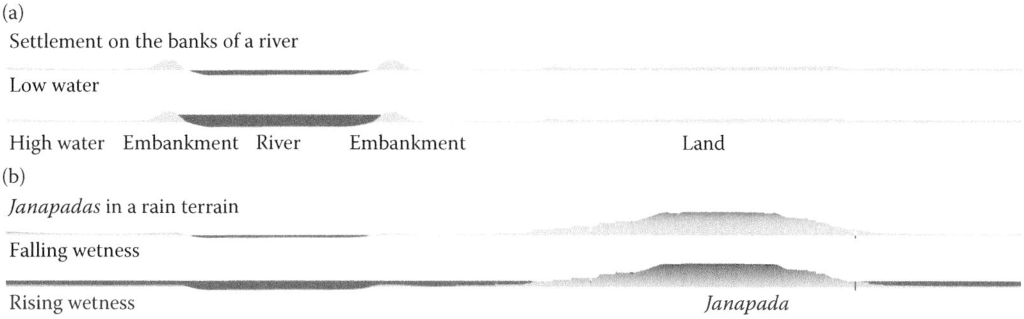

(a)
Settlement on the banks of a river

Low water

High water   Embankment   River   Embankment   Land
(b)
Janapadas in a rain terrain

Falling wetness

Rising wetness   *Janapada*

FIGURE 1.8
(a) Human habitation in a river landscape such as Ganges occurs in settlements across the line of a riverbank which more often than not is raised in an embankment. (b) Human habitation in a rain terrain such as Ganga occurs on the slopes of a *janapada*, literally, a foothold. (Images courtesy of the author.)

The Case of the Chesapeake Bay

The Coastline of the Tidewater Region

The Chesapeake Bay is, without a doubt, an estuary: a number of rivers meet the sea through it. Its edge is complex, and its varying salinity does not just reach up rivers, it reaches into the creeks that run into the rivers and the rills that run into the creeks. The bay forms an intricate coast that harbors unique ecologies and species, many of which move between freshwater and saltwater. When Europeans arrived on this coast with ambitions for settlement, they brought with them the idea of a coastline. One of the earliest drawings of this line was by the artist John White in 1585, *La Virginea Pars*. In 1587, White returned at the behest of Sir Walter Raleigh as leader of a group of over 100 colonists. But the colony that he was instructed to begin, called Virginia, on Roanoke Island, did not take root. However, two decades later the English set a firm foot on land that they separated from water with a clear and distinct line, a line respected not just for the separation it made, but also the property that it marked. It was the making of place famously at odds with the "native peoples" encountered by Europeans post-1492. These people were seen to lack the concept of possession (and civilization), a lacking that would inspire the legal concept of *Terra Nullius*, meaning land not belonging to anyone. The concept was used to justify European occupation in Australia, Africa, and the Americas until 1973, when the International Court of Justice ruled it indefensible and recognized the peoples of these places as "first inhabitants." But does the concept of *Terra Nullius* really stem from the idea of possession, or does it stem from a preferred moment of the water cycle when Europeans could claim possession, most effectively by means of the map? This is a moment probably as unfamiliar to the natives as maps were, or perhaps it was ephemeral to them because they chose to anchor their time in the moment of precipitation, a moment exemplified by the hurricane that prevented John White from looking for his famous "lost colony."

A month after setting up his colony on Roanoke, White had to return to England to get supplies for a colony unable to live off the land (and water). He could not come back for three years, and when he did, the colony was no more. He was not able to search for survivors; a hurricane was bearing down on them and his captain refused to stay. White never returned but others would; in 1607, the settlement of Jamestown was founded. These settlers drew the coast with a firmer line, doing so when it was not raining. This line has been threatened and often erased by hurricanes, and each time, after some debate, dissent, and marginal change, it has been redrawn. Hurricane Sandy in 2012 is the most recent hurricane to erase the coastline not merely in Virginia, but up and down the East Coast. Its destruction was tremendous. The debate that it has instigated is seen in a different light than previous hurricanes. It is framed by climate change, which is discernible in global warming and sea level rise with consequences expected in the increasing frequency of hurricanes and larger amounts of rain. The Lower Chesapeake has already experienced a rise of 38 centimeters in the last eight decades. One option being discussed is to defend the coastline against an advancing sea with seawalls and gates, including one across the entrance to the Chesapeake Bay. This option is not just difficult to implement because of the intricate nature of the coastline; it also forces a reliance on pumps, besides being detrimental to species that live and move between freshwater and saltwater. A second option being considered is to retreat from the coastline, allowing it to be erased and replaced by a "higher" line. In the face of these two alternatives, we ask: What if we see rain and hurricanes not as outsiders but as events that make us reevaluate the very idea of a coastline? What if we see the coast in a moment of rain?

"Fingers of High Ground" in a Rain Terrain

There is another reading of the coast in the tidewater region of Virginia. It is not divided between land and water but structured by fingers of high ground. These grounds grow from higher fingers and extend into lower ones. Their fall is complemented by the rise of the webs between them, webs that in a moment of flow formation have been characterized hierarchically as rills flowing into creeks, creeks flowing into rivers, and rivers flowing into the bay. However, in a moment of precipitation, fingers and webs are merely two different "directions" of the same ground that operates by a dynamic gradient of wetness—high to low, low to high.

Fingers and webs are not just another reading of "nature's ground"; they are also another reading of existing settlements with an eye and an imagination driven to appropriate and make gradients of wetness. Design here is necessarily opportunistic, seeking out places that allow a meaningful complementarity of falling rain and rising tide. Unlike a ground enclosed by seawalls and gates that requires "completion" in order to be effective, fingers and webs enjoy an autonomy that allows their conception and construction to be open to time, growth, and replication with a sense of experiment, adaptability, learning, and strategy. They protect by accommodating rising and falling tides rather than confronting the sea, avoiding the catastrophes that can result from the failures of walls, gates, and pumps.

The design of fingers of high ground drives us to rework the way we understand "land use," which is typically seen in terms of exclusive areas of the earth surface. In the moment of rain, land use can be infused with an appreciation of section, direction, and time, such that occupancies on high ground can extend opportunistically and temporally to low ground; and those on low ground can extend in the same way to high ground. Here, humans are not the only beneficiaries; organisms that occupy gradients of salinity stand to gain as well, particularly in the face of rising seas when entire ecological gradients are shifting up. Add to this the possibility that fingers of high ground can be an emergent process both in their construction and in their accommodation of risk, and design can become a powerful means of turning the coast from a continuous line to one of cumulative gradients.

It is often assumed that design is led by a site analysis followed by a proposal that responds to that analysis. In the face of storm events there is the added challenge of identifying, mapping, and modeling sites based on their vulnerability to surge and flooding. This approach front-loads the design process with measures of a problem that design must address and to which design is held responsible. Sea level rise, however, poses a challenge of a different order for designers. It calls for addressing more than the vulnerability of sites; it calls for revisualizing places in terms that do not make them vulnerable to begin with. Visualizing the coast in terms of fingers of high ground does this. It affords a new beginning for tidewater Virginia, positioning sites on a gradient between rain and tide rather than in a confrontation between land and sea.

In Norfolk, we had the opportunity to demonstrate, at least on paper, the possibilities that follow from seeing a coast in terms of fingers of high ground rather than a line. At times, these fingers were obvious in ridges, hummocks, highways, and other infrastructural corridors; at other times, history, particularly in the form historical maps and oral recordings, led us to their traces in grounds that were flattened in the process of development in the last century; and at yet other times, we found opportunity in post-industrial and superfund sites that could be raised or needed to be capped in some way. Whether found, recovered, or opportunistically raised, fingers of high ground operate in an open field. Process, strategy, and participation are key to their realization, as are the extensive dredging operations in the waterways and bay adjoining Norfolk that provide the material for raising ground.

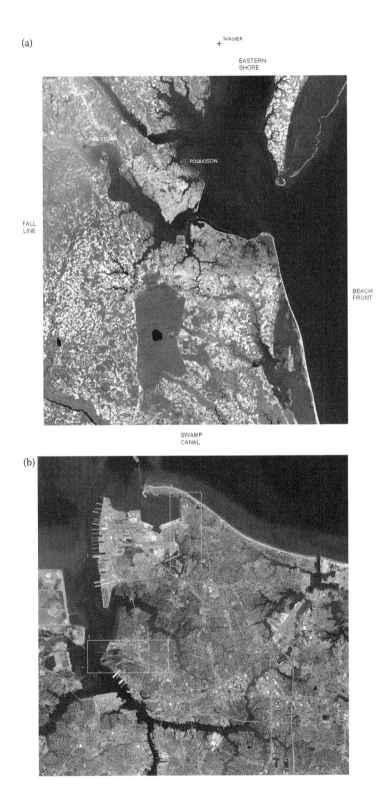

FIGURE 1.9
Location of Lambert's Point proposal (1) in Norfolk, tidewater Virginia. (Images courtesy of the author.)

FIGURE 1.10

Lambert's Point proposal. (Images courtesy of the author.)

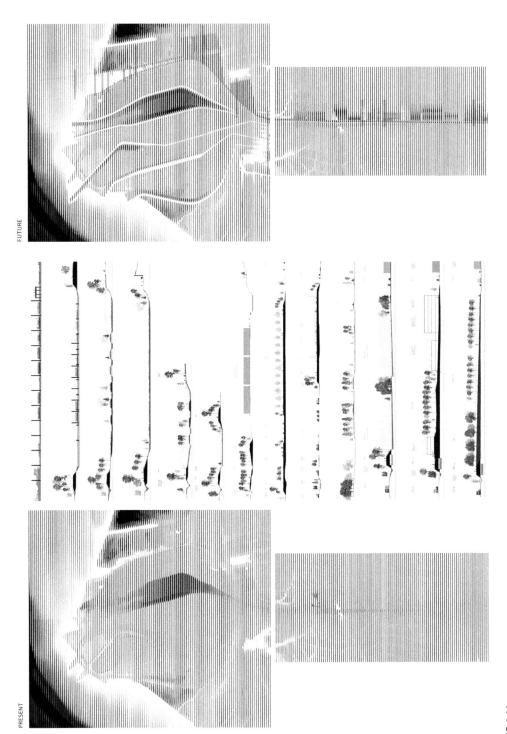

FIGURE 1.11
Lambert's Point proposal. (Images courtesy of the author.)

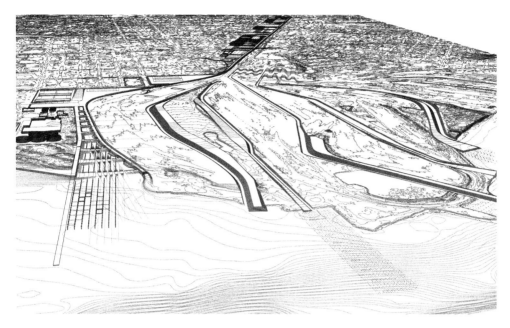

FIGURE 1.12
Lambert's Point proposal. (Images courtesy of the author.)

In this essay, we demonstrate one starting point going forward in Norfolk. It involves the revisualizing of Lambert's Point in terms of fingers of high ground, a place where rail lines and shipping piers meet the Elizabeth River. It is home to a coal-exporting facility that extends inland along rail lines that run on a ridge flanked by warehouses that are currently in transition to post-industrial uses. The ridge presents itself as a finger of high ground that divides via sidings into several smaller fingers to eventually reach piers on the Elizabeth River. The transitions from ridges to piers, webs (between fingers) to wetlands, storm drains to rivers, afford diverse gradients and consequently varied opportunities and programs. The project is best told through drawings, some of which are shown in Figures 1.9 through 1.12. It is important to the larger aspiration of coastal resilience that fingers of high ground do more than raise land; they must perform richly as the future DNA of the coast.

Over four decades ago, architect Kevin Lynch asked, "what time is this place?" The evidence of time, he writes, "is embodied in the physical world." And the image of time "is crucial for individual well-being and also for our success in managing environmental change, and that the external physical environment plays a role in building and supporting that image of time" (Lynch, 1972). We have suggested that this image of time, exemplified in features, such as the coastline and the riverbank, is conceived and constructed in a particular moment of the hydrologic cycle when it can be believed that we have the capacity to keep land and water apart. This choice has made events, such as the monsoon and the hurricane outsiders. The possibility of anchoring in another moment of water and, as such, conceiving and constructing "another time and another place" where the monsoon and hurricane are insiders, opens new horizons for designing the coast. Both the janapada and finger of high ground exemplify this possibility. The more we study them, the more their infrastructure potential, stemming as it does from accommodating rather than confronting wetness, makes sense in the face of rising seas. At the very least, they expand the conversation on coastal resilience.

References

Amos, William. 1971. *The Infinite River: A Biologist's Vision of the World of Water*. 244. New York, NY: Random House.

Eck, Diana L. 2004. *India: A Sacred Geography*. 131. New York, NY: Harmony Books.

Encyclopaedia Britannica, 15th ed., s.v. "river."

Ghosh, Amitav. 2004. *The Hungry Tide*. 6–7. New Delhi: HarperCollins Publishers.

Herodotus. 2003. *The Histories*, bk. 2, 97, trans. Aubrey de Sélincourt. London: Penguin Books.

Horton, Robert E. 1931. "The Field, Scope, and Status of the Science of Hydrology." In *Transactions of the Geophysical Union* 12(1931):192.

Hunter, W. W. 1875. *A Statistical Account of Bengal: District of the 24 Parganas and Sundarbans, vol. 1.* xiii–xiv. London: Trubner & Co.

Kandel, Robert. 2003. *Water from Heaven: The Story of Water from the Big Bang to the Rise of Civilization, and Beyond*. 6–7. New York, NY: Columbia University Press.

Lynch, Kevin. 1972. *What Time Is This Place?* 1. Cambridge, MA: MIT Press.

Oldfather, C.H. trans. 1933. *Diodorus of Sicily*, vol. 1, bk. 1. 135. New York, NY: G. P. Putnam's Sons.

Singh, Khushwant. 1987. "The Indian Monsoon in Literature." In *Monsoons*, ed. Jay S. Fein and Pamela L. Stephens, 37. New York, NY: John Wiley & Sons.

2

Fraying at the Edges: On Coastal Life and Rising Seas

Abbas El-Zein

CONTENTS

In 2007, Jill Walker, a resident of Sandon Point—about 100 kilometers south of Sydney, along the Illawarra coast—took the`then Minister for Planning of the State of New South Wales, Frank Sartor, to court over his decision to approve the development of a large retirement village close to where she lived. As one of numerous disputes over development projects heard in Australian courts every year, the case would have been unremarkable had it not been for the novelty of the central argument put forward by the plaintiff—and the fact that the minister was using special powers recently invested in him to override decisions by local councils for "state-significant" projects. The site in question was likely to experience more intense and frequent flooding over the course of this century, as a result of sea level rise driven by anthropogenic climate change. The partial drying of three creeks on the site, envisaged in the planned development, was likely to exacerbate the site's vulnerability to future flooding. Lawyers acting on Walker's behalf argued that the minister failed to take this into consideration before granting the approval, hence the decision should be invalidated.

The New South Wales Land and Environment court ruled in Walker's favor, based on the precautionary principle—the notion that where there is uncertainty about outcomes, but the stakes are high, decisions must err on the side of caution.[1] The precautionary principle, however, is notoriously elastic, and much depends on how much uncertainty is judged acceptable and exactly how high the stakes must be before action is taken. The Court of Appeal later reversed the decision, and Walker failed in her bid to have the case heard by the Supreme Court. The development went ahead.

In the Court of Appeal, the case essentially came down to whether the minister should have taken the public interest into account in his decision—to which the Court gave an affirmative answer—and whether principles of ecologically sustainable development and, specifically, considerations of climate change are parts of the public interest. The Court found that knowledge about climate change, at the time of the decision by the minister in 2006, was not sufficiently advanced to be considered as part of the public interest, hence the Court's decision. However, the judgment further asserted that this was "likely to change" and that ecologically sustainable development was likely to become an element of public interest to be considered in such decisions in the future.[2]

Hence, while Jill Walker and the residents of Sandon Point lost their battle, this was a Pyrrhic victory for developers, specifically those who were arguing against any attempts at limiting development in anticipation of future climate change. Indeed, for the last 10 years or so, municipal councils, state governments and the Courts of Justice in Australia have been grappling with, not so much whether, but how to respond to sea level rise and the changes to our coasts it will inevitably bring about.

Humans have always had a mixed fascination for water—as a vital resource without which they would die of thirst or hunger and as a threat through the prospect of flooding. Australia is no exception, and the contrast between the great deserts of the interior and the oceans surrounding our continent is a singularly antipodean take on this drama. The image in the Australian psyche of a vast, terrifying, largely inhospitable landscape lives side-by-side with that of a "lucky country" that yields riches and carefree, beachside lifestyles. Enter climate change and sea level rise.

Sea level rise is believed to have started accelerating in the middle of the nineteenth century, with a total of 15–20 centimeters (cm) rise level now reached relative to pre-industrial times. The global mean rate of rise currently stands at 3.3 millimeters (mm)/year and is increasing. This is compared to a rate of a few tenths of a millimeter before 1800. Nor is the rise the same everywhere—variations in ocean currents cause significant fluctuations around the mean. For example, along the north coast of Australia, sea levels have been rising at more than 7 mm/year. All in all, by the end of this century, sea levels are projected to increase by between 25 cm to 1 meter, relative to pre-industrial times, according to the latest Intergovernmental Panel on Climate Change (IPCC) report to policy makers, released in 2014. The IPCC calls this range a "medium-confidence" estimate, and the actual rise could be higher. Much hinges on whether the Arctic and Antarctic glaciers will melt and add to thermal expansion. Most troubling is that we are most likely locked into a long-term rise in sea levels which, climate scientists say, will continue for the next few hundred years and millennia. And it is precisely the escalating nature of this risk that gives sea level rise and its problems much of their complexities.

One meter of sea level rise by 2100 might not sound like much. After all, waves many times bigger hit coastlines around the world every day. However, two aspects of the problem in particular make it dangerous. First, land adjacent to marine or riverine water would be inundated, especially at high tide, and the cost of damage and/or protection can easily run into billions of dollars in dense urban centers (Figure 2.1a and b). Second, the rise in sea levels is accompanied by an increase in the frequency and intensity of extreme weather events. For instance, a study by the federal Australian research organization CSIRO (Commonwealth Scientific and Industrial Research Organisation) found that a rise of 90 cm would be accompanied by a 10,000-fold increase in the frequency of storm surges and flooding events at some locations along the eastern coast of Australia.[3] This means that an event that takes place once every hundred years today would occur once every few days by 2100.

Loss of land to the sea in partial or total inundation, flooding and beach erosion as a result of storm surges, and sea water intrusion into groundwater are all likely consequences (Figure 2.2a–c). The so-called king-tide storm that hit the New South Wales coast in June 2016 gave us a taste of what this might mean, with 50 m of beach lost in some places, houses and infrastructures damaged, and power supply interrupted.[4] There is no doubt that the projected change in sea levels over the course of this century would make for a very different coastline to the one we know today.

On the other hand, lives and livelihoods in Australia have evolved in close proximity to the coastline, especially since the beginning of white settlement, and 85% of Australians today live within 50 kms of the coast. Having to give up parts of our coasts to the sea—or at least rearrange the way we live to accommodate higher sea levels—is almost too painful to contemplate. Fortunately, this is precisely what many are doing; and not just contemplating. Adapting the coastline to sea level rise is a question that is looming large

(a)

(b)

FIGURE 2.1
Projected inundation around Sydney's Parramatta River basin, in 2100 under high tide and a 74 cm sea level rise scenario. *Author note: Photo (b) is a Zoom-in on Drummoyne area to illustrate impact on properties.* (Coastal Risk Australia, www.coastalrisk.com.au, accessed April 18, 2016.)

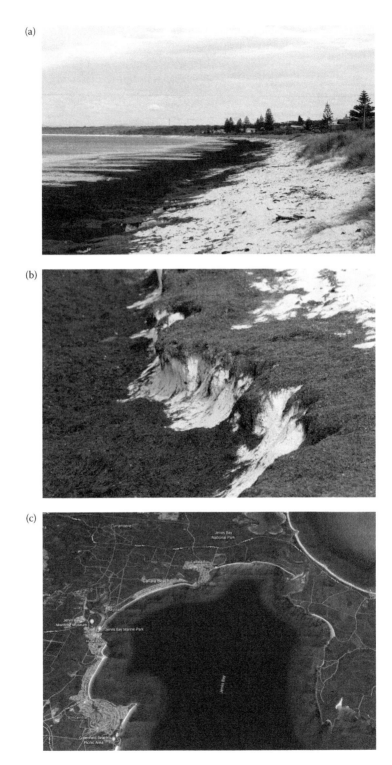

FIGURE 2.2
Loss of beach sand after a storm at Callala Beach, Jervis Bay, Shoalhaven, a three-hour drive south of Sydney.
(Photos by author on September 9, 2014. Map from Google Earth.)

over coastal councils, communities, and government agencies, even if concrete answers are yet to emerge about how exactly we should take up this challenge—with the thorny politics attached to it.

A small community of about 2,000 people has grown around Mollymook Beach, a four-hour drive south of Sydney, in the region of Shoalhaven. More than a third of Mollymook residents are over 65 years of age, and around a fifth are under the age of 19. In summer, the beach area swells with visitors, serviced by an entertainment center (the Golf Club), beach rental apartments, and a surf club (Figure 2.3a–c).

A set of properties has been built on the sand dunes adjacent to the beach. Almost all development at Mollymook has taken place after World War II, when Australia witnessed a period of sustained economic and population growth (Figure 2.4a and b).

Mollymook Beach is one of three communities in Shoalhaven identified as especially exposed to the erosional effects of sea level rise. The other two are Callala (see Figure 2.2a–c) and Vincentia's Collingwood (see Figure 2.5a–c), both of which are located in Jervis Bay.

In Mollymook, the Golf Club, the Surf Club, and a number of private properties are at risk of periodic flooding and, more insidiously, gradual weakening of foundations which may make them structurally unsafe. To make matters worse, the beach harbors key water mains and a wastewater station that is connected to a bigger sewerage network. Mitchell Road, running alongside the beach, services significant traffic and is also at risk. Damage to the sewerage station would affect services to beach residents and beyond—more than 5,000 households according to one estimate.[5] Some assets are already at risk from a storm surge. A 50 cm rise scenario for 2050 estimates the public and private properties affected to be valued at AU\$17 million and AU\$25 million, respectively.

None of the traditional options of coastal management—protect, accommodate, relocate—are particularly suitable. Protecting the beach by seawalls has been found, in detailed engineering studies, to cost AU\$32 million, beyond the capacity of the local council. And, even if money is found in the coffers of the state or federal governments, 1% of the national GDP would be spent before this treatment is extended to 300 beaches—less than 3% of the 10,000+ beaches found in Australia. And that's only beaches; hardly a blueprint for a national adaptation strategy.

What's more, while seawalls are effective in protecting property, they can interfere with beach access and amenity. They also tend to diminish the natural ability of the dunes to replenish the beach because they reflect rather than absorb the wave energy—hence accelerating erosion. Building protective groins or setting up beach nourishment are just as financially and environmentally problematic. Globally, a study by researchers at the Global Climate Forum and Potsdam Institute for Climate Impact Research[6] found that protecting coasts around the world with dikes and seawalls could cost up to US\$71 billion by 2100. This is much less than the cost of damages from floods, as estimated by the same authors, but would increase the risk of catastrophic failures, such as those witnessed in New Orleans when flood levees failed.

Relocating the public infrastructure systems away from Mollymook Beach, to avoid repeated damage from storms, is easier said than done. For example, the Shoalhaven City Council is monitoring the risk to the sewerage station and, if a risk threshold is exceeded, plans to relocate it. However, finding a suitable alternative location is proving difficult, especially because the station is part of a bigger sewerage network. As for Mitchell Road, relocating it sounds like a contradiction in terms since the whole purpose of a local road is to be where it is! The larger point here is that the provision of infrastructure services at Mollymook, as in other coastal communities around Australia, will become increasingly difficult and more expensive as sea levels continue to rise.

(a)

(b)

(c)

FIGURE 2.3
Mollymook Beach, four-hour drive south of Sydney. (Photos by author on January 12, 2014.)

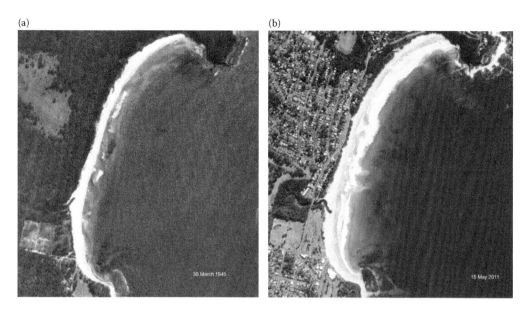

FIGURE 2.4
Post-World War II development as revealed by two aerial photos of Mollymook Beach. (Photos provided by the Shoalhaven City Council and the Office of Environment and Heritage of the New South Wales Government.)

FIGURE 2.5
Collingwood Beach, Vincentia, Jervis Bay, Shoalhaven. Photos show dune erosion and residential properties along the beach, three hours south of Sydney. (Photos by author on January 7, 2014.)

Shoalhaven City Council (SCC)—which is responsible for 165 kms of coastline including Mollymook, Callala, and Collingwood beaches—has been attempting to develop long-term planning for sea level rise for a number of years now. However, local and state politics have made the task harder.

In 2009, the then Labor government of New South Wales mandated that local councils should use a benchmark of 40 and 90 cm rise, relative to the 1990s, by 2050 and 2100, respectively. Based on these benchmarks, hazard lines for the Shoalhaven coastline were generated that identified properties and public assets at risk at present and in the future. Inevitably, when such information was published, many property owners objected, not unreasonably, that the values of their assets would go down while their insurance premiums would increase, all as a result of a set of hazard lines which many saw as speculative.

A new Liberal Coalition government (on the conservative side of politics) was elected in 2011 and was quick to slash the benchmarks. However, rather than proposing new ones, the government directed local councils to develop their own, from locally specific risk assessments and risk tolerances. The decision made little sense since benchmarks for future sea level rise could only be based on interpretations of Intergovernmental Panel on Climate Change (IPCC) reports, simulations of global circulation models, and nationwide readings of past and current rise—a task much more within the remit and resources of state and federal government agencies than those of municipal councils. Only when the implications of these benchmarks are to be translated into risk assessments and hazard lines do local factors usually become important.

In reality, the new government was buying time while it developed a new statewide policy. In 2013, it put to Parliament a white paper[7] that set up a new legal framework for coastal development. However, the bill sank in the Upper House and it was back to the drawing board. A new Coastal Management Bill[8] was finally made into law in May 2016, and it does a far better job of addressing some of the conflicts between private and public interests than previous legislation. However, it says almost nothing about how the long-term effects of sea level rise should be tackled.

When the benchmarks were slashed, Shoalhaven City Council teamed up with the neighboring local Council of Eurobodalla, and together they commissioned a consultant, Whitehead and Associates, to conduct a scientific and legal study and advise them on which benchmarks to use. The report recommended the adoption of higher-end predictions given in the latest IPCC report,[9] under the representative concentration pathway 8.5 (RCP8.5) worst-case scenario of intensive CO_2 emissions continuing through this century. This translated into a 28 cm rise by 2050 and 98 cm by 2100, relative to 2015. The recommended benchmarks were therefore 7 cm lower than the discarded ones for 2050 but 13 cm higher for 2100.[3] (The lower 2050 benchmark reflected updated scientific projections in the fifth IPCC report.) Predictably, the report came in for criticism from local communities, with 67 out of 79 submissions, in response to an exhibited draft, disagreeing with its conclusions,[10] according to one source. The report even earned the disapproving attention of the Nongovernmental International Panel on Climate Change (NIPCC), a climate-sceptic organization financed by the conservative Chicago-based Heartland Institute!

Elected local councilors are understandably more open to pressures from community groups, and councilors in Shoalhaven and Eurobodalla ultimately adopted a more conciliatory path, opting for the more optimistic IPCC RCP6.0 scenario which assumes some success at CO_2 emission reduction by 2100. Hence, both councils adopted 23 cm for 2050, the higher-end projection from RCP6.0. For 2100, Shoalhaven adopted 36 cm at the lower end of RCP6.0, while Eurobodalla opted for twice that value at the higher end of RCP6.0, in stormy meetings in which the threat of legal action loomed large.[11]

Local councils in Australia have always found themselves between a rock and a hard place when it comes to planning for sea level rise—legally required to put in place long-term coastal planning and protect public amenities and assets, while catering for strong private property rights enshrined in Australian statutes. Sea level rise can bring these two spheres of coastal management into serious conflict. The effects of publication of hazard lines on property values, mentioned earlier, is such an instance. Another example is given by cases where beachside properties are at risk from storm surges. Does past development approval provided by the local council imply that the council is legally bound to protect that property from past and/or future hazards? If not, should property owners be allowed by the council to build such protection themselves, even if it interferes with beach amenity? These questions have been tested in the courts over the last decade or so. What local councils need—and what repeated state governments have ultimately failed to provide—is clear, legislatively based guidance that would protect them from a morass of legal liabilities.

Underlying all the issues raised previously are two overarching question which, 20 years ago, might have come across as purely academic but have now acquired a level of practical urgency, thanks to sea level rise. First, what is it that we value most about our coastlines and our beaches? In other words, which element of the coastline and the lives with which it intersects, are we keen to preserve and, conversely, which are we willing to forsake?

Second, should adaptation be small, piecemeal, and highly democratic—the outcome of a multitude of local negotiations and adjustments—or should it take the form of a grand scheme in which higher-level government agencies play a central planning role and which local councils merely implement? Or can we work something out in between these two extremes?

The two questions are of course related. Not just because different arms of government need to be brought into play, depending on which values are to be preserved. It works the other way as well. How one chooses to approach the problem might quietly favor some values and interests, while pushing aside others. And therein lies the political rub of the problem.

At stake are competing visions of coastal settlements—living behind seawalls and dikes in a modern incarnation of medieval towns, letting the water conquer part of the coast, either by retreating or, more daringly, by choosing to live on water, or, most likely, some combination of these extremes. However, regardless of which of these "futures" might ultimately prevail, a coastal management approach that is driven primarily by present-day conflicts around coastal properties is unwise. This is because rising coastal property prices and economic and population growth almost inevitably create political dynamics which, unless countered, make the "stay-and-protect" option the only viable one no matter how costly and irrational it might be in the long run. This is tantamount to walking half-blindly into the future, head down, far too focused on the present.

How far into the future should our planning extend, given the long-term nature of climate change? Our late-modern scientific rationality is accustomed to working within time frames of years and decades, up to half a century at most. The machines and artifacts we make typically last from a few years—stationery, clothes, computers—to a decade or two—white goods, furniture, motor vehicles. Those who design our buildings, public parks, and cities do consider the future contexts in which their creations will function, but seldom more than 25 to 50 years ahead. Even our identification with our progeny is time-limited, and most of us would find it difficult to empathize with our descendants beyond grandchildren.

And yet, if we dare to look to 2100 and beyond, into the twenty-second and twenty-third centuries, sea level rise becomes even more troubling. The oceans have a high capacity for storing carbon dioxide and heat, hence acting as a dampener on global warming. In other words, were it not for the oceans, we would have experienced far more warming, far earlier. The corollary to this, however, is that the thermal expansion of oceans is likely to go well into this millennium, possibly even the next. Should polar ice sheets melt—and there remains significant uncertainty as to whether or not we have already crossed the point of no return on this—the magnitude and extent in time of sea level rise would increase further. Projections of more than 5 m rise by 2300 have appeared for the first time in the latest IPCC report, a level that would no doubt transform our cities well beyond recognition.

The most potent argument against attempting to plan for 2100 and beyond lies in the large uncertainties inherent in not just climate projections, but the technological, social, and institutional drivers that shape our cities. Why bother, one could ask, when we know so little about 2100, let alone the twenty-second or twenty-third century?

There are three possible answers to the question—complementary, rather than mutually exclusive. First, it is possible to reverse the question and ask how many of the chronic dysfunctionalities found in our cities today have been at least partly caused by decisions made, say 70 or more years ago. Think of today's blighted transport networks and traffic jams around the world, from Sao Paolo and Los Angeles to Tehran, Cairo, and Beijing. Arguably, the roots of our current transport problems can be found in specific policies favoring car ownership and highway building on a large scale, which started in the early and mid-twentieth century in the United States and then spread to the rest of the world. This is a reminder that, while our planning imagination may not extend beyond 50 years, the impacts of our planning decisions most certainly do.

Second, the uncertain nature of climate change inevitably draws attention to the institutions managing our cities and the extent to which they are able to cope with the sudden shocks that the climate system may bring to them. A group of researchers from the United Kingdom[12] conducted one-on-one interviews and a one-day policy exercise among decision makers responsible for managing the Thames Estuary, including London, in order to study a scenario under which the Western Arctic shelf begins to collapse in 2030, leading to a sea level rise of 5 m over the following 100 years. They found that, given the significant disagreements among decision makers, the most likely response was policy paralysis, at least for a while. In other words, attempting to understand the implications for our cities of higher sea level rise is important not just in spite of, but because of the uncertainty.

Finally, and perhaps most importantly, exercises investigating strategic coastal-planning directions for our cities[13] can be useful in developing a clearer vision of the kinds of futures available to us, how desirable they might be, and what is it that might prevent us from achieving them. In the words of Karen O'Brien, a scholar of climate change adaptation at the University of Oslo, "[deliberate transformations are] not about social engineering or 'designing' the future, but rather about recognizing that some fundamental shifts are necessary to enable desirable futures to emerge."[14]

The British who invaded and settled the Australian continent in the eighteenth and nineteenth centuries, famously came up with the *Terra Nullius* fiction to justify land seizure from traditional indigenous owners of the land. This is the legal doctrine according to which, since Aboriginal Australians had no written code of property allocation, the land did not belong to anyone and the European settlers were free to seize it. There is a cruel irony, a reverse *Terra Nullius* as it were, in the way we, descendants of white settlers and/ or beneficiaries of their conquests, are now having to deal with another threat coming at us from the ocean. This represents a threat that, much like white Europeans 250 years ago, is entirely blind to our way of living, undermines our sense of ownership of the land, and insists that we change the way we do things in order to accommodate it. What is especially cruel about the irony is that today's indigenous Australians living close to the coast will suffer the consequences of sea level rise, just as much as the rest of the Australian population—if not more, given persisting disadvantage.

The analogy between white settlement and sea level rise has its limits of course, not least in that the price indigenous Australians paid for white settlement is of a much higher order, no matter how dire climate change projections might be. However, it is not beyond a stretch of imagination that a historian, in 300 years from now, looking back at the way sea level rise will have transformed our coastline might see it as a driver of change that is comparable to white settlement or post-World War II economic and population boom. Conflicts such as those described previously—about development and assets along the coastlines of Illawarra, Shoalhaven, and Eurobodalla—are likely just the first stirrings of a problem that will grow and stay with us for a long time.

Aboriginal Australians who live near the sea refer to their coastal living environment as *Sea Country*. What better way of highlighting the intimate connections between two parts of our world whose strict separation we are now called upon to reexamine? Some have even suggested that *Sea Country* may be a way in which land inundated in the past, and now under water, is reclaimed, culturally and linguistically, by indigenous Australians. But there is another Aboriginal word that may speak to us even more. *Ganma*—in the languages of the Yolngu indigenous Australian people of North-Eastern Arnhem land—denotes "the meeting and mixing of [sea] water … and fresh water from the land"[15] but also means "two-way learning." Whatever happens to our coasts and whichever way we end up adapting to change, there is a chance that, in a century from now, this word will have found its way into Australian English.

Arguably, this is the first time in our history that we have had to deal with a problem of this nature and of such magnitude. Whether our institutions can rise to the challenge or not, and what kind of transformation they must undergo over the next few decades in order to be able to tackle the problem effectively, is one of the big questions of adaptation (though seldom explicitly stated). On the other hand, it is important to keep reminding ourselves—amid all the gloomy climate projections and all the talk about vulnerability, inaction, and climate skepticism—that it's *not* all bad news. Indeed, some aspects of the problem are tilted in our favor and three in particular spring to mind.

First, although it would have been undoubtedly better if we had acted earlier to prevent climate change or adapt to it, it is still likely the case that, thanks to climate science, no threat in human history has been foreseen so far ahead in time. The ancient Mayans and Egyptians, some of whose civilizations had to endure climatic havoc and likely died out as a result, would no doubt be envious of our ability to anticipate threats whose symptoms and causes are only discernible through sophisticated science and technology. Will the story of climate change be one in which the worst is averted and foresight is converted into sensible actions? Or will it be a chronicle of a death foretold? We don't know the answers yet, but this is where the other two positive aspects come into the picture.

Unlike riverine and coastal communities in other parts of the world, such as the Nile Delta, Bangladesh, Cambodia, and small-island states, we are not fighting for basic survival. Fifteen of the 20 megacities of the world[16] are highly exposed to the effects of sea level rise and most of these are in the Global South where high levels of poverty persist and many live in slum areas. The kind of changes that might be required of us over the next 100 years or so will no doubt be painful and difficult. However, we are not under pressure to abandon our country—as citizens of the Maldives and Marshall Islands are—nor will we have to cope with hunger, disease, death, and injury on a large scale. At least not in the twenty-first century.

More importantly, we are part of the bigger European history of industrialization of the last 300 years and as such, we have at least reaped some benefits from the fossil fuels that have been silently cooking the planet's atmosphere and warming its oceans and landmass. One of these benefits is the creation, over the relatively short history of this continent's European colonization, of governance institutions, vigorous civil associations, strong scientific research organizations, and so-called "adaptive capacity" that may yet allow us to reconfigure the way we occupy the land so as to coexist in some harmony with higher sea levels. We cannot be certain that our institutions will rise to the challenge, but at least we know that they probably have what it takes.

This is not self-congratulation or a rehash of the "lucky-country" trope: luck has little to do with it, and indigenous Australians have paid a horrendous price for it. It is, among other things, a reiteration of the most glaring injustice at the heart of climate change—that those living in affluent countries and who, historically, have contributed most to the greenhouse gas problem, will likely suffer far less from its consequences than those living in the Global South, who have contributed least.

The Paris Agreement on Climate Change[17]—signed in December 2015 and entered into force just under a year later—brought about a wave of optimism, despite its rather modest ambition. However, as these hopes begin to fade now that a climate change denier has been elected as President of the United States and has pulled his country out of the Paris Agreement, it appears, more than ever, that the more extreme predictions of sea level rise may need to be taken more seriously.

The problems created by sea level rise are undoubtedly "wicked,"[18] as system scientists might say, that is, they are complex, dogged with uncertainty and unpredictability, subject to fundamental disagreements about their nature and causes, involving competing and often irreconcilable interests, and are without obviously optimal solutions. In addressing them, we need all the creative energy and resources we can channel into them.

Although many individuals and organizations are attempting to tackle the problem, judged against the scale of the coastal changes awaiting us, our collective efforts remain paltry, nationally and globally. And, given how high the stakes are for our children, the earlier we rise to the challenge, the better.

Acknowledgment

Several students of mine have contributed, over the years, to the research underpinning this article, including Dr Fahim Tonmoy, Mr Tanvir Ahmad, Mr Matt Brown and Ms Peta Papadoropoulos. I am grateful to the Shoalhaven City and Inner West Councils for their collaboration on parts of this research.

Endnotes

1. http://d3n8a8pro7vhmx.cloudfront.net/edonsw/pages/466/attachments/original/1387246030/sandon_point_casenote.pdf?1387246030.

2. http://www.lec.justice.nsw.gov.au/Documents/preston_influence%20of%20climate%20change%20litigation.pdf.

3. http://www.cmar.csiro.au/sealevel/downloads/SLR_PA.pdf.

4. http://www.abc.net.au/news/2016-06-04/sydney-and-nsw-coast-set-for-wild-weather-all-weekend/7476948.

5. Fahim Tonmoy and Abbas El-Zein. 2018. Vulnerability to Sea Level Rise: A Novel Local-Scale Assessment Methodology and Application to Eight Beaches in Shoalhaven, Australia. *Ecological Indicators*, 85:295–307.

6. Jochen Hinkel, Daniel Lincke, Athanasios T. Vafeidis et al. 2016. Coastal flood damage and adaptation costs under 21st century sea-level rise. *PNAS*, 111(9):3292–3297.

7. https://www.edonsw.org.au/planning_reforms.

8. https://www.parliament.nsw.gov.au/bills/Pages/bill-details.aspx?pk=3291.

9. https://www.ipcc.ch/report/ar5/.

10. https://www.batemansbaypost.com.au/story/2721187/worst-case-benchmark-recommended/.

11. https://www.batemansbaypost.com.au/story/3471945/sea-level-motion-back-on-agenda-for-eurobodalla/.

12. Richards S. J. Tol, Maria Bohn, Thomas E. Downing. 2006. Adaptation to Five Metres of Sea Level Rise. *Journal of Risk Research* 9(5):467–482.

13. https://www.ice.org.uk/getattachment/media-and-policy/policy/facing-up-to-rising-sea-levels/Facing-Up-to-Rising-Sea-Levels-Document-Final.pdf.aspx.

14. Karen O'Brien. 2012. Global Environmental Change II: from adaptation to deliberate transformation. *Progress in Human Geography* 36(5): 667–676.

15. http://livingknowledge.anu.edu.au/learningsites/seacountry/11_gapu.htm.

16. http://siteresources.worldbank.org/INTUWM/Resources/340232-1205330656272/CitiesandClimateChange.pdf.

17. https://unfccc.int/process/the-paris-agreement/what-is-the-paris-agreement.

18. http://www.enablingchange.com.au/wickedproblems.pdf.

3

(Re)Think (Re)Design for Resilience

Nina-Marie Lister

CONTENTS

The summer of 2017 was one of the worst on record for flooding anywhere. From North America to Europe to South East Asia, record flood events were posted in coastal and inland areas alike. Torrential rainstorms, record-setting hurricanes, and surging Great Lakes levels combined to raise insurance premiums, public awareness, and political promises for flood protection and stormwater strategies. At a time of unprecedented global urbanization, both in scale and population, urban and urbanizing regions suffer disproportionate effects of flooding: in hard-surfaced cities, with more built form than vegetation, from rainfall to snowmelt, urban stormwater declines rapidly in quality and often has nowhere to flow but overland, bringing urban contaminants into swelling lakes, rising rivers, and subsiding coasts. In the transition to an urban world, the Anthropocene has ushered in a triple-threat: it's becoming clear that the profound social, technological, and built-form changes of urbanization are being exacerbated, tied inexorably through positive feedback to large-scale ecosystem and climate change. In short, the urban world is in urgent need of new modes of planning and design for resilience.

In the face of a changing climate, increasing vulnerability to extreme weather takes many forms. On December 21, 2013, for example, the city of Toronto and its metropolitan area of 5,000,000 inhabitants—along with a sizeable portion of southern Ontario and northern New York—experienced an unseasonably warm winter storm. The storm dropped more than 30 millimeters of freezing rain on the city. Temperatures hovered around freezing for almost 36 hours and then rapidly plummeted to −25°C and stayed there, locking the city under a blanket of ice for almost two weeks and leaving more than half a million residents in the frozen dark following the Winter Solstice. Under the weight of the ice, more than 20% of the city's 10,000,000 trees were felled, bringing down power lines and cables in the process and leaving thousands of homes without power, heat, or light through Christmas and the holiday season. With an estimated cost of CDN$106,000,000 to the city of Toronto alone in clean-up and emergency services, the eastern North American ice storm of 2013 is recorded as one of the worst natural disasters in Canadian history.[1] Yet, notably, this figure does not account for the loss of the green infrastructural value and the attendant ecosystem services of the loss of one-fifth of the city's mature tree canopy. The city will continue to suffer long-term related impacts of the ice storm through increased soil erosion, decreased flood protection, carbon sequestration, urban heat mitigation, and so on.

The ice storm, however, was not an isolated incident. In February 1998, a similar ice storm caused a massive power outage throughout Québec that lasted more than two weeks, affecting more than 50,000 homes in the middle of a deep freeze. The Red River floods of 1998 and 2012 crippled the cities of Winnipeg, Minneapolis, and St. Paul, while Alberta's Bow River flood of 2012 virtually shut down the city of Calgary and the Trans-Canada highway for more than a month. These are but a few of many recent, locally catastrophic storm events. The better-known "monster storms," such as Hurricane Katrina which devastated New Orleans in 2005, and Superstorm Sandy in 2011, which left half of midtown Manhattan without power for more than a week, are globally significant events. By virtue of their reach and effect in major urban centers, these storms have catalyzed a new wave of research into urban environmental planning, coastal defense, urban vulnerability, and related policy responses that link urbanism, planning, and ecology.

In addition to the economic, social, and environmental costs of such storms, there is growing recognition that these events pose significant challenges to the world's urbanizing areas and their largely outdated systems of governance and planning. Cities across the globe are facing the reality that the increasing magnitude and frequency of major storm events are evidence of human-induced global climate change, and with this reality has come a range of increasing challenges to our systems of survival, including a need for new design approaches to cope with ecological change and vulnerability.[2] Identified as a global threat by the *Intergovernmental Panel on Climate Change* (IPCC) and grounded in a wide range of policy-related research linked to long-term sustainability, climate change is now an accepted phenomenon for which adaptation strategies must be developed and implemented from municipal to national scales.[3]

Long-term environmental sustainability demands the capacity for resilience—the ability to recover from a disturbance, to accommodate change, and to function in a state of health. In this sense, sustainability refers to the inherent and dynamic balance between social-cultural, economic, and ecological domains of human behavior that is necessary for humankind's long-term surviving and thriving. Ann Dale has described this dynamic balance as a necessary act of reconciliation between personal, economic, and ecological imperatives that underlie the primordial natural and cultural capitals on Earth.[4] With this departure from conventional "sustainable development," Dale has set the responsibility for long-term sustainability squarely in the domain of human activity and appropriately removed it from the ultimately impossible realm of managing "the environment" as an object separate from human action.

A growing response to the increasing prevalence of major storm events has been the development of political rhetoric around the need for long-term sustainability and, specifically, resilience in the face of vulnerability. As a heuristic concept, resilience refers to the ability of an ecosystem to withstand and absorb change to prevailing environmental conditions. In an empirical sense, resilience is the amount of change or disruption an ecosystem can absorb, by which, following these change-inducing events, there is a return to a recognizable steady state in which the system retains most of its structures, functions, and feedbacks.[5] In both contexts, resilience is a well-established concept in ecological systems research, with a robust literature related to resource management, governance, and strategic planning. Yet, despite more than two decades of this research, the development of policy strategies and planning applications related to resilience is relatively recent. While there was a significant political call for resilience planning following Superstorm Sandy in 2011 and the ice storm of 2013, there remains a widespread lack of coordinated governance, established benchmarks, implemented policy applications, and few (if any) empirical measures of success related to climate change adaptation.[6] In this context, there has been little critical analysis of and reflection on the need to understand, unpack, and cultivate resilience beyond the rhetoric. In this essay, I argue that concomitant with the language of resilience is the need to develop nuanced, contextual, and critical analyses coupled with a scientific, evidence-based understanding of resilience; that is, we need an evidence-based approach that contributes to adaptive and ecologically responsive design in the face of complexity, uncertainty, and vulnerability. Put simply: What does a resilient world *look* like, how does it *behave*, and how do we plan and design for resilience in an urban world?

Why Resilience? Why Now?

The emergence of resilience as a rhetorical idea is tied not only to the emerging reality of climate change, but to an important and growing synergy between research and policy responses in the fields of ecology, landscape architecture, and urbanism—a synergy that is powerfully influenced by several remarkable and coincidental shifts since the turn of the second millennium. Most notable is the global population shift, in which our contemporary patterns of settlement are tending toward large-scale urbanization, a hallmark of the Anthropocene. The last century has been characterized by mass migration to ever-larger urban regions, resulting in the rise of the "megacity" and its attendant forms of suburbia, exurbia, and associated phenomena of the modern metropolitan landscape.[7] For most of the world's population, the city is fast becoming the singular landscape experience.[8]

In North America, and the United States in particular, this shift in urbanism has come, paradoxically, with a widespread decline in the quality and performance of the physical infrastructure of the city. The roads, bridges, tunnels, and sewers that were built during the late nineteenth and early twentieth centuries to service major urban centers are now aging and crumbling, in some cases, while both the political will and the public funds to rebuild this outdated but essential public infrastructure are disappearing. More significantly, these infrastructures continue to decay, and they are increasingly vulnerable to catastrophic failure in the face of more frequent and severe storm events, thus compounding the cost of their loss and the extent of impact (Figure 3.1).

The emergence of a new direction and emphasis in ecology represents another significant and concomitant shift with the change in urbanism and the reality of climate change. During the last few decades, the field of ecology has moved from a classical, reductionist concern with stability, certainty, predictability, and order, in favor of more contemporary understandings of dynamic, systemic change and the related phenomena of uncertainty, adaptability, and resilience. Increasingly, these concepts in ecological theory and complex systems research are found useful as heuristics for decision making generally and, with empirical evidence, for landscape design in particular.[9] This offers a powerful new disciplinary and practical space; one that is informed by ecological knowledge both as an applied science and as a construct for managing change within the context of sustainability. As a practice of planning *for and with* change, resilience is, in itself, a conceptual model for design.[10]

With this new ecological approach has come another important shift in creating the synergy necessary for resilience-thinking: the renaissance of landscape as both a discipline and praxis throughout the last two decades and its (re)integration with planning and architecture in both academic and applied professional domains. Landscape scholars have identified the rise of post-industrial urban landscapes coupled with a focus on indeterminacy and ecological processes as catalysts for the reemergence in landscape

FIGURE 3.1
Four views of a washed-out section of a major arterial roadway in Toronto after heavy rain and flooding of the Don River followed Hurricane Katrina, which was downgraded to a tropical storm when it hit Toronto on August 29, 2005. (Photo collage by Carmela Liggio and Nina-Marie Lister, 2005.)

theory and praxis.[11] Understood today as an interdisciplinary field, linking art, design, and the material science of ecology, landscape scholarship, and application now includes a renewed professional field of practice within the space of the city.[12]

Considered together in the era of climate change and vulnerability, these shifts in our collective understanding of urbanism, landscape, and ecology have created a powerful synergy for new approaches in planning and design to the contemporary metropolitan region. This synergy has been an important catalyst for the emergence of resilience as a rhetorical idea, but much work remains to be done to move toward evidence-based implementation of strategies, plans, and designs for resilience. The scale and impact of North American megastorms, such as Hurricane Katrina in 2005 and Superstorm Sandy in 2011 have been effective triggers for a new breed of policy and planning, initiatives in disaster preparedness in general, and flood management plans in particular. Conventional policy and planning approaches to natural disasters have long been rooted in the language of *resistance* and *control*, referencing coastal defense strategies, such as fortification, armoring, and "shoring up" by using brute-force engineering responses designed to do battle with natural forces.[13] By contrast, emerging approaches in design and planning reference the language of *resilience* and *adaptive management*, terms associated with elasticity and flexibility, leading to the use of hybrid engineering of constructed and ecological materials that adapt to dynamic conditions and natural forces.[14] Recent coastal management policies and flood management plans following the major storm events abound in this language of resilience, including New Orleans's *Water Management Strategy*, Louisiana's *Coastal Management Plan*, New York City's *Rebuild by Design* program, and Toronto's *Wet Weather Flow Master Plan*. These examples are notable as responses (reactive and proactive) to catalytic storm events and climate change, yet they remain, for the most part, speculative, untested, and unimplemented, relying on a language of resilience that is heuristic and conceptual rather than experiential, contextual, or scientifically derived.

The general concept of resilience has origins across at least four disciplines of research and application: psychology, disaster relief and military defense, engineering, and ecology. A scan of resilience policies reveals that the concept is widely and generally defined with reference to several of the original fields and is universally focused on the psychological trait of being flexible and adaptable. Examples are having the capacity to deal with pressure and the ability to "bounce back" to a known normal condition following periods of stress, maintaining well-being under stress, and being adaptable when faced with change or challenges.[15] The use of resilience in this generalized context, however, begs important operational questions of how much change is tolerable, which state of "normal" is desirable and achievable, and under what conditions is it possible to return to a known "normal" state. In policies that hinge on these broadly defined, psychosocial aspects of resilience, there is little or no explicit recognition that adaptation and flexibility may result in transformation and, thus, require the *transformative capacity* that is ultimately necessary at some scale in the face of radical, large-scale, and sudden systemic change. Using sea level as an example, if we accept that waters naturally rise and fall within a range of seasonal norms, we might be better off to embrace a gradient of acceptable "normal" conditions rather than a single static, and ultimately brittle, state that is unsustainable (Figure 3.2). A more critical and robust systems-oriented discussion of resilience will force all concerned to confront a difficult but

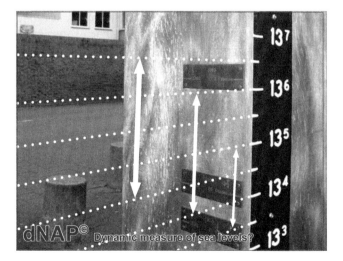

FIGURE 3.2

The Normaal Amsterdams Peil (NAP) is a measure used to gauge the rise in sea level and to establish national policies, laws, and regulations on the basis of a fixed, "normal" water level. In contrast, the Dynamic Normaal Amsterdams Peil or d(NAP), shown here, is a proposed measure of sea level for the Netherlands Delta Region that acknowledges dynamic water levels to address better changing hydrological regimes; for example, to reflect seasonal flooding. (Diagram courtesy of Kimberly Garza and Sarah Thomas, 2010.)

essential question: How much can a person, a community, or an ecosystem change before it becomes something unrecognizable and functions as an altogether different entity?[16] If resilience is to be a useful concept in application and, in particular, to inform design and planning strategies, it must ultimately instruct us *how* to change safely rather than how to resist change completely. Current policies and eventual design strategies will risk the potential power of resilience by emphasizing a misguided focus on "bouncing back" to a normal state that is, ultimately, impossible to sustain.

Unpacking Resilience

Before one can implement applied strategies and associated indicators for resilience in design and planning, it is useful and, arguably, necessary to unpack the history, theory, and conceptual development of resilience as it emerged in ecology. We can do so critically with reference to a well-established social-scientific literature derived principally from ecosystem ecology and, in particular, with research applications in natural resource management. Decades of research related to complex systems ecology, and thinking about and practice of social-ecological systems offers both broad heuristic and empirical contexts for the study and application of resilience. As such, both the construct and measures of resilience are important to embed, apply, and test within policies and designs for long-term sustainability. As an essential capacity for sustainability, applications of resilience are derived from research in complex systems ecology, first published by the American ecologist Howard T. Odum and later developed by the Canadian ecologist, Crawford Stanley ("Buzz") Holling.[17] Yet it should be noted that the foundations of resilience thinking were laid much earlier. Well before the language of complex systems was embraced within ecological science, the early twentieth century conservation movement was already concerned with the health of natural systems, which was conceptualized variously, from self-renewal to healing and balance, with implications for management practices. For example, Aldo Leopold used the concept of "land health" to refer to the land's capacity for self-renewal—essentially resilience—which he saw as threatened by and at odds with unchecked exploitation of land and resources for economic growth.[18] Similarly, Gifford Pinchot's perspectives on the need for cautious resource extraction, however utilitarian, gave rise to an early version of adaptive management to accommodate changes in nature and the landscape.[19] By the 1960s, with the birth of modern environmentalism, there were more urgent calls for caution. Notable among these was Rachel Carson's characterization of nature as resilient, changeable, and unpredictable: "… the fabric of life … on the one hand delicate and destructible, on the other miraculously tough and resilient, capable of striking back in unexpected ways."[20]

The late 1970s and early 1980s marked the beginning of a significant theoretical shift in the evolving discipline of ecology. In general, ecological research at all scales has moved toward a more organic model of open-endedness, indeterminacy, flexibility, adaptation, and resilience and away from a deterministic and predictive model of stability and control, based on engineering models for closed (usually mechanical) systems. Ecosystems are now understood to be open, self-organizing systems that are inherently diverse and complex and behave in ways that are, to some extent, unpredictable.

This shift, influenced by the early ecosystem analyses of the Odum brothers (Eugene P. and Howard T.), followed the rise in complexity science and the groundbreaking work of Ilya Prigogine, Ludwig von Bertalanffy, C. West Churchman, Peter Checkland, and other systems scholars throughout the latter half of the twentieth century. Ecological research came into its own discipline, distinct from biology and zoology, by focusing on large-scale and cross-scale (connected) functions and processes of an ecosystem. As an outgrowth of research in complex systems coupled with the emerging new discipline of landscape ecology and associated spatial analyses—made possible by new tools, such as high-resolution satellite imagery— ecosystem ecology led to multi-scaled, cross-disciplinary, and integrated approaches in land use planning. Beginning in the 1970s with F. Herbert Bormann's and Gene Likens' first ecosystem-based study of the Hubbard Brook watershed, long-term ecological research programs (known as LTERPs) became established, influencing, throughout the 1980s and 1990s, a growing recognition of the dynamic processes inherent and essential to living, layered landscapes, and the understanding of ecosystems as open, complex systems within which structure and function are interrelated and scale-dependent.[21]

The dynamic ecosystem model has been an important development in ecology and a significant departure from the conventional, linear model of ecosystems that dominated scholarly twentieth-century thought. Resilience is an important concept that emerged from this development. Defined by the process of ecological succession, the linear model held that ecosystems gradually and steadily succeed into stable climax states from which they will not routinely move unless disturbed by a force external to that system.[22] An old-growth forest is the typical example, in which a forest matures and then remains in that state permanently such that any disturbance from that state is considered an aberration. Yet we now know that not only is change built into these systems, but, in some cases, ecosystems are dependent on change for growth and renewal. For example, fire-dependent forests contain tree species that require the extreme heat of fire to release and disperse seeds and to facilitate a forest's renewal and, sometimes, a shift in the complement of a species following a major fire. The dynamic ecosystem model, based on long-term research in a variety of global contexts, asserts that all ecosystems go through recurring cycles with four common phases: rapid growth, conservation, release, and reorganization. Known as the adaptive cycle, or the Holling Figure Eight, this generalized pattern is a useful conceptual description of how ecosystems organize themselves over time and respond to change.[23] The adaptive cycle of every ecosystem is different and contextual; how each system behaves from one phase to the next depends on the scale, context, internal connections, flexibility, and resilience of that system (Figure 3.3).

Ecosystems are constantly evolving, often in ways that are discontinuous and uneven, with slow and fast changes at small and large scales. While some ecosystem states appear to be stable, stability is not equated in a mathematical sense but rather in a human-scale or time-limited perception of stasis. C. S. Holling pioneered this concept in application to resource management, in which he described ecosystems as "shifting steady-state mosaics," implying that stability is patchy and scale-dependent and is neither a constant nor a phenomenon that defines a whole system at any one point in time or space.[24] The key point is that ecosystems operate at many scales, some of which are loosely and others tightly connected, but all subject to change at different rates and under different conditions. An ecosystem we perceive as stable in a human lifetime may, at a longer scale, be ephemeral, and this realization has profound implications for how we choose to manage, plan, or design for that system (Figure 3.4).

There is an important connection between stability, change, and resilience—a property internal to any living system and a function of the unique adaptive cycle of that system. Resilience has both heuristic and empirical dimensions, arising from its origins in psychology, ecology, and engineering. As a heuristic or guiding concept, resilience refers to the *ability* of an ecosystem to withstand and absorb change to prevailing environmental conditions and, following these change events, to return to a recognizable steady state (or

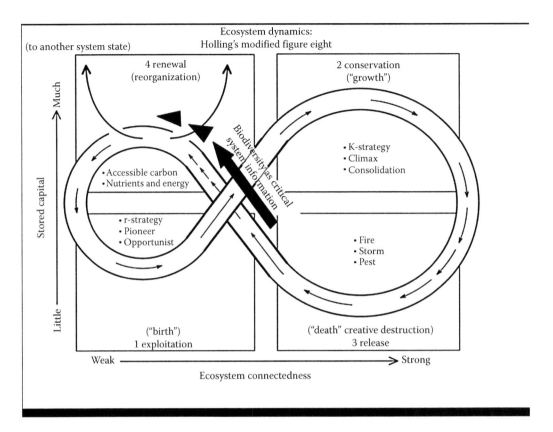

FIGURE 3.3
Ecosystem Dynamics and the Adaptive Cycle: Holling's Modified Figure Eight. Ecologist C. S. Holling's dynamic cycle of ecosystem development is the foundation of a complex systems perspective in ecology. (Diagram courtesy of Waltner-Toews, David, James J. Kay, and Nina-Marie E. Lister, eds. 2008. *The Ecosystem Approach: Complexity, Uncertainty, and Managing for Sustainability.* New York, NY: Columbia University Press. 97; modified from Holling, C. S. 2001. "Understanding the Complexity of Economic, Ecological, and Social Systems." *Ecosystems* 4(5): (August 2001): 390–405.)

a routinely cyclic set of states) in which the system retains most of its structures, functions, and feedbacks. As an empirical construct in engineering, resilience is the *rate* at which an ecosystem (usually at a small scale, with known variables) returns to a known and recognizable state, including its structures and functions, following change events. Such events, considered disturbances—which C. S. Holling strategically referred to in the vernacular as "surprises"—are usually part of normal ecosystem dynamics, yet they are also unpredictable, in that they cause sudden disruption to a system.[25] These can include, for example, forest fires, floods, pest outbreaks, and seasonal storm events.

The ability of a system to withstand sudden change at one scale assumes that the behavior of the system remains within a stable regime that contains this steady state in the first place. However, when an ecosystem suddenly shifts from one stable regime to another (in the reorganization phase, via a flip between system states or what is called a "regime shift"), a more specific assessment of ecosystem dynamics is needed. In this context, *ecological resilience* is a measure of the *amount of change* or disruption that is required to move a system from one state to another and, thus, to a different state of being maintained by a different set of functions and structures than the former (Figures 3.5 through 3.7).[26] Each of

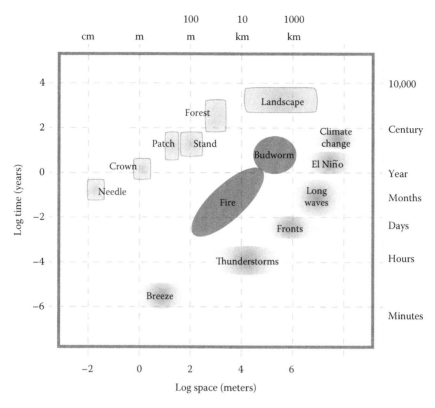

FIGURE 3.4
Ecosystem dynamics are observed here across multiple scales of time and space. (Redrawn by Marta Brocki and adapted from Holling, C. S. 2001. "Understanding the Complexity of Economic, Ecological, and Social Systems." *Ecosystems* 4(5): (August 2001): 390–405 [393].)

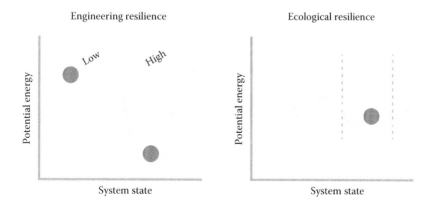

FIGURE 3.5
Shown here are two contrasting perspectives on resilience: (left) Engineering Resilience in closed systems (limited uncertainty and known variables) versus (right) Ecological Resilience in open systems (inherent uncertainty and infinite variables). (Redrawn by Nina-Marie Lister and Marta Brocki and adapted from Holling, C. S. 1996. "Engineering Resilience versus Ecological Resilience." In Schulze, P. C., ed. *Engineering within Ecological Constraints.* 31–44 [35]. Washington, DC: National Academy Press.)

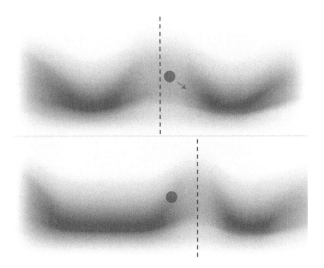

FIGURE 3.6
Resilience, seen here as a function of social-ecological system conditions, is described metaphorically as a ball in a changing basin. The basin represents a set of states that share similar functions, structures, and feedbacks. Though the location of the ball remains the same, changes in the surrounding conditions bring about a shift in state. (Redrawn by Marta Brocki and adapted from Walker, Brian, C. S. Holling, Stephen R. Carpenter, and Ann Kinzig. 2004. "Resilience, Adaptability and Transformability in Social–ecological Systems." *Ecology and Society* 9(2): (December 2004): 4. http://www.ecologyandsociety.org/vol9/iss2/art5.)

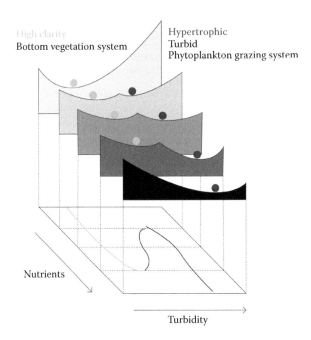

FIGURE 3.7
In this early schematic of a complex systems perspective in ecology, we visualize multiple states—all possible—in a freshwater ecosystem. (Courtesy of James J. Kay, as sketched in lectures from a course, "Systems Design Engineering," at the University of Waterloo, 1994, in which the author was a student. Redrawn by Marta Brocki and adapted from Kay, James J., and Eric Schneider. 1994. "Embracing Complexity: The Challenge of the Ecosystem Approach." *Alternatives Journal* 20(3):(July 1994): 32.)

these nuanced aspects of resilience is important. They underscore the social-cultural and economic challenges inherent in defining what "normal" conditions are and, in turn, how much change is acceptable at what scale.

It becomes critical to understand the ecological systems in which we live, and, given their inherent uncertainty, we ought to do so through a combination of ways of knowing: experiential, observational, and empirical. Indeed, if there are multiple possible states for any ecosystem, there can be no single "correct" state—only those we choose to encourage or discourage. Notably, these are not questions of science but of social, cultural, economic, and political dimensions—they are also questions of design and planning. The trajectory of research in resilience has been instrumental in exploring the paradoxes inherent within living systems—the tensions between stability and perturbation, constancy and change, predictability and unpredictability—and the implications of these for management, planning, and design of the land. Resilience, in short, as Brian Walker declares, "is largely about learning *how* to change in order not to *be* changed."[27]

From Rhetoric to Tactic: Toward Resilient Design

More recently, applied ecology has been focused on trying to understand what the ecosystem states are that we perceive to be stable, at what scales they operate, and how they are useful to us. It is important to recognize that stability can be positive or negative, just as change is neither universally good nor bad. Thus, while designers want to encourage a desirable stability (such as access to affordable food or a state of health for a majority of citizens), they also wish to avoid pathological stability (such as chronic unemployment, a state of war, or a dictatorship). This approach has significant implications for management, planning, and design, as it rests on the recognition that humans are not outsiders to any ecosystem but, rather, participants in its unfolding and agents of its design.

In this context, the sub-science of urban ecology developed during the 1990s has created a new niche for resilience.[28] Related practices of urban design, environmental planning, and landscape architecture have cross-pollinated in the service of design and planning for healthier cities within which connected vestiges of natural landscapes might thrive. The work of environmental scholars (such as William Cronin, Carolyn Merchant, and David Orr), together with the practice of landscape architects (such as Anne Whiston Spirn, Frederick R. Steiner, and James Corner) effectively brought nature into the embrace of the city, challenging the hierarchical dualism of humans versus nature.[29] The once-discrete concepts of "city" and "country" grew tangled and hybridized and the boundaries between the urban and the wild blurred. This blurring of boundaries, coupled with the contemporary ecological paradigm of nature as a complex, dynamic open system in which diversity is essential and uncertainty the norm, represented a significant break from ecological determinism and its slavish pursuit of perpetual stability underpinned by the illusion of the balance of nature.[30] The increasing hybridization of cultural and natural ecologies has created a powerful aperture for the development of resilience in thought and practice—and with it a new realm for design developed formatively through the interdisciplinary study of social-ecological systems science, in which coupled systems of humans *within* nature are the norm.[31]

What does design for resilience look like? What tactics do urban planners and designers need to engage in for attaining resilience? To activate such a model for design, one can summarize key principles of adaptive complex systems, generally, and of resilience, specifically.[32] First, change can be slow and fast, at multiple scales. This means that it is essential to look beyond one scale in both space and time and to use various tools to understand the ecological system. Slow variables are arguably more important to understand than fast ones, as they provide necessary stability from which to study change at a safe distance. Yet there can be no universal point of access or ideal vantage point. Mapping, describing, and analyzing the system from multiple perspectives, using different ways of knowing and with a diversity of tools, is critical. If uncertainty is irreducible and predictability is limited, then the role of the traditional expert is also limited—and the role of designer is more akin to a facilitator or curator.

Second, some connectedness or modularity across scales is important, and feedback loops should be both tight and loose. Resilient systems are not so tightly coupled that they can't survive a shock throughout the system that moves rapidly and destructively. For example, children need some limited exposure to viruses to develop immunities but at not too large a scale of impact so as to endanger long-term health. In the same way, design strategies for resilience must consider novelty and redundancy in terms of structures and functions. A useful example is a trail system in a park, which is somewhat connected using a hierarchy of paths that is legible and efficient and yet not so tightly connected that it compromises habitat, folds in on itself, or prohibits spontaneous exploration.

Third, even as there are multiple states in which an ecosystem can function, there is no single correct state. It is important to determine where, in the adaptive cycle, the system of interest is, such that decision makers and designers can learn patterns and anticipate change (if not predict it). Eventually, perceived stability in any phase will end, and the system will move to a new phase in its adaptive cycle. A non-linear approach to design that encompasses oscillating or changing states within various phases of a system's development will help facilitate change. For example, it may be desirable to design for seasonally flooded landscapes or along a gradient of water that changes rapidly in a short period of time.

Finally, resilient systems are defined by diversity and by inherent but irreducible uncertainty. Successful strategies for resilient design should use a diversity of tactics through *in situ* experimental and ecologically responsive approaches that are safe-to-fail, while avoiding those erroneously assumed to be fail-safe.[33] This distinction is important, for conventional engineering relies on prediction and certainty to assume an idealized condition of fail-safe design. Yet this is impossible under dynamic conditions of ecological and social complexity in which predictability is limited at best to one scale of focus. Even knowing one scale exhaustively and managing for it specifically and exclusively may compromise a system's overall function and resilience. The reductionist caveat of "scaling up," using knowledge gained at one scale and applying it to the whole system, cannot work in complex systems in which scales are nested. Design strategies that support and facilitate resilience should, for example, model its attributes, using living infrastructures that mimic ecological structures and their functions, and to design them to be tested and monitored, from which learning and adaptation to changing conditions are built into the design. When design experiments fail, they should fail safely, at a scale small enough not to compromise long-term health.

These and other emerging approaches to design for resilience tend to reflect the characteristics of the theoretical paradigm shifts that have laid its foundation. They are often interdisciplinary, integrating architecture, engineering, and ecology, specifically, and art and science, broadly. They cross-pollinate freely across scales and hybridize in surprisingly novel ways.[34] The growing use of living "blue" and "green" infrastructures[35] to soften seawalls, anchor soils, provide rooftop habitats, clean stormwater, soak and hold floodwater, and move animals safely across highways[36] are a collective and optimistic testament to the emergence of a new breed of urban and landscape designers whose creative work mimics, models, and manifests the living systems that inspire and sustain us. Yet activating resilience requires a subtle and careful approach to design: one that is contextual, legible, nuanced, and responsive, one that is small in scale but large in cumulative impact. In (re)thinking design, and in (re)designing for change with this sensibility, we have begun to cultivate a culture of resilience and the adaptive, transformative capacity for long-term sustainability—thriving beyond merely surviving—with change in the urbanizing landscapes that now define us.

Acknowledgments

This chapter has been adapted from the original which appears as "Resilience Beyond Rhetoric" (Chapter 13) in *Nature and Cities: The Ecological Imperative in Urban Design and Planning* (F. Steiner, G. Thomson, and A. Carbonell, eds., 2016, Cambridge, MA: Lincoln Institute of Land Policy). I am grateful to Marta Brocki for assistance with research and the collection of illustrations in this essay, and to my landscape architecture colleagues for discussion and creative design work that helped shape these ideas.

Endnotes

1. City of Toronto. 2014. *Impacts from the December 2013 Extreme Winter Storm Event.* Staff Report to City Council (January 8, 2014): 2. www.toronto.ca/legdocs/mmis/2014/cc/bgrd/backgroundfile-65676.pdf.
2. See, for example, Steiner, Frederick R. 2011. *Design for a Vulnerable Planet.* Austin, TX: University of Texas Press.
3. Intergovernmental Panel on Climate Change. 2013. *IPCC Fifth Assessment Report (AR5).* Geneva, Switzerland: IPCC. http://www.ipcc.ch/report/ar5/mindex.shtml. Corroborating evidence is published by an independent association of insurance industries in Canada's *Institute for Catastrophic Loss Reduction:* www.iclr.org. Municipal strategies for climate change are evaluated in Robinson, Pamela, and Chris Gore. "Municipal Climate Reporting: Gaps in Monitoring and Implications for Governance and Action." *Environment and Planning C: Government and Policy* 33(5):1058–1075.
4. Dale, Ann. 2001. *At the Edge: Sustainable Development in the 21st Century.* Vancouver: University of British Columbia Press. In this essay, I use the term "management" in the context of Dale's definition of sustainability; that is, in the context of managing *human activities* within the environment, rather than regarding the environment as an object.
5. Holling, C. S. 1973. "Resilience and Stability of Ecological Systems." *Annual Review of Ecology and Systematics* 4:1–23.
6. See, for example, *The Post-Sandy Initiative: Building Better, Building Smarter—Opportunities for Design and Development* (May 2013), initiated and undertaken by the American Institute of Architects, New York Chapter (AIANY), and the AIANY's Design for Risk and Reconstruction Committee (DfRR), available at http://postsandyinitiative.org.
7. The United Nations projects that, in 2030, there will be 5,000,000,000 urbanites with three-quarters of them in the world's poorest countries. See United Nations. 2011. *World Urbanization Prospects: 2011 Revision.* http://esa.un.org. In 1950, only New York City and London had more

than 8,000,000 residents, yet today there are more than 20 megalopoli, most in Asia. See Chandler, Tetris. 1987. *Four Thousand Years of Urban Growth: An Historical Census*. Lewiston, NY: St. David's University Press; and Yvonne Rydin and Karolina Kendall-Bush. 2009. *Megalopolises and Sustainability*. London, UK: University College London Environment Institute. http://www.ucl.ac.uk/btg/downloads/Megalopolises_and_Sustainability_Report.pdf.

8. According to the World Health Organization, the percentage of people living in cities is expected to increase from less than 40% in 1990 to 70% in 2050. See "Global Health Observatory: Urban Population Growth," World Health Organization, available at http://www.who.int/gho/urban_health/situation_trends/urban_population_growth_text/en/.

9. Lister, Nina-Marie. 2008. "Sustainable Large Parks: Ecological Design or Designer Ecology?" In Czerniak, Julia, and George Hargreaves, eds., *Large Parks*. 31–51. Princeton, NJ: Princeton Architectural Press.

10. See, for example, Reed, Chris, and Nina-Marie Lister, eds., 2014. *Projective Ecologies*. New York, NY: Actar, in association with the Harvard University Graduate School of Design.

11. As articulated and elaborated in Corner, James. 1997. "Ecology and Landscape as Agents of Creativity." In Thompson, George F., and Frederick R. Steiner, eds., *Ecological Design and Planning* 80–108. New York, NY: John Wiley & Sons; Corner, James. 1999. "Recovering Landscape as a Critical Cultural Practice." In Corner, James, ed., *Recovering Landscape* 1–26. Princeton, NJ: Princeton Architectural Press; and Waldheim, Charles. ed., 2006. *The Landscape Urbanism Reader*. Princeton, NJ: Princeton Architectural Press.

12. Reed and Lister, *Projective Ecologies*.

13. This phenomenon is well articulated by Mathur, Anuradha, and Dilip da Cunha. 2009. *Soak: Mumbai in an Estuary*. Mumbai, India: Rupa & Co.

14. See, for example, Lister, Nina-Marie. 2009. *"Water/Front,"* *Places*. Design Observer Online. http://places.designobserver.com/feature/water-front/10227/.

15. See, for example, a variety of North American and international examples of resilience policies at http://resilient-cities.iclei.org/resilient-cities-hub-site/resilience-resource-point/resilience-library/examples-of-urban-adaptation-strategies/. The U.S. Department of State's *Deployment Stress Management Program* (http://www.state.gov/m/med/dsmp/c44950.htm) defines resilience in a psychosocial context, and the same language of resilience is often used in policy documents referencing resilience.

16. Brian Walker, Chair of the Resilience Alliance and research fellow at the Stockholm Resilience Centre, provides an excellent overview of this aspect of resilience in https://www.project-syndicate.org/commentary/what-is-resilience-by-brian-walker (accessed July 5, 2013).

17. Seminal references are Odum, Howard T. 1983. *Systems Ecology: An Introduction*. New York, NY: John Wiley & Sons; and Holling, "Resilience and Stability of Ecological Systems."

18. Discussed in F. Berkes, N. C. Doubleday, and G. S. Cumming. 2012. "Aldo Leopold's Land Health from a Resilience Point of View: Self-Renewal Capacity of Social–Ecological Systems." *Eco Health* 9(3):278–287.

19. Discussed in A. R. Johnson. 2012. "Avoiding Environmental Catastrophes: Varieties of Principled Precaution." *Ecology and Society* 17(3):9. http://dx.doi.org/10.5751/ES-04827-170309.

20. Carson, Rachel. 1962. *Silent Spring*. 297. New York, NY: Houghton Mifflin; reprinted in 2002.

21. Bormann, F. Herbert, and Gene Likens. 1979. *Pattern and Process in a Forested Ecosystem*. New York, NY: Springer-Verlag. For continuing work based on this pioneering study, see http://www.hubbardbrook.org.

22. Succession is a process by which one ecosystem's community is gradually replaced by another.

23. The adaptive cycle was first described by C. S. Holling. 1986. In "Resilience of Ecosystems: Local Surprise and Global Change," in Clark, W. C., and Edward (Ted) Munn, eds., *Sustainable Development of the Biosphere*. Cambridge, UK: Cambridge University Press; modified in Gunderson, Lance, and C. S. Holling, eds. 2002. *Panarchy: Understanding Transformations in Human and Natural Systems*. Washington, DC: Island Press; and, more recently, by Reed and Lister. In *Projective Ecologies*.

24. Holling, C. S. 1992. "Cross-scale Morphology, Geometry and Dynamics of Ecosystems." *Ecological Monographs* 62(4): December 447–502.

25. Holling. "Resilience of Ecosystems."

26. Holling, C. S. 1996. "Engineering Resilience versus Ecological Resilience." In Schulze, P. C., ed. *Engineering within Ecological Constraints* 51–66. Washington, DC: National Academy Press; further developed in Walker, Brian, C. S. Holling, Stephen R. Carpenter, and Ann Kinzig. 2004. "Resilience, Adaptability and Transformability in Social–ecological Systems." *Ecology and Society* 9(2): December 5. http://www.ecologyandsociety.org/vol9/iss2/art5.

27. For Brian Walker's view on resilience, see https://www.project-syndicate.org/commentary/what-is-resilience-by-brian-walker (July 5, 2013).

28. See, for example, Pickett, S. T. A, M. L. Cadenasso, and J. M. Grove. 2004. "Resilient Cities: Meaning, Models, and Metaphor for Integrating the Ecological, Socio-economic, and Planning Realms." *Landscape and Urban Planning* 69(4): October 369–84.

29. Cronin, William, ed. 1980. *Uncommon Ground: Rethinking the Human Place in Nature.* New York, NY: W. W. Norton; Merchant, Carolyn. 1980. *The Death of Nature.* San Francisco, CA: Harper & Row; Orr, David. 1992. *Ecological Literacy: Education and the Transition to a Postmodern World.* Albany, NY: State University of New York Press; Spirn, Ann Whiston. 1984. *The Granite Garden: Urban Nature and Human Design.* New York, NY: Basic Books; Steiner, Frederick R. 1990. *The Living Landscape.* New York, NY: McGraw-Hill; and Corner. "Landscape and Ecology as Agents of Creativity." In Thompson, George F. and Steiner, Frederick R., eds. Ecological Design and Planning. New York: John Wiley & Sons.

30. As discussed by Ellison, Aaron. 2013. "The Suffocating Embrace of Landscape and the Picturesque Conditioning of Ecology." *Landscape Journal* 32(1): September 79–94.

31. The development of social-ecological systems science, supported by case study analyses, can be followed in: Gunderson and Holling, *Panarchy*; Berkes, Fikret, Johan Colding, and Carl Folke, eds. 2002. *Navigating Social-Ecological Systems: Building Resilience for Complexity and Change.* New York, NY: Cambridge University Press; and Waltner-Toews, David, James J. Kay, and Nina-Marie Lister, eds. 2008. *The Ecosystem Approach: Complexity, Uncertainty, and Managing for Sustainability.* New York, NY: Columbia University Press.

32. Related versions of these principles—described variously as system attributes, tenets, and characteristics—are elaborated in Gunderson and Holling, ibid.; Waltner-Toews, Kay, and Lister, ibid.; and, more recently, in Walker, Brian, and David Salt. 2012. *Resilience Practice: Building Capacity to Absorb Disturbance and Maintain Function.* Washington, DC: Island Press.

33. Lister. "Sustainable Large Parks."

34. See a variety of designed examples in Steiner, Frederick, George Thomson and Armando Carbonell (eds). 2016. *Nature and Cities: The Ecological Imperative in Urban Design and Planning.* Cambridge, MA: Lincoln Institute of Land Policy.

35. See for example Green, Jared. 2015. *Designed for the Future: 80 Practical Ideas for a Sustainable World.* New York, NY: Princeton Architectural Press.

36. A diversity of examples of wildlife crossing infrastructure is available at https://arc-solutions.org/.

4

Resilience and the Translation of Expertise

Jane Wolff

CONTENTS

In the children's novel *Hans Brinker, or the Silver Skates,* a little boy saves the Netherlands by blocking a hole in a dike with his fingers until grownups come to repair it permanently (Dodge, 1896). Written by a native of the United States, that story about the survival of a landscape under threat from water has more to do with American culture than with Dutch infrastructure, physical or social: the resolution of the problem depends on the action of an individual at a single, pivotal place and time rather than on the constant, collective maintenance of a system distributed across the country. The story is a useful reminder that people tend to evaluate problems and propose solutions based on what they know. That tendency is important to keep in mind as the professional globalization of design, planning, and engineering and the ubiquitous hazards of climate change make it more and more tempting to lump landscapes of the same broad type together. Similar physiographic circumstances can give rise to cultural landscapes that are quite different, even if they are comparable, and culture gaps make designing for resilience much more than a technical problem.

In the wake of climate-related disasters like Hurricane Katrina and Superstorm Sandy, the Dutch landscape has returned to American consciousness. Dutch experts have worked on water management strategies in Louisiana, New York, Florida, and California, and the advanced measures that the Netherlands is taking to protect itself from climate change and sea level rise have been offered as models by the *New York Times* and on public radio.[1] The media coverage and the clientele of Dutch water managers are right about technical expertise. But the conversation about importing Dutch knowledge does not fully engage the particular historical, political, and social context that produced such expertise, and it does not reckon with profound contextual differences between the Netherlands and the United States, or, for that matter, most of the rest of the world.

The recent histories of the Netherlands and New Orleans comprise an allegory about the importance of cultural translation in the transfer of specialized knowledge. The landscapes make close analogues. They are both urbanized deltas in the developed world; both have been subject to the same unexpected consequences of land reclamation; and both have been canaries in the coal mine of climate change. Their deeply different responses to that threat—despite the substantial presence of Dutch advisors in New Orleans—demonstrate the importance of local history, politics, and consciousness to the success of proposals that might seem purely technical.

Defining Resilience in the Delta of the Netherlands

Many of the Dutch experts who arrived in New Orleans in the aftermath of Hurricane Katrina were also involved in an ambitious new project to address the impacts of climate change in the Netherlands. In the winter of 1994 and 1995, after 40 years without significant flooding, the watery Dutch landscape had faced a new threat. Historically, floods had come from the west, from the sea, but climate change dispatched a harbinger in the form of more water from upstream. Storms in France and Germany and melting snow in the Alps sent unusually high volumes of water into the Maas and Waal rivers, whose courses run close together through the Eastern Netherlands. More than 200,000 people had to be evacuated, and the nation was shaken out of complacency about its water management policies.

The events brought on a paradigm shift in landscape planning, design, and policy, one conceived for resilience. After the catastrophic Walcheren floods of 1953, which killed nearly 2,000 people in the province of Zeeland, the Dutch government launched an ambitious program to keep the country dry at any cost. This endeavor produced the Delta Works, a series of dynamic barriers that hold the North Sea at bay. Completed in 1972, the system had worked well as coastal defense, but it was not designed to address the problems that now faced the country's river landscapes. The floods on the Maas and Waal made it clear that policy makers, designers, and planners needed to attend to the rivers, and the character of the threat suggested that maintenance of dry feet at all times and in all places was no longer a realizable goal. Instead, over the next 20 years, management strategies shifted toward the creation of zones for controlled inundation and water storage.

The future of the landscape became a subject of both popular and professional discussion. In 1998, the theme of the annual nationwide Book Week was "Panorama of the Netherlands: City and Land in Prose and Poetry;" in the essay published as part of the event, the noted author Geert Mak wrote about the history and present circumstances of the Dutch landscape (Mak, 2001). The same year, journalist and cultural critic Tracy Metz published *New Nature*, a collection of reportages on the revival of ideas of nature in the highly controlled Netherlands (Metz, 1998). The National Planning Service, a government agency, sponsored *Nederland 2030*, a series of workshops that involved a broad cross-section of designers, planners, and policy experts in scenario planning for 2030, and published the results (Ministerie van Volkshuisvesting, ruimtelijke ordening en milieubeheer, 1998). Even language in newspaper articles about water policy began to shift; the word *beheren*, to manage, began to appear in contexts where *beheersen*, to control, had previously been used. In 2001, the Ministry of Environmental Planning released its policy document *The 5th Note on Spatial Planning* and put forward an agenda that made more room for water:

> The traditional approach of 'diking, pumping, and channelization' ... is coming to the limits of its possibilities. The new policy philosophy is 'moving with water.' In accord with that, the *5th Note* treats questions of safety, flooding, and fresh water provision as spatial topics and puts water to the forefront as an organizing principle ... the uncertainties that are connected to water demand the return of space to the water system (especially the great rivers) and the reduction of vulnerability of uses to flooding in spatial planning ... The *5th Note* speaks in this connection of a 'water test' as part of the existing procedures conforming to the Spatial Planning Law, in all phases of the plan development (design phase, elaboration and plan testing).

Sociaal-Economische Raad, 2001

This mandate began to take shape in 2004 with the planning of an initiative called Room for the River. The initiative identified 30 projects to increase floodplains and create areas for controlled flooding in the eastern part of the Netherlands. Undertaken in 2007, the program was declared successful and substantially complete by the Dutch parliament in 2018 (Ruimte voor de Rivier, 2018). Its final components will be completed in 2022 (Dutch Water Sector, 2017). A collaborative effort among local, regional, and national water management agencies, the endeavor is described in detail for the public on a website provided by the national government (Ruimte voor de Rivier, 2017). The site explains all aspects of the initiative: the threat that it addresses; its strategies for change; and its component projects and their trajectories, tactics, and degree of completion. It links to all of the agencies involved in each project and provides a way into detailed information about the nation's system for managing water.

The paradigm shift in the Netherlands was made possible by a deep cultural understanding of the need for landscape management. Centuries of struggle with water had produced not only strong institutions for governance, administration, and maintenance but also widespread public awareness, at least in general terms, of the national necessity of such efforts. Responding to the crisis of the 1990s was a question of changing course rather than building culture and consciousness. Those had developed over the past millennium.[2]

Landscape practices and processes began to shape Dutch society about 1,200 years ago, when farmers with shovels first dug drainage ditches through the soggy moors and bogs in the delta of the Rhine, the Maas, and the Schelde rivers. Over centuries, as the landscape evolved and technology developed, small-scale, individual efforts became collective, large-scale endeavors by capitalists and then massive public works projects by the national government. Driven by need for flood control and demand for arable land, the landscape's transformation from a wet, windy coastal plain into one of the most intensely urbanized and cultivated places in the world produced institutions, artefacts, and language that permeate Dutch society. The Netherlands today cannot be separated from the work of land reclamation. Its four biggest cities and significant portions of its territory and population occupy land that people have claimed from the fluctuating ground of the delta.

The landscape's evolution took place through four distinct methods of reclaiming land, each developed in response to local conditions and available technology. All depend on the creation of *polders*, areas enclosed by dikes that separate the water within them from water outside. The surface drainage and groundwater levels in these polders are controlled by pumps: rain and high groundwater are carried up, out, and away to the sea through a system of ditches that operate from local to regional scales. The first polders were constructed from the peat lands of what is now North Holland, South Holland, and Groningen. In the Middle Ages, farmers trying to make land dry enough to cultivate dug ditches through waterlogged peat, and water flowed to natural drainage courses by gravity. To protect reclaimed land from water flowing from higher areas, a boundary ditch was dug at the border of the reclaimed area and a low dike was built on the reclamation side of the boundary ditch. But reclamation had an unexpected consequence. The highly organic soil of the peat polders oxidized, and as the temperature of the soil rose with cultivation, the ground subsided. The moors fell from their original levels of one to three meters above sea level to between one and two meters below sea level. As the land sank it became less suitable for cultivation; land further back from the original drainage course had to be reclaimed; and the polders required complete enclosure by perimeter dikes.

The subsidence of the peat polders made large areas vulnerable to storm surges, especially in the region north of Amsterdam, and by AD 1250, flooding had created an inland arm of the North Sea, the Zuiderzee, and a series of lakes connected to it. The technique of *droogmakerij* was developed to combat the flood hazard from these lakes and to create agricultural land for sale. Usually funded by groups of investors, workers dug a ring canal around a lake, built a dike between the lake and the canal with the displaced earth, and used windmills to pump the lake water into the canal. From the ring canal, the water went to a regional canal system, the *boezem*, where it was transported to the sea. The clay soil of the lake bottom, meters below sea level, was arable as long as rain and groundwater were pumped away. As improvements in windmill technology after the beginning of the seventeenth century made paddle wheels more efficient, mills ganged together could pump large quantities of water to significant heights. The most famous of the large projects enabled by this technology, the Beemster polder (1612), was 7,100 hectares. Like the peat polders, the *droogmakerij* polders had unexpected consequences for the natural system they changed. Before reclamation, the lakes had functioned as part of the *boezem*. Their drainage reduced its surface area and volume significantly, and the limitation of its storage capacity became a new cause of flooding. This problem was not resolved until the development of steam pumps in the nineteenth century made constant and consistent pumping possible.

In the region between Rotterdam and Antwerp, a third method of reclamation, *bedijking*, or diking, reclaimed land at the edges of existing areas along the coast and on the islands in the delta of the Schelde River. Dikes were extended from high ground into adjacent tidal marshes and water was pumped out from behind. Like the peat polders, polders reclaimed by this method grew by accretion. Newer polders were added to older ones over the course of centuries; changing technology during that period of time produced different scales in adjacent parcels.

In the twentieth century, advances in pumping technology enabled a vast reclamation project called the IJsselmeer polders. The technique, called *nieuwe droogmakerij,* reclaimed land from the Zuiderzee, which was subject to wind-driven flooding and had been a significant hazard since its formation. The process occurred in two stages. In 1932, the Zuiderzee was closed off from the North Sea by the *Afsluitdijk* to shorten the coastline and reduce tidal flooding. The enclosed water body, now called the IJsselmeer, turned fresh and became both a water storage area and a source of freshwater for agricultural and urban uses. Over the next 40 years, four polders were reclaimed by pumping water from behind closed dikes built out into the water. The project, which comprised 165,000 hectares at elevations approaching seven meters below sea level, required infrastructure at a monumental scale and would not have been possible before the development of steam and electric pumps. Conceivable only as a project of the national government, the IJsselmeer polders required an unprecedented degree of planning, calculation, testing, and documentation in every aspect of their design and function. The level of management exercised here and across the country is almost inconceivable in North America.

Managing and maintaining the watery landscape has been both one of the great efforts of Dutch society and one of the great forces shaping it. When land is reclaimed, local situations have dramatic implications for larger ones: if the ditches that drain a field or polder are not maintained and water accumulates in an area, neighboring land will be subject to flooding. The integrity of the system depends on the participation of all members, and certain individual rights, like the right of a landowner not to maintain his or her ditches, must be subordinated to the good of the collective. However, because any member of the group can jeopardize the security of the whole effort, every landowner has power. This prisoner's dilemma has produced a society that emphasizes cooperation and consensus. The Dutch call their characteristic way of running things the "polder model."

The institutions that developed to manage the drainage system were always democratic in character. Known as water boards, they arose to enforce compliance with community maintenance standards and to manage drainage problems and structures that involved more than one community. Because the borders of drainage areas were rarely contiguous with town, village, or provincial boundaries, water boards developed at a local and regional level parallel to and in some cases even before the system of civil government. They acquired the authority to summon inhabitants of drained land for dike maintenance and watching, to tax them for the upkeep of infrastructure, to control pumping from drained land if the level of the *boezem* became dangerously high, and to authorize reclamation of new polders. Rijkswaterstaat, the national agency for water management (and later, road and rail infrastructure) dates to 1798 (Rijkswaterstaat, 2017). In a way, the institutions that manage water are the true power structures in the Netherlands: they control the land itself. But complete centralization of control remains impossible. According to the Rijkswaterstaat's textbook on the history of land reclamation in the Netherlands,

> in practice, the constant maintenance of the intricate pattern of drainage cannot be executed from a central point. So, to this day, field drains are maintained by those directly involved; without their constant care the greater part of the Netherlands would turn into a swampy area.

> **van de Ven, 2004**

The management demands of the Dutch landscape—and the need for widespread support of management practices—have produced highly developed information for both technical and popular consumption. The problem of controlling complicated geographies has had a powerful impact on Dutch cartography. The widespread reclamation projects of the seventeenth century provided a strong economic incentive for improvements in technology and techniques for surveying and mapmaking: before lakes could be made into polders, their boundaries had to be known.[3] As the water management systems grew and the interdependence of its elements increased, the need to see relationships among parts produced cartographic methods and maps unique to the Netherlands. Since the 1860s, the Rijkswaterstaat has published the *Waterstaatskaarten*, a special series of maps that document, describe, and explain drainage structures and systems. Drawn in reference to the national 1:25,000 topographic survey and distributed under government restrictions, these maps are the code to the landscape's operation. They show each polder, sluice, lock, weir, and pump in the Netherlands—and the height of the water table, which is managed within tight tolerances[4] (Figures 4.1 through 4.3). Most of the time, this information allows water agencies to keep land dry, but systematic inundation has been used as a tool in the eighteenth, nineteenth, and twentieth centuries.[5]

Even widely available, non-specialist maps like the ANWB's recreation series[6] or the *Topographic Atlas of the Netherlands* (Topografische Dienst, 1987) provide clues to the landscape and its history. Polders from each type and era of reclamation display characteristic patterns,[7] and to a person who knows the vocabulary of form, the sequence of transformation is clear. The landscape was assembled over time: the land that was most easily made was occupied first, and the more difficult places in between were filled in as need warranted and technology permitted. The resulting patchwork manifests reclamation by different methods at different scales[8] in both rural and urban landscapes (Figures 4.4 through 4.6). In Amsterdam, water is present in a range of different ways: as the IJ, an arm of the former Zuiderzee; along the River Amstel; at the Dam and in the narrow canals of the medieval city; through the structure of the seventeenth-century canal ring; in the neighborhood of the Jordaan, whose subdivision clearly derives from the medieval parcellization of the peat polders on which the city was built; and at the Singel, the fortified canal built at the edge of the city center for defense. In Rotterdam, the structure of the city follows the pattern of

FIGURE 4.1
Water channels draining part of the Beemster polder, the greatest reclamation project of the Golden Age.

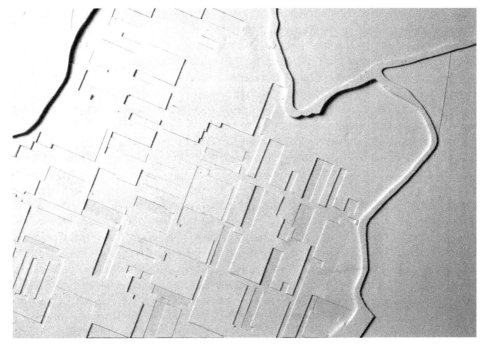

FIGURE 4.2
Physical model of the water table for the same part of the Beemster. Groundwater levels are precisely controlled by pumping and can be adjusted field by field. Water is pumped up from the center of the polder to the ring canal, where it travels to the *boezem*.

FIGURE 4.3
The Beemster polder.

bedijking; small islands enlarged by polders at their edges grew together to form the urban landscape.[9] The traces of reclamation comprise a rich physical vocabulary in everyday places and in heritage landscapes: five of the Netherlands' nine UNESCO World Heritage Sites are related to water infrastructure (Figures 4.4 through 4.6).

Landscape consciousness in the Netherlands begins in childhood. The ubiquitous presence of water has made swimming lessons a basic part of primary education. They are regarded as an academic subject, not a sport. For decades, the public schools of Amsterdam, Rotterdam, The Hague, and Utrecht all taught swimming, and proficiency was required to finish elementary school. Now children take their lessons at certified, universally accessible swim schools, and a child must have a swimming diploma to use a public pool. Breaststroke—called school stroke in Dutch—is a requirement. Children make acquaintance with landscape systems and infrastructure through *Droppie Water,* a cartoon series originally published in print by a union of regional water boards and now available on the Internet as a collection of stories, games, and school lessons. The cartoons describe adventures in the landscape among a group of Dutch types, all recast as giant water drops. The hero, the eponymous Droppie Water, is a little boy who lives in a village on a dike, and other characters include the boy's kindly grandfather, his schoolmates, his geography teacher, a water engineer, and a family friend who runs a pumping station. The boy learns not to be afraid of rain and thunder as he makes a series of forays into the polders to learn how they work. Each adventure ends with a note of caution, even though the dikes are safe for the moment. In a strange conjunction of people with the material they are trying to control, the characters are drawn as oversized water drops.[10]

With climate change and the Internet, more and more clear, accessible, engaging information is being put forward by the government. One, entitled "If Our Water Rises," allows people to enter their postal codes to learn whether their houses and neighborhoods

FIGURE 4.4
Water channels in and around the Vinkeveense Plassen, 1849.

FIGURE 4.5
Water channels in and around the Vinkeveense Plassen, 1994. Some areas that were wet in the middle nineteenth century have been pumped dry; some areas that were dry are now inundated.

FIGURE 4.6
A canal in the city of Edam, North Holland, the Netherlands.

could flood in case of disaster, to what depth, and with what implications for basic services; how to decide whether to stay home or evacuate; what kinds of larger disruptions might occur; what advance measures can be taken at home; and whom to contact in case of emergency (Als ons water stijgt, 2017). The site "Current Elevations in the Netherlands" is a digital topographic map with an average of eight elevation points per square meter. Used by water boards, provinces, and Rijkswaterstaat to monitor land and water elevations across the country, the data are also available to researchers, cities, and businesses that work with water and the general public (Actueel Hoogtebestand Nederland, 2017). The most expansive site, "Our Water in the Netherlands" (subtitled "Clean, Safe, Sufficient Water Isn't Self-evident") includes an array of topics: "Working on Our Water," "Trouble from Our Water," "If Our Water Rises," "Our Water in the Sewer," "Our Water from the Tap," "The Mineral Content of Our Water," "Our Water in Images," "Creating Wetlands from Open Water," "Experiencing Our Water," "Education for Water-Related Work," a version of the game "Memory" with water, "Learning about Our Water" (a link to Droppie Water's own site), "Bringing Our Water into Balance," and "Who Manages Our Water?" (Ons water in Nederland, 2017). No wonder the mayor of Rotterdam told the *New York Times* that ordinary citizens would understand if he told them they all needed boats (Kimmelman, 2017).

The website includes another extremely important section, "The State of Our Water," which discusses the cost and specific allocation of spending related to water infrastructure nationally. The Dutch have a robust physical and managerial system because they pay for it. Between 1975 and 2000—in other words, for the 35 years following the end of the Vietnam War—the proportion of GDP devoted to infrastructure spending in the Netherlands approached, matched, or exceeded the proportion of GDP devoted to military spending in the United States.[11] And the polder model has produced a culture in which government policy about landscape issues has real power. An article in the *New York Times* about Room for the River and other demonstrations of Dutch expertise in relation to New Orleans, New York, and New Jersey cites the example of farmers who have been moved to enable the creation of controlled flood zones in the eastern part of the country (Kimmelman, 2013). This is actually the second large-scale relocation of farmers and reallocation of land in fairly recent memory: in the 1950s and 1960s, to rationalize agricultural production, the government instituted a widespread policy of land swaps to consolidate parcels and allow the modernization of farming methods. The prisoner's dilemma of the Dutch landscape has created a culture in which planning and its institutions have real power. The society's mind-set, customs, and policies—in sum, its cultural landscape—are products of a long-term relationship between people and place.

From One Delta to Another

The cultural landscape of New Orleans, in the Mississippi Delta, is another world. Though its physical geography compares closely to the western Netherlands, its traditions and institutions are deeply different, and the reclamation of New Orleans took place in a society poorly equipped to understand or negotiate its dilemmas.

The city was founded by Europeans who came to the New World in search of resources, and its site made it an invaluable *entrepôt*. Located at a portage route between Lake Pontchartrain and the Mississippi river, the spot was ideally suited to export raw materials from the interior of North America to Europe. But inhabiting the watery, fluctuating landscape at the bottom of the Mississippi Delta was not easy. The river's active patterns of flooding, erosion, and deposition and the soggy, mucky character of the area between the river and the lake meant that for almost 200 years, urban development was confined to the high ground along the Mississippi's natural levee, where flood processes had deposited gravel and other stable soils over time.

Initial efforts to drain the back-of-town cypress swamps began in the late nineteenth century. In the late 1890s, the State of Louisiana chartered two agencies to manage water and reclaim land in New Orleans, the New Orleans Drainage Commission and the New Orleans Sewerage and Water Board.[12] In 1913, ten years after those agencies merged, A. Baldwin Wood's screw pump made their work technically feasible (Sewerage and Water Board of New Orleans, 2012). Two years later, 11 wood screw pumps had been installed, and the city spread rapidly: by 1925, the drainage system network was 560 miles long and served 30,000 acres (Campanella, 2002, 60). Thirty years later, all of the swampy land in central Orleans Parish had been drained and developed. The new urban landscape left no room for water except in the narrow canals that led from the city's pumping station to Lake Pontchartrain.

Just as in the Netherlands, the processes of reclamation and drainage caused subsidence. The soils in the back-of-town swamps oxidized and disappeared. Pumping pulled water from the pores in the earth, and soil particles collapsed onto each other. The removal of the swamps meant no new soil was created through decomposition. By the turn of the twenty-first century, central New Orleans had assumed the shape of a giant bowl: land reclaimed over the course of 90 years had fallen to as low as 3.6 meters below sea level, and the city's boundaries had to be protected from the river and the lake by constructed levees.[13]

Within that bowl lie a series of smaller drainage units, or polders, all kept dry by constant pumping. Groundwater must be removed continuously from subsided ground, even on dry days. On wet days, rain falling anywhere in the bowl—even on land above sea level—runs to the center and has to be evacuated mechanically. When the pumping stations were built, they lay above sea level and pushed water horizontally to Lake Pontchartrain. Now, after a century of subsidence, they have to pump rain and groundwater up to lake level; then the water travels through canals whose high walls hide their contents and divide the city (Figure 4.7). System capacity is limited: the pipes and canals within New Orleans can store 12 millimeters (mm) (½ inch) of rainfall, and the pumps can remove 12 mm of rain in one hour. Rain that falls any faster backs up into catch basins and out into streets, and in the city's subtropical climate, low-level flooding is a constant problem.[14] Management is made even more difficult by institutional fragmentation. The Department of Public Works is responsible for water that runs in gutters and in pipes less than 914 mm (36 inches) in diameter; the Sewerage and Water Board controls pipes 914 mm or bigger and pumps and

FIGURE 4.7
The Orleans Avenue canal, New Orleans.

canals inside the city; and the Corps of Engineers is in charge of new gates and booster pumps where the canals enter Lake Pontchartrain. No agency has official responsibility for managing groundwater (Waggonner & Ball, 2013).

New Orleans's dilemmas are compounded by two more factors: regional development and a lack of public awareness about the processes of the delta. Southern Louisiana's wetlands used to protect the city from flooding by dampening the force of storms and surges from the Gulf of Mexico. The culture of resource extraction turned those marshes into oil and gas fields, and over the course of the twentieth century, excavation for pipelines and access channels diminished their capacity substantially. The Mississippi River-Gulf Outlet (MR-GO), a federally funded shipping channel that reduced the distance from the Gulf to the Port of New Orleans, has also been a major instigator of wetland loss. In 1965, when it was completed, the channel measured 198 meters (m) across; erosion increased its width to more than 457 m and destroyed or damaged at least 8,094 hectares of the wetlands that comprised southern Louisiana's delta. However, the state did not establish agencies for the integrated governance, management, or regulation of regional wetlands until after Hurricane Katrina.

As for public awareness, New Orleans's situation demonstrates the dark side of twentieth-century engineering successes: the mechanics of drainage made landscape processes opaque rather than legible. Hidden behind high levee walls, water was invisible. The incremental pace of subsistence led to the widespread acceptance of unstable ground and low-level flooding as ordinary nuisances.[15] Faith in technological solutions led to the collective disavowal of risk. Until it failed catastrophically, the system worked well enough that most residents of New Orleans lost track of the landscape's fundamental sogginess. At the national level, the city's ecology and infrastructure were unique—and isolated—and there was almost no discussion about ways in which development upstream and Corps of Engineers policies along the Mississippi created continental-scale impacts on a local situation.

In 2005, when Hurricane Katrina struck, New Orleans and the United States faced a crisis of similar proportion to the Dutch floods 10 years earlier. The storm surge that Katrina created in the Gulf of Mexico hit New Orleans through two routes. First, southern Louisiana's diminished wetlands were unable to dampen the surge significantly enough to prevent high water from reaching Lake Pontchartrain. The lake backed up into the outlet canals that drained central New Orleans. Two were breached, and large parts of the twentieth-century city were inundated. Second, the MR-GO provided a straight path through which the surge travelled directly to the Industrial Canal. High water breached and overtopped levees, and the Lower Ninth Ward and parts of the Upper Ninth Ward flooded. Restoring the pumping system took time, and some areas remained under water for weeks. The ensuing destruction and the failure of local, state, and federal government agencies to respond in timely and sufficient ways made international news.

The justified expressions of public outrage after Katrina tended to focus on narrow questions about the engineering of physical infrastructure: why, for instance, had a levee been built as an I-wall rather than a T-wall, or had the Corps of Engineers done its calculations properly? There was very little sense that the state of the physical landscape reflected broad cultural conditions. For most citizens of New Orleans, Louisiana, and the United States, the landscape's fundamental characteristics were invisible or illegible. The institutions and agencies whose work affected its function were not well known, and their mandates were fragmented, uncoordinated, and underfunded. The government was unable to provide clear information about topics fundamental to the urban landscape: topography, soils, tree cover, and most important, the structure and mechanics of the drainage system.

The public documentation and discussion of metropolitan New Orleans as a dynamic landscape emerged not through official channels but from a range of grassroots efforts, including collaborations between neighborhood organizations, local non-profits and universities; research and teaching endeavors by academics in landscape architecture; and a series of workshops entitled "Dutch Dialogues," funded by the Embassy of the Netherlands in the United States, the American Planning Association, and Waggonner & Ball, a local architecture firm, to bring together Dutch and North American water experts for conversations about the future.[16] Through the work of committed and energetic volunteers, these ad hoc endeavors made a compelling case for urban landscape planning that would address the ecology and hydrology of the Mississippi Delta. Most important, they offered clear, widely accessible explanations of the city's hydrological and hydraulic history and present circumstances and offered conceptual models about what might be possible in the future.

Twelve years after Katrina, there have been important steps toward a more resilient plan for New Orleans, including the city's comprehensive zoning ordinance, which offers incentives for landscape management of stormwater; work by the New Orleans Redevelopment Authority to manage water on vacant lots; and the *Greater New Orleans Water Plan*, which reconvened many of the participants from "Dutch Dialogues" in an extensive consultancy project.[17] The *Greater New Orleans Water Plan*, completed in 2013, was funded by Greater New Orleans, Inc., a non-governmental economic development organization, through community development and disaster recovery grants it received from the U.S. Department of Housing and Urban Development. The large team, which was led by Waggonner & Ball and involved a significant number of Dutch water experts and landscape architects, has proposed a water management strategy for the metropolitan area of New Orleans (Waggoner & Ball, 2013). Its strategies and goals—developing urban landscape spaces and water circulation procedures to increase storage capacity, slow subsidence, and create public amenities—are sound. The city's Sewerage and Water Board has committed US$500,000 a year to supporting this work (O'Neil, 2017), and a federal grant of US$141,000,000 will implement a series of pilot projects in one of the city's lowest-lying districts over the next five years (LaRose, 2016). However, despite the support it has received from the city and the U.S. Department of Housing and Urban Development, the *Greater New Orleans Urban Water Plan* is not an official policy document. It exists outside city, state, and federal planning authorities, and the enactment of its recommendations relies on persuasion rather than the force of law. And the grants for pilot projects, while substantial in an American context, do not begin to approach the level of financial support for infrastructure and public information in the Netherlands. The signs of progress are real, but they still depend on individual initiatives without a comprehensive governmental, administrative, or funding structure. The plan itself acknowledges these obstacles to change; in the section entitled "Vision," it discusses gaps in the responsibilities and relationships of the several agencies that manage water in New Orleans and proposes that

> The best way to realize the *Urban Water Plan* is to convene a regional water management authority with adequate resources to guide implementation in the coming years. This authority can coordinate water-related initiatives at system and district scales, conduct research on regional issues, and represent the region's interests and capabilities in a national and international context. The authority will also need to work with municipalities and local agencies to coordinate groundwater management; provide technical support and regional data; initiate a storm water retention credits trading

system, wetlands mitigation bank, or other market-based funding strategy; and establish ongoing regional monitoring and hydraulic modeling efforts.

Waggonner & Ball, 2013

In other words, the success of the plan depends on an institutional framework that does not exist at present and may be difficult to establish in the current American political climate. The *Urban Water Plan* has made excellent use of technical knowledge from the Netherlands, but what is even more necessary is guidance about the development of institutions for the governance and administration of the landscape—and the cultural consciousness and public funding to support them. This is a question at regional and national scales as well as locally. Louisiana's Coastal Protection and Restoration Authority, founded in 2005 in the wake of Katrina, has prepared a series of plans to increase the integrity and resilience of the state's wetlands but does not link directly to city efforts. Elsewhere in the United States, New Orleans's dilemmas are regarded as unique and local rather than as a test case for climate change adaptation or as a manifestation of the continental-scale transformation of the Mississippi River's drainage basin.

The Moral of these Stories?

These accounts of the Netherlands and New Orleans are not intended to valorize one society and villainize another. They are offered to say that a landscape's physical resilience depends on a citizenry that understands its dilemmas, at least in broad terms, and that is willing to make comprehensive, long-term investments in the establishment and maintenance of physical and institutional systems for its management. American society has tended to view the landscapes of the United States as either sublime visions or resource stockpiles; individual rights to property are among the strongest of national values; and taxation and the role of government in environmental regulation have been divisive, contested issues since the Reagan era. Attitudes and public finances vary widely among the American landscapes under threat from climate change, and in the absence of national consensus about the need for climate change adaptation, it seems almost inevitable that regional differences in culture, policy, and funding will reflect and even intensify the country's current fragmentation.

As for the Dutch at work on water landscapes across the United States—or any experts providing technical knowledge in faraway places—climate change is an urgent and terrible threat, and it is essential to be able to learn from expertise that originates elsewhere. But it is also essential to recognize that the implementation of complex technical strategies in different places is a project in translation. Landscapes are not only physical and technical: they are dynamic artefacts of culture and politics. The possibilities for design in similar places depend as much on their incongruities as their resemblances, and the knowledge of consistent typological tendencies is not enough to produce meaningful proposals. Design for resilience has to be inflected, case by case, by the recognition and understanding of public consciousness, institutional structure, historic patterns, and local idiosyncrasy. And those in search of advice from afar must remember that they are not just installing hardware. Their most complicated task is to change the hearts, minds, and habits of the landscape's citizens.

Appendix A

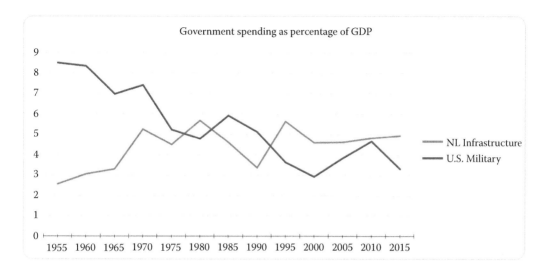

Source: Centraal Bureau Voor de Statistiek. 1954. *Jaarcifers voor Nederland 1947–1950.* Utrecht: Uitgeversmaatschappij W. de Haan N.V.; Netherlands Central Bureau of Statistics. 1966. *Statistical Yearbook of the Netherlands 1963-1964.* Hilversum: Uitgeversmaatschappij W. de Haan; Netherlands Central Bureau of Statistics. 1970. *Statistical Yearbook of the Netherlands 1967–1968.* The Hague: Staatsuitgeverij; Netherlands Central Bureau of Statistics. 1972. *Statistical Yearbook of the Netherlands 1972.* The Hague: Staatsuitgeverij; Netherlands Central Bureau of Statistics. 1978. *Statistical Yearbook of the Netherlands 1977.* The Hague: Staatsuitgeverij; Netherlands Central Bureau of Statistics. 1983. *Statistical Yearbook of the Netherlands 1982.* The Hague: Staatsuitgeverij; Netherlands Central Bureau of Statistics. 1988. *Statistical Yearbook of the Netherlands 1987.* The Hague: Staatsuitgeverij; Netherlands Central Bureau of Statistics. 1993. *Statistical Yearbook of the Netherlands 1993.* The Hague: Staatsuitgeverij/CBS-publications; Statistics Netherlands. 1997. *Statistical Yearbook of the Netherlands 1997.* Voorburg: Statistics Netherlands; Statistics Netherlands. 2003. *Statistical Yearbook of the Netherlands 2003.* Voorburg: Statistics Netherlands; Eurostat, last updated April 24, 2017. General government expenditure by function (COFOG). http://ex.europa.eu/eurostat/web/products-datasets/-/gov_10a_exp (accessed July 27, 2017); The World Bank. Military expenditure (% of GDP). http://data.worldbank.org/indicator/MS.MIL.XPND>GD.ZS?locations (accessed July 27, 2017); The World Bank. GDP (current US$). http://data.worldbank.org/indicator/NY.GDP.MKTP.CD?locations=NL (accessed July 27, 2017).

Endnotes

1. Michael Kimmelman, "The Dutch Have Solutions to Rising Seas. The World Is Watching," *New York Times*, June 15, 2017, accessed July 24, 2017, https://www.nytimes.com/interactive/2017/06/15/world/europe/climate-change-rotterdam.html. Chris Bentley, "As sea levels rise, Rotterdam floats to the top as an example of how to live with water," *PRI*, June 20, 2016, accessed July 27, 2017, https://www.pri.org/stories/2016-06-20/sea-levels-rise-rotterdam-floats-top-example-how-live-water. Chris Bentley, "Holland is relocating homes to make more room for high water," *PRI*, June 22, 2016, accessed July 27, 2017, https://www.pri.org/stories/2016-06-22/holland-relocating-homes-make-more-room-high-water. Adam Wernick, "The Netherlands, always vulnerable to floods, has a new approach to water management," *PRI*, July 16, 2017, accessed July

27, 2017, https://www.pri.org/stories/2017-07-16/netherlands-always-vulnerable-floods-has-new-approach-water-management. Eleanor Beardsley, "Netherlands Shares Climate Change Advice With Poorer Countries," *NPR*, December 4, 2015, accessed July 27, 2017, http://www.npr.org/2015/12/04/458427019/netherlands-shares-climate-change-advice-with-poorer-countries.

2. My understanding of the Dutch landscape and its relationship to national culture emerged from time spent in the Netherlands through two research grants, a Charles Eliot traveling fellowship funded by the Harvard Graduate School of Design (1994) and a Fulbright scholarship funded by the Council for the International Exchange of Scholars (1998).

3. For reproductions of maps important to the reclamation projects of the Golden Age and a history of Dutch cartographic technology, please see Jan Werner, Chris Streefkerk and Frouke Wieringa, *Perfect Gemeten: landmeters in Holland Noorderkwartier ca. 1550–1700* (Holland: Stichting Uitgeverij Noord-Holland, 1994).

4. These groundwater elevations are compared to Dutch zero, Normaal Amsterdams Peil (NAP), the mean level of the River IJ at the Amsterdam harbor when it was still in open connection with the North Sea.

5. The Holland Defense Line, a string of fortifications along a swath of land that could be flooded to create a cross-country barrier from north to south, was developed in the eighteenth and nineteenth centuries to stop invaders. During the Second World War, the newly reclaimed Wieringermeer polder was flooded to foil the Germans.

6. The ANWB, the Royal Dutch Touring Club, is the Dutch equivalent to North American national automobile associations.

7. The physical structure of the polders has three elements: the organization of dikes, the parcellization of land by ditches and the three-dimensional organization of the water table. Each of the four basic methods for making polders has produced a characteristic arrangement of dikes, parcellization, and water table specific to a historical period, a technological capability, and a geographic condition.

8. For instance, the heritage landscape of the Vinkeveense Plassen was originally reclaimed as peat polders. When the polders subsided and became less valuable for agriculture, people manipulated the land in a second way: peat soil was harvested for fuel. The peat was cut in lines from the surface of the ground to below the level of the ground water and dried on the strips of earth left in between. Then natural process gained the upper hand. Wind washed away the strips of land between dredged lines. Broad shallow lakes were created, and the initial pattern of reclamation was erased. These lakes became flood hazards and were reclaimed again in the nineteenth century using the method of *droogmakerij*, which filled in the voids with a new pattern.

9. For a detailed account of Rotterdam's morphological development, see Frits Palmboom's *Rotterdam: verstedelijkt landschap* (Rotterdam: Uitgeverij 010, 1990).

10. "Welkom bij Droppie Water: Ontdek hier alles over water," *Droppie Water*, accessed July 26, 2017, http://www.droppiewater.nl/home/desktop.php. Droppie Water's forays into the landscape are described in a series of comic books drawn by Robert van der Kroft and published by the Union of Water Boards. The trip to the dike takes place in Robert van der Kroft, *Droppie Water 3: Op de Dijk*. (The Hague: Unie van Waterschappen, 1992). In other issues, Droppie visits the sewer system, follows the course of rainwater, visits a nature area, and learns about the climate.

11. Please see Appendix A.

12. Craig E. Colton's *An Unnatural Metropolis: Wresting New Orleans from Nature* (Baton Rouge: Louisiana State University Press, 2005) and Richard Campanella's *Time and Place in New Orleans: Past Geographies in the Present Day* (Gretna: Pelican Pub. Co., 2002) provide detailed accounts of the development of the city's drainage infrastructure.

13. The levees north and south of central New Orleans follow the edges of the Mississippi River and Lake Ponchartrain. The third, to the east, borders the Industrial Canal, a shipping channel excavated in the 1920s. The fourth, at the city's boundary with the suburb of Metairie, belongs to the 17th Street Canal, the largest of the canals that carries pumped water out of the city and into the lake. Levees have had their own unexpected consequence: the construction of higher

and higher levees along the Mississippi has caused the water flowing past the city to rise. At flood stages, it grew higher because levees kept it within narrower boundaries than before; at ordinary stages, it grew higher because the sediment trapped in the river's main channel by the levees fell to and raised the channel bottom.

14. For a more detailed explanation of New Orleans's drainage system, see Jane Wolff, Derek Hoeferlin, and Elise Shelley, *Gutter to Gulf: Legible Water Infrastructure for New Orleans*, http://www.guttertogulf.com/. For a discussion of New Orleans in relation to current ecological thinking and management strategies, see Jane Wolff, "Cultural Landscapes and Dynamic Ecologies: Lessons from New Orleans," in *Projective Ecologies*, ed. Chris Reed & Nina-Marie Lister (Cambridge: Harvard University Graduate School of Design; New York: Actar Publishers, 2014), 184–203.

15. John McPhee's essay "Atchafalaya" offers a beautiful portrait of ad hoc solutions to subsidence in the domestic landscapes of New Orleans. In *The Control of Nature* (New York: Farrar, Straus, & Giroux, 1989), 60.

16. I was involved in several efforts of this kind. I organized Longue Vue House and Gardens' work with the Pontilly Disaster Collaborative to develop rehabilitation strategies for the neighbourhoods of Pontchartrain Park and Gentilly Woods. This endeavor (which expanded to include Julie Bargmann, Elizabeth Meyer, and William Morrish and their students at the University of Virginia, Elizabeth Mossop and her students at Louisiana State University, and Mia Lehrer & Associates) was documented in the *Pontchartrain Park & Gentilly Woods Landscape Manual* (New Orleans: Longue Vue House & Gardens, 2009) and in a chapter by the same name in the book *New Orleans Under Reconstruction* (New York: Verso, 2014). I participated in the Dutch Dialogues workshops 1 and 3 in 2008 and 2010. In 2011, I served as a technical advisor to the consultant and advocacy team that instigated and prepared the New Orleans Water Plan. Most important, from 2009 to 2013, I led a teaching and research initiative with Elise Shelley and Derek Hoeferlin at the University of Toronto and Washington University to document and describe New Orleans's water infrastructure. This work, published as *Gutter to Gulf* (www.guttertogulf.com) served as base information for the Dutch Dialogues and the *Greater New Orleans Urban Water Plan*. In 2009, I worked with Austin Allen, Walter Hood, and Elizabeth Mossop on a landscape proposal to the Make It Right Foundation for the renewal of the Lower Ninth Ward.

17. For an extended discussion of these and other recent landscape initiatives in New Orleans, please see Elizabeth Mossop's article "New Orleans Owns Its Water: Rethinking the Urban Landscape After Katrina," *Landscape Architecture Magazine*, October 2016, 110–119.

References

Actueel Hoogtebestand Nederland. http://www.ahn.nl/index.html (accessed July 26, 2017).

Als ons water stijgt. Woon jij in een risicogebied? http://www.overstroomik.nl (accessed July 26, 2017).

Beardsley, E. 2015. NPR, December 4. "Netherlands shares climate change advice with poorer countries." http://www.npr.org/2015/12/04/458427019/netherlands-shares-climate-change-advice-with-poorer-countries (accessed July 27, 2017).

Bentley, C. 2016a. PRI, June 20. "As sea levels rise, Rotterdam floats to the top as an example of how to live with water." https://www.pri.org/stories/2016-06-20/sea-levels-rise-rotterdam-floats-top-example-how-live-water (accessed July 27, 2017).

Bentley, C. 2016b. PRI, June 22. "Holland is relocating homes to make more room for high water." https://www.pri.org/stories/2016-06-22/holland-relocating-homes-make-more-room-high-water (accessed July 27, 2017).

Campanella, R. 2002. *Time and Place in New Orleans: Past Geographies in the Present Day*. Gretna: Pelican Pub. Co.

Central Bureau of Statistics. 1993. *Statistical Yearbook 1993 of the Netherlands*. The Hague: Staatsuitgeverij/ CBS-publications.

Centraal Burean Voor de Statistiek. 1954. *Jaarcifers voor Nederland 1947–1950*. Utretch: Uitgeversmaatschappy W. de Haan N.V.

Colton, C. E. 2005. *An Unnatural Metropolis: Wresting New Orleans from Nature*. Baton Rouge: Louisiana State University Press.

Dodge, M. M. 1896. *Hans Brinker, or the Silver Skates*. London: Sampson Low, Marston, and Company.

Droppie Water. Welkom bij Droppie Water: Ontdek hier alles over water. http://www.droppiewater. nl/home/desktop.php (accessed July 26, 2017).

Dutch Water Sector. 2017. "Room for the River programme nears completion with new by-pass of Ijssel river, the Netherlands." 23 February. https://www.dutchwatersector.com/news-events/ news/23707-room-for-the-river-programme-nears-completion-with-new-by-pass-on-ijssel-river.html (accessed June 8, 2018).

Eurostat. 2017. General government expenditure by function (COFOG). http://ec.europa.eu/eurostat/ web/products-datasets/-/gov_10a_exp (accessed July 27, 2017).

Kimmelman, M. 2013. Going with the flow. *New York Times*, February 13. http://www.nytimes. com/2013/02/17/arts/design/flood-control-in-the-netherlands-now-allows-sea-water-in.html (accessed May 24, 2017).

Kimmelman, M. 2017. The Dutch have solutions to rising seas. The world is watching. *New York Times*, June 15. Accessed July 24, 2017. https://www.nytimes.com/interactive/2017/06/15/world/ europe/climate-change-rotterdam.html (accessed July 24, 2017).

Kroft, R. van der. 1992. *Droppie Water 3: Op de Dijk*. The Hague: Unie van Waterschappen.

LaRose, G. 2016. *New Orleans Times-Picayune*, January 26. "Gentilly resiliency district goes against flow of how New Orleans handles stormwater." http://www.nola.com/politics/index.ssf/2016/01/ gentilly_resiliency_district_g.html (accessed July 25, 2016).

Mak, G. 2001. *Het Ontsnapte Land*. Amsterdam: Atlas.

McPhee, J. 1989. "Atchafalaya." In *The Control of Nature*. New York: Farrar, Straus, & Giroux.

Metz, T. 1998. *Niewe Natuur: Reportages Ovewr Veranderend Landschap*. Amsterdam: Ambo, cop.

Ministerie van Volkshuisvesting, ruimtelijke ordening en milieubeheer. 1998. *Nederland 2030*. Den Haag: Ministerie van Volkshuisvesting, ruimtelijke ordening en milieubeheer.

Mossop, E. 2016. New Orleans owns its water: Rethinking the Urban Landscape after Katrina. *Landscape Architecture Magazine*. October 2016. 110–119.

Netherlands Central Bureau of Statistics. 1966. *Statistical Yearbook of the Netherlands 1963–1964*. Hilversum: Uitgeversmaatschappij W. de Haan.

Netherlands Central Bureau of Statistics. 1970. *Statistical Yearbook of the Netherlands 1967–1968*. The Hague: Staatsuitgeverij.

Netherlands Central Bureau of Statistics. 1972. *Statistical Yearbook of the Netherlands 1972*. The Hague: Staatsuitgeverij.

Netherlands Central Bureau of Statistics. 1978. *Statistical Yearbook of the Netherlands 1977*. The Hague: Staatsuitgeverij.

Netherlands Central Bureau of Statistics. 1983. *Statistical Yearbook of the Netherlands 1982*. The Hague: Staatsuitgeverij.

Netherlands Central Bureau of Statistics. 1988. *Statistical Yearbook of the Netherlands 1987*. The Hague: Staatsuitgeverij.

O'Neil, L. 2017. *The Atlantic*, September 2. "Why doesn't New Orleans look more like Amsterdam?" https://www.theatlantic.com/technology/archive/2015/09/why-doesnt-new-orleans-look-like-amsterdam/402322/ (accessed July 26, 2017).

Ons water in Nederland. Schoon, veilig en voldoende water is niet vanzelfsprekend. Accessed July 26, 2017. https://www.onswater.nl/ (accessed July 26, 2017).

Palmboom, F. 1990. *Rotterdam: verstedelijkt landschap*. Rotterdam: Uitgeverij 010.

Rijkswaterstaat: Ministerie van Infrastructuur en Milieu. Onze historie. https://www.rijkswaterstaat. nl/over-ons/onze-organisatie/onze-historie.aspx (accessed January 11, 2017).

Ruimte voor de Rivier. https://www.ruimtevoorderivier.nl/ (accessed July 27, 2017).

Ruimte voor de Rivier. 2018. "Eindevaluatie Ruimte voor de Rivier aangeboden aan Tweede Kamer." 22 March 2018. https://www.ruimtevoorderivier.nl/eindevaluatie-ruimte-rivier-aangeboden-aan-tweede-kamer/ (accessed June 8, 2018).

Sewerage and Water Board of New Orleans. A Historical Look at the Sewerage and Water Board. http://www.swbno.org/history_history.asp (accessed July 7, 2017).

Sociaal-Economische Raad. 2001. *Vijfde Nota Ruimtelijke Ordening: Advies Over de Vijfde Nota de Ruimtelijke Ordening 2000–2020*. Den Haag: Social-Economische Raad. https://www.ser.nl/~/media/db_adviezen/2000_2009/2001/b19865.ashx (accessed May 25, 2017).

Statistics Netherlands. 1997. *Statistical Yearbook 1997 of the Netherlands*. Voorburg: Statistics Netherlands.

Statistics Netherlands. 2003. *Statistical Yearbook of the Netherlands 2003*. Voorburg: Statistics Netherlands.

The World Bank. GDP (current US$). http://data.worldbank.org/indicator/NY.GDP.MKTP. CD?locations=NL (accessed July 27, 2017).

The World Bank. Military expenditure (% of GDP). http://data.worldbank.org/indicator/MS.MIL. XPND.GD.ZS?locations (accessed July 27, 2017).

Topografische Dienst. 1987. *Grote Topografische Atlas van Nederland 1:50,000*. Groningen: Wolters-Noordhoff Atlasprodukties.

van de Ven, G. P. (ed.) 2004. *Man-Made Lowlands: History of Water Management and Land Reclamation in the Netherlands*. Utrecht: Uitgeverij Matrijs.

Waggoner & Ball, Architects. 2013. *Greater New Orleans Urban Water Plan: Vision*. New Orleans: Greater New Orleans, Inc. http://livingwithwater.com/blog/urban_water_plan/reports/ (accessed May 25, 2017).

Werner, J., C. Streefkerk and F. Wieringa. 1994. *Perfect Gemeten: Landmeters in Holland Noorderkwartier ca. 1550–1700*. Holland: Stichting Uitgeverij Noord-Holland.

Wernick, A. 2017. PRI, July 16. "The Netherlands, always vulnerable to floods, has a new approach to water management." https://www.pri.org/stories/2017-07-16/netherlands-always-vulnerable-floods-has-a-new-approach-water-management (accessed July 27, 2017).

Wolff, J. 2014. "Cultural Landscapes and Dynamic Ecologies: Lessons from New Orleans." In *Projective Ecologies*, edited by Chris Reed & Nina-Marie Lister. Cambridge: Harvard University Graduate School of Design; New York: Actar Publishers.

Wolff, J. and C. McMichael Reese. 2014. "Pontchartrain Park + Gentilly Woods Landscape Manual." In *New Orleans under Reconstruction*, edited by Carol McMichael Reese, Michael Sorkin and Anthony Fontenot, 381–392. New York: Verso.

Wolff, J. and C. Reese (eds.) 2009. *Pontchartrain Park & Gentilly Woods Landscape Manual*. New Orleans: Longue Vue House & Gardens.

Wolff, J., E. Shelley and D. Hoeferlin. *Gutter to Gulf: Legible Water Infrastructure for New Orleans*. http://www.guttertogulf.com/ (accessed July 26, 2017).

Section II

Methods and Practices

The complexity of dynamic coastal environments calls for new methods in planning and design across disciplines and cultures.

5

The Joy of Counterintuitivity

Rob Roggema

CONTENTS

Introduction

The Dutch have a long tradition of coastal engineering and management. The designs made for closure dams and dikes and land reclamation were functional, engineering wonders. With the changes in climate and the increasing attention for ecological systems approaches, the attitude toward coastal design is changing. The engineered plan is no longer seen as the ultimate goal; the new paradigm is how the power of natural processes could help to create a stronger coast. In this chapter, the role of complex systems is further elaborated to understand the contribution these systems may deliver to further enhance coastal resilience. The focus is on designing the intervention, rather than controlling the entire plan, and trusting that the powers of nature will take the landscape to a more resilient level. Before this approach is presented, a brief overview of Dutch coastal design, planning, and policy is given, after which the paradigm shift from an engineering approach to a more organic worldview is described. This chapter concludes by introducing Swarm Planning, a methodology to design more adaptive landscapes.

A Brief History of Dutch Coastal Planning

For many years the Dutch have fought against water. Smart as they were, they combined fighting with a trade-off. With every battle, they reclaimed a little piece of land and, after the battle was won, the Dutch built a dike to protect the new land from flooding. This reclamation process went on for centuries and is one of the reasons a substantial part of the country lies below sea level. It is also the reason the Netherlands is one of the most vulnerable countries to sea level rise and storm surges (Climate Central, 2014). Within the tradition of water-managing the Dutch landscape, engineers in the early twentieth century began to show serious land reclamation projects on the map of the Netherlands. The whole of the then-called Zuiderzee was supposed to be turned into land. There have been plans to do so for centuries, the plan of Hendrik Stevin being the first in 1667. In 1891, Cornelis Lely drew the plan (see Figure 5.1) that was eventually partly realized (Nieuw Land, undated), and there were even plans to reclaim the whole Wadden Sea (Walsmit et al., 2009), today one of the most valuable wetlands in Europe.

Starting from the idea that nature could be controlled, large-scale reclamation started with building the Closure Dam, separating the Zuiderzee from the open ocean. This happened in 1932, and the Zuiderzee became the IJsselmeer (IJsselmeer). At the time there were serious doubts, mainly of a financial nature, whether Lely's plan to transform the IJsselmeer into land was the right thing to do (Civiele Techniek, 2007). This changed after the national flood in 1916. The threat of a new flood combined with the need to produce food suddenly made the fertile soils at the bottom of the IJsselmeer of great interest (Boerma, 2016). The flood of 1953 again accelerated thinking around protecting the country. Twenty days after the flood, a Delta Works committee was installed (Deltawerken online, undated) to give advice about the execution of the Delta Plan, which had already been made in 1937 by the Dutch Ministry of Public Works. The Delta Plan would finally be realized in 1997 (Deltawerken online, undated).

In the 1960s and 1970s plans for dikes and dams (Brouwersdam, Grevelingendam, Philipsdam, Oosterscheldekering and other dams) were made and realized for closing off the estuary of Zeeland, and at the same time the polders in the IJsselmeer, Northern and Southern Flevoland were pumped dry. Huge areas of new land were added to the map of the Netherlands, which was substantially transformed during these two decades (Figure 5.2a and b).

The saying "God created the World, but the Dutch created the Netherlands" became true more than ever. In the 1970s, however, with the emergence of a strong environmental movement, there were questions as to the need for, and ecological impacts of, building a dam in the Oosterschelde. The environmental movement in collaboration with the fishermen's union was successful in achieving a lower impact solution, as the government decided to create an open dam (Dekkers and de Jong, 2002) that can be closed during a storm surge, the so-called Oosterscheldekering. The environmentalists also prevented the Markerwaard, which was supposed to be the finishing piece in reclaiming the entire IJsselmeer and turning it into land (Huitema, 2004).

Today, the Dutch continue with reclaiming land all around the world. However, current projects work with nature to a greater degree than historically, in order to compensate or even increase the quality of the ecological environment. The second Maasvlakte, allowing the harbor of Rotterdam to move seaward (Verweij et al., 2014); IJburg, a new urban precinct near Amsterdam in the IJssel Lake (Kinder, 2011; Savini, 2017); and the Palmeira Island in Dubai (Gibling, 2013); though not all extremely eco-friendly (Cressey, 2011), are only a few of the examples the Dutch have created in recent history.

All this engineering power, however, could only reduce the flood risk to a certain extent. As climate change became a more serious issue, causing increasing sea level rise and the accompanying increased risk of storm surges, there were several near misses of river flooding (Engel, 1997; Tol and Langen, 2000) in the Netherlands. People began to understand that an even greater risk was posed by the possible breach of even the strongest dike or dam.

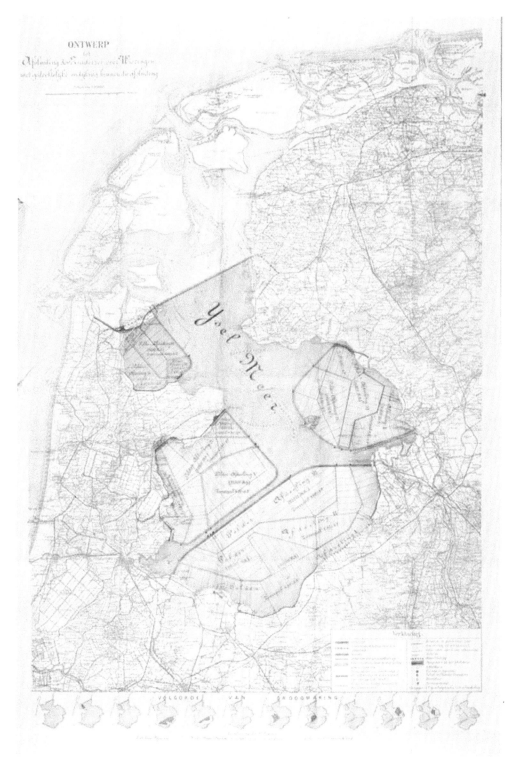

FIGURE 5.1
Plan for reclaiming the entire Zuiderzee, Cornelis Lely, 1891. (Collectie Nieuw Land, Zuiderzeecollectie.)

(b)

(a)

FIGURE 5.2
Comparing the Netherlands before (a) and after (b) the Delta Program. (https://upload.wikimedia.org/wikipedia/commons/c/c9/Carte_Hollande_1843.jpg; https://visibleearth.nasa.gov/view.php?id=54661.)

The river system and the coast showed themselves to be increasingly vulnerable to storms and flooding. The "Room for the River" program was initiated (Climate Wire, 2012; Nijssen and Schouten, 2012) and designs for the weak points (so-called zwakke schakels) along the Dutch coast began to be developed (Ministerie van Verkeer en Waterstaat, 2003) (see Figure 5.3a and b).

(a)

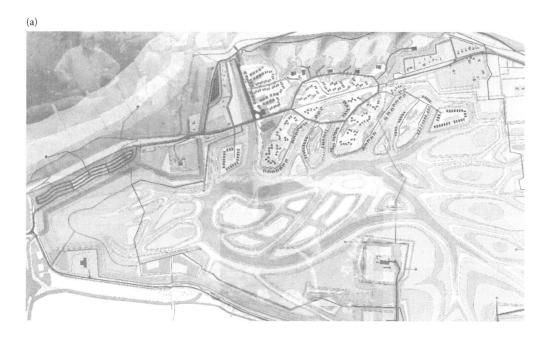

(b)

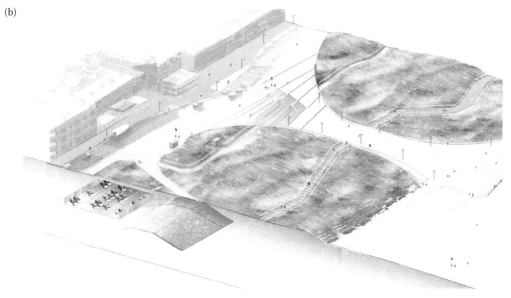

FIGURE 5.3
The weak points (zwakke schakels) of Waterdunen in the province of Zeeland (a), and with a parking building under the new dunes in Katwijk (b). (From Provincie Zeeland and https://www.deingenieur.nl/artikel/katwijk-heeft-dijk-en-parkeergarage-in-duin.)

In 2008, a second Delta Committee was instituted to develop a vision of how the Netherlands could be protected in the long term (Deltacommissie, 2008), eventually leading to the Delta Program in 2010 (Ministeries van V&W, LNV en VROM, 2010), to be followed by seven successive programs, which currently implement the Delta Plan.

State of the Art

The current approach to designing urban landscapes can be characterized by a transition from the short-term belief that engineering a solution for a threat or problem could solve the problem, toward a more long-term organic point of view, which believes in the strengths of systems to overcome problems (Brouwer, 2011; Roggema and Broess, 2014). Brouwer identifies four worldviews: the mythic, the static, the mechanic, and the organic. As an example, the way we deal with water problems is illustrative. In the mythic worldview building small artificial hills on top of which people could live dealt with the flooding. In the static worldview, the problem was dealt with by building a dike. The mechanic (or engineering) worldview solved the problem through pumping the water out, and the organic worldview solved the problem by introducing a self-regulating system of water absorption and discharge, the WADI (Beenen and Boogaard, 2007).

Introducing self-regulating systems, for instance when the problem of sea level rise and storm surges must be dealt with, translates in the Dutch context as "building with nature" (Waterman, 2010). The idea is that using the powers of nature to create the system allows a resilient system that is capable of dealing with the changes. This is a form of adaptive planning (Schuetze and Chelleri, 2011; Wirkus, 2016). Adaptive planning is the "joint capability to create and revise plans rapidly and systematically, as circumstances require" (definedterm.com). Applying this type of planning implies that the governing agency takes a step back and allows the system to adapt by itself. From this starting point, the adaptation of the system means it is capable of adjusting the design, planning, and management of a plan or region down the line. The Delta Program (Ministeries van IenM and EZ, 2016) has taken adaptive planning as its leitmotiv, treating the unpredictable future as changes and disruption occur. This includes disruption in content, such as uncertainties around sea level rise, superstorms or the perfect storm, and disruption by the complex participatory approach chosen as the way to implement the program. The latter sees regional agencies, such as provinces, water boards, and municipalities being responsible for developing the designs together with local stakeholders and residents, but in close collaboration with the program directorate. Recently, several examples based on this thinking have been realized.

The sand engine (Stive et al., 2013) is a project off the Western Coast of the Netherlands that saw a one-off deposit of 21.5 million cubic meters of sand that will stay in the system for about 20 years. It will allow the development of a permanent ecosystem. The sand is redistributed along the coast by currents and wind, natural forces that deposit sand in the right places to build up coastal protection and ecological systems, providing an environment for recreational use (De Vlieger, 2017).

Located on the Western Coast of the Netherlands near Camperduin and Petten, the Hondsbossche Dunes (www.west8.com/projects/hondsbossche_and_pettemer_zeewering/)

FIGURE 5.4
The Hondsbossche Dunes in the making. (http://www.hoogwaterbeschermingsprogramma.nl/Communities/
Common/Images/Hoogwaterbeschermingsprogramma/Zwakke%20Schakels%20NH.jpg.)

replace the old dike with a dune landscape surrounding a lagoon in the middle. This forms
a sandy and dynamic solution as opposed to a hardscape. It simultaneously strengthens the
region, enhancing and embracing the qualities of the natural environment, while bolstering
recreational activities.

The tranquility, nature, and clear access to the beach are defining features of this place,
and the Hondsbossche Dunes begin a new chapter in the narrative of this landscape. The
longest uninterrupted dunescape is created with varied topography, overgrown with native
vegetation, complete with young drifting dunes, dune valleys, and a beach that varies in
width. The project meets the dual goals of safety and environmental quality (Figures 5.4
and 5.5).

Restoring the natural habitat could help build resilience in the coastal protection
system, as shown in these examples. However, this will only partly increase safety. More
effort is needed to deal with unprecedented impacts, in turn requiring the allowance of
future dynamics in the current design. The first step in this process is to initialize the
spatial transformation that could accommodate the unprecedented changes. Second, once
the spatial intervention has initialized system change, the landscape needs to be (re)
designed to spatially accommodate the impacts of the unknown dynamics of the future.
Finally, the actual space must be created to allow these dynamics to form an emergent
new landscape.

FIGURE 5.5
The Hondsbossche Dunes after realization. (www.west8.com/projects/hondsbossche_and_pettemer_zeewering/.)

Problem Definition

Despite these recent initiatives, the inherent uncertainty of the climate system and its impacts could bear down on the landscape, for instance when a superstorm (Hansen et al., 2016), in combination with accelerated sea level rise (Overpeck et al., 2006; Chen et al., 2017) and normal king tide, hits the unprepared hinterland. Exactly this scenario was broadcast on the Dutch-Flemish television in the TV series "When the Dikes Breach" (NPO, 2016), which explored the debate around whether environmental disasters can happen even if we think we have all coastal defenses in place. In her comments on that TV series, the Minister for Water Management admitted that coastal disasters could happen but reasoned that as all coastal safety measures have been put in place, it will not ever happen. This illustrates a way of thinking whereby we "engineer our way out of the problem." This approach attempts to control nature, designing the future in great technical detail. Logic and rationale prevail as the dominant way of thinking, and there is no reliance on the natural systems' ability to form a resilient system. Hence the landscape and urban systems are not designed in a way that adeptly or thoroughly accommodate the dynamics of natural systems. With the ever-increasing complexity of climate change impacts, developments have become non-linear. A core characteristic of these wicked problems (Rittel and Webber, 1973) is that they cannot be ultimately solved. The common responses do not work, and a linear solution to the problem is not possible. While experiments such as the sand engine have inbuilt adaptive management principles, the magnitude of future change and the interrelatedness of systems will require further advanced approaches, in which uncertainty, adaptivity, and counterintuitivity will need to have a prominent position.

Counterintuitivity

Instead of thinking in ways we have always done, the solution to these wicked problems might lie in counterintuitive thinking (Concevitsch, 2007; Lane and Harris, 2014; Srivastava, 2014). If we would like to give space to the dynamic of the landscape to create a resilient self-organizing system strong enough to deal with sudden change, an unprecedented intervention could enable this. This approach, Swarm Planning (Roggema, 2012), entails looking at the change in the landscape systems at a different pace and finding the unique spot to trigger modification that can accommodate the wicked and interrelated problems such as those posed by climate change. Non-linearity is a key concept that applies to these problems and is the reason why Swarm Planning seeks spatial interventions that are novel and counterintuitive. A counterintuitive thought is one that creates new conditions and crosses the boundaries of existing frameworks. Instead of creating a presumed increase in safety, as we do by raising the heights of dikes, a counterintuitive intervention could allow nature to form the landscape and regain safety in a natural way. This way the design goes beyond restoring habitats to create new ones in a self-responsive way. Another aspect of counterintuitive thinking relates to the design profession itself. Designs are often seen as a masterpiece of how the world would look after the design has been realized. In the context of coastal resilience, this approach is counterproductive as a dynamic equilibrium is sought. If the intervention is designed only, the space for an emergent system and room for dynamics are enhanced. Once the impacts of the intervention are clear, and the dynamic equilibrium is reached, then the landscape can be designed.

The landscape is a series of self-organizing systems: responsive, adaptive, and complex. Both urban environments (Allen, 1996; Portugali, 2000, 2008; Batty, 2005, 2013; Bettencourt, 2013, 2015) and landscapes are seen as complex systems (Favis-Mortlock and de Boer, 2003; Roggema, 2013). There are several concepts useful when designing for a complex spatial system.

Self-organization of a system is "the potential to spontaneously and unpredictably develop new forms and structures by itself out of chaos" (Merry and Kassavin, 1995). It is also the ability to spontaneously arrange its components or elements in a purposeful (non-random) manner, under appropriate conditions but without the help of an external agency (Kauffmann, 1995; Krugman, 1996). This implies that design should create the conditions for a spontaneous process, that is, a landscape able to provide its own foundational elements, such as sand and water, if the coast needs to develop.

Emergence occurs when a system moves from a certain state into another. The system is not, but rather becomes. This will happen if certain elements together form new entities. Emergence cannot be planned, only stimulated by creating the starting conditions (Krugman, 1996; Goldstein, 1999). For the design of the landscape, this means that certain elements need to be put together, not that the end state of the landscape must be designed but solely considered to be the point of departure.

Adaptive capacity is the ability of a system to adjust to climate change (including climate variability and extremes), to moderate potential damages, to take advantage of opportunities, or to cope with the consequences (Brooks et al., 2005). In design terms, adaptive capacity can be enhanced if a landscape can be easily adjusted or make use of opportunities. The spatial elements in the landscape should therefore not be fixed but should be afforded their own innate capacity to change over time.

Fitness landscapes are complex adaptive systems that self-organize and adapt, in order to remain within their current state. The system only shifts to other attractors (alternative states) after a shock that drives the system out of its current state (e.g., due to significant change). Major adjustments are needed and, after the initial shock, the system will self-organize to achieve those adjustments. The process this system goes through can be represented in the form of a fitness landscape (Langton et al., 1992; Mitchell Waldrop, 1992; Cohen and Stewart, 1994). This fitness landscape includes positions considered favorable (the peaks) and less favorable (the valleys). A complex system tends to move across less favorable valleys to the highest possible position in the landscape, the attractor. At the peak, the adaptive capacity is highest, which allows the system to adapt more easily to changes in its environment. The design of a landscape should freely allow the dynamic play of elements that will enable that landscape's potential to re-shape and re-organize itself and seek a new stable situation.

The following process illustrates how a dynamic and resilient landscape can be designed using these concepts. The example of the Floodable Landscape (Roggema, 2012) is used to illustrate the different design steps.

First, the characteristics of the system are analyzed to evaluate the elevation and topography of the landscape, the potential sea level rise, eventual weaknesses in the coastal protection, and cultural-historic values of the landscape.

Second, the counterintuitive intervention is proposed. In this case, the proposition does not comprise a strengthening of the dams and dikes along the coast but the opposite: through a hole in the dike the seawater is let in the hinterland, on purpose. At first this only happens virtually during the design process to examine how different levels of sea level rise impact the landscape (see Figure 5.6a–d).

Third, different design options can be tested to respond to the impacts of the intervention. New inland dams can be built, complete floating villages established, or a natural systems process instigated. In this Research by Design process (Roggema, 2016) the merits of different design solutions are tested and judged by their impact on the quality of the landscape.

In the fourth stage, the landscape interventions that support building with nature processes are designed. The proposition here is to create wooden structures (see Figure 5.7a) that will capture sand and clay with the tidal process of seawater coming in and will leave the ground level of the landscape slowly rising. This is a process very familiar to the people in this region who have reclaimed land from the sea by exactly the same approach, only at the outer side of the dikes. It has worked in the past by raising the ground level until it emerges above sea level and creating new land, called kwelders.

Based on this intervention, a detailed urban design can be founded on the new features of the landscape. The landscape will eventually flood and will become a dynamic landscape. This means that new nature and water will be all around the region, and the fluctuating conditions will accordingly demand responsive housing. The houses will be attached to poles to prevent them from floating away and to ensure they can move up and down with the tide. The houses will be simultaneously fixed and floating, innately well adapted to future conditions of a rising sea level (Figure 5.7b).

Current Developments, Future Outlook

Following the dynamic equilibrium approach, the initial designs need to inform policy debate and development and be integrated in implementation frameworks. The concept of an inundated landscape that is responsive to sea water and raises the ground level of the landscape has been taken up in the national Delta Program (see Figure 5.8) and the

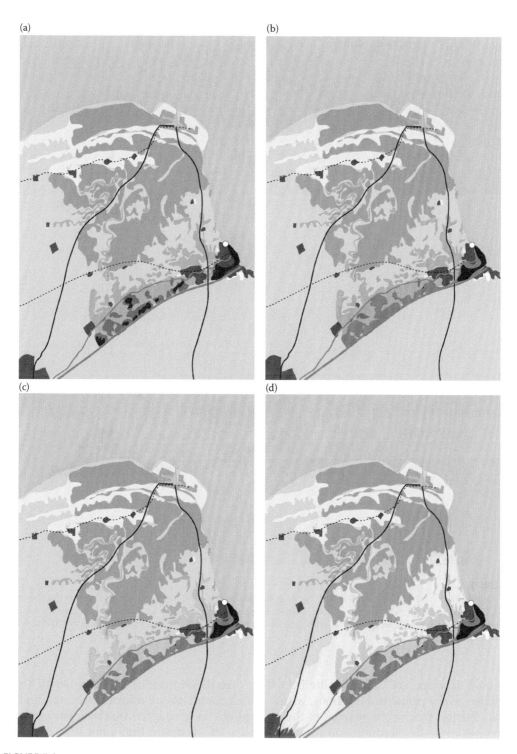

FIGURE 5.6
The impact of different levels of sea level rise: 0.3 m (a), 0.6 m (b), 0.9 m (c), and 1.2 m (d); on the landscape after creating a gap in the dike. (From Roggema, R. 2012. "Swarm Planning: The development of a spatial planning methodology to deal with climate adaptation." PhD thesis. Delft University of Technology, Delft.)

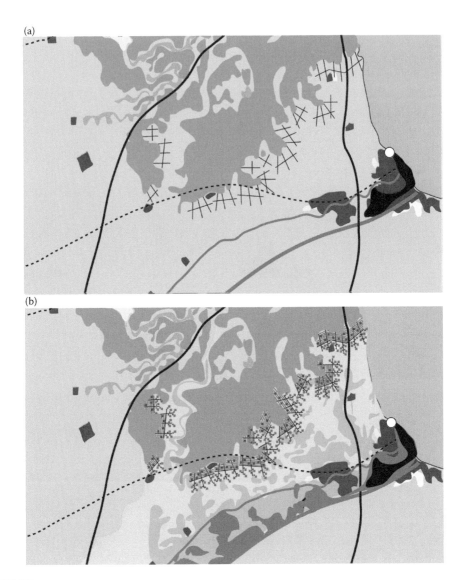

FIGURE 5.7
(a) The design of a system of wooden structures in the landscape, ahead of eventual water entering the area. (b) A flooded landscape with floating houses. (From Roggema, R. 2012. "Swarm Planning: the development of a spatial planning methodology to deal with climate adaptation." PhD thesis. Delft University of Technology, Delft.)

regional development vision for the Eemsdelta area (Groningen, 2013). These small areas function as the prime test locations, which, if successful, can be implemented at a higher level. The fact that, seven years after their inception, these counterintuitive ideas are now taken seriously by planning and regulatory agencies illustrates that innovative concepts can not only influence policy development, but that they are critically needed to challenge and transform regular policy in a time of significant environmental change.

The urgency is genuine. We know unparalleled climatic impacts will hammer the coast for many years to come, but regular policy affords little more than a standstill, static

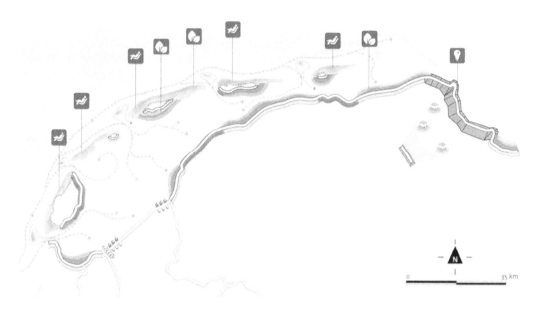

FIGURE 5.8
The inundation strategy as part of the national Delta Program. (Ministry of Infrastructure and Environment, Delta Program, 2014.)

approach unable to respond to the challenges ahead. Therefore counterintuitivity is a way of approaching unprecedented problems and can be used to invent solutions that were once deemed impossible. The dynamic coast can be a clear beneficiary of this thinking. Working with interventions that start a process of coastal landscape forming could create stronger and safer coasts than those forged through damming and engineering. Two plans that have recently been released in Australia illustrate how counterintuitive thinking could lead to resilient coasts: A Sydney Barrier Reef and Badgerys Island.

A Sydney Barrier Reef

Climate change is influencing the temperatures in the Pacific (Australian Bureau of Meteorology and CSIRO, 2011). This leads to a warming ocean both in the area of the Great Barrier Reef and also in front of Sydney's coast. In recent years the rise in temperature has caused bleaching in up to 75% of the Barrier Reef. If the ocean temperature continues to rise, the reef will eventually die. Assuming this will happen, and climate predictions show this, even if global warming is stopped immediately, the warming will continue for several decades or even longer (Frölicher et al., 2014). The choice is to live without a Great Barrier Reef or design an intervention that could serve as a next best option. Regular policy will emphasize measures to save the Great Barrier Reef, and many actions and regulations have been put in place that aim to reach this result (Commonwealth of Australia, 2015). A counterintuitive approach looks at the potential of re-creating a Barrier Reef. As the Pacific Ocean near Sydney is warming too, the optimal conditions for reef ecology will slowly move southward and reach the Sydney coastal zone. Therefore, this is the ideal place to create a Sydney Barrier Reef (Roggema, 2017). This reef would be constructed as an artificial reef, using old oilrigs that create the conditions for reef growth. The inhabitation of these steel bases would take a while, but tests have shown the corals would grow rapidly and faster than in a natural condition (Forsman et al., 2015). As temperatures of the southern Pacific rise, the chance that Sydney will be hit by a cyclone also increases. The reef offers a protective shield for the built-up value of real estate and the significant population living in metropolitan Sydney. At the same time, the design of a Sydney Barrier Reef offers funnels where the remarkable surf beaches are. Through these funnels, the wave activity is increased, which leads to better surf conditions, including higher waves and longer breaks. During storms these funnels will be closed off by the wave-plants that generate energy out of the waves. These pontoons that capture the wave activity are normally loosely secured but in case of a storm they would be tied very strongly and so dampen the waves. The reef would potentially operate as a refuge for the Great Barrier Reef ecosystem. Once climate change is under control, the water flora and fauna can return to their northern origin. In the meantime, and perhaps forever, the Sydney Barrier Reef could become a tourist attraction and valuable ecosystem at the same time.

Badgerys Island

A second example of counterintuitive design thinking is the concept for Sydney's Badgerys Island. If global warming is not controlled to reasonable levels, eventually all ice on Earth will melt. This will lead to a sea level rise of 70 or even 80 meters (Williams and Hall, 1993; Poore et al., 2011; Williams and Ferrigno, 2012). In these circumstances, it is inevitable to think about the retreat of urban population to inland areas where the landscape elevation is higher. For the landscape around Sydney, this rise in sea level means that a large portion of the current city will drown, and only a modest piece of inhabitable land will remain. The new coastline will form where the Blue Mountains currently are, some 60 kilometers west of Sydney's beaches, and the majority of Sydney residents would be forced to move to this area. An interesting coastal zone will remain, comprising an island and several peninsulas, where new activities and living can be located. A new harbor area will be distinguished, as this large habitable island emerges. This island, located in the Badgerys Creek area, could host the new Badgerys Creek airport and provide home to approximately one million people. Because the island would be surrounded by the ocean, thorough coastal protection is required, consisting of reefs, dunes, and cliffs that together will form a responsive system against storm surges. In keeping with Sydney's tradition, the island could only be reached by ferries that would serve as the people movers but would also bring supplies to Badgerys Island. The internal transportation would utilize smaller barges, bringing people to their houses and around the island. On Badgerys Island, there would be a substantial amount of jobs for the inhabitants. Tourism would be a large employer as natural bays and boat-life culture are extremely suitable for yachting and culinary exquisiteness. Some of the island's food would be produced on the island itself, in small farms dotted along the coast, and in the aquaponic systems that could be integrated in the apartment buildings.

This plan illustrates a counterintuitive approach to huge sea level rise. The approach of regular policies would be to adjust the existing coastal defenses to a level that eventually proves to be neither strong nor high enough. Instead of waiting for this disaster to occur, the counterintuitive option to retreat implies the sea will occupy what we now know as Sydney. Understanding this, the ultra-long-term vision shows the new coastline and its habitable places. The lead-up to this ultimate situation affords time and capacity for residential development, businesses, and valuable goods to be slowly relocated to higher places (Figure 5.9).

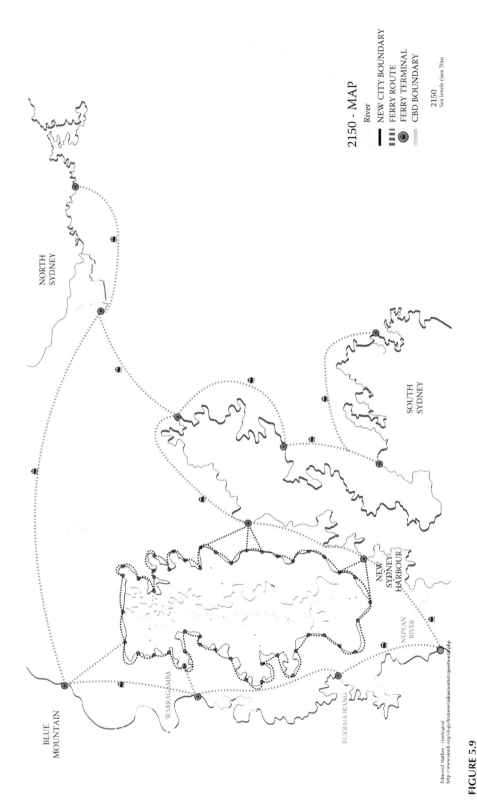

FIGURE 5.9

Badgerys Island. (Drawing by Kanha Keat, Design studio GO WEST! UTS Masters of Architecture, Autumn 2017, Studio leaders: Rob Roggema & Craig Allchin.)

Conclusion

The examples shown in this chapter underwrite the necessity for counterintuitive thinking in order to find solutions for serious changes that are ahead of us. It is clear the regular policies we already know will fall critically short and prove insufficient to deal with future disruptive events on the scale science is predicting.

An approach that opts for a designed end result will not work, as the processes in the coastal zones are by definition dynamic. Therefore, a systemic approach that builds on the core qualities of the complex coastal system and the qualities that natural dynamics bring is required.

In our designs, we can learn from complexity concepts such as emergence, self-organization, adaptive capacity, and the fitness landscape. Using these concepts in design means we are translating our understanding of how a swarm operates to the design of resilient and adaptable landscapes. As a swarm adjusts its shape whenever new impacts, interventions, or incentives occur, similarly, a single intervention in the design of a landscape will start the dynamic process of landscape forming. The swarming landscape is then adjusted to respond to a future change in the best possible way. It creates a resilient landscape and delivers new qualities, not after a disaster has taken place, but beforehand: this is what Swarm Planning does (Roggema, 2012).

References

Allen, P. M. 1996. *Cities and Regions as Self-Organizing Systems: Models of Complexity*. London and New York, NY: Taylor and Francis.

Australian Bureau of Meteorology and CSIRO. 2011. Climate Change in the Pacific: Scientific Assessment and New Research. Volume 1: Regional Overview. Volume 2: Country Reports.

Batty, M. 2005. *Cities and Complexity: Understanding Cities with Cellular Automata, Agent-Based Models, and Fractals*. Cambridge, MA and London: MIT Press.

Batty, M. 2013. *The New Science of Cities*. Cambridge, MA: MIT Press.

Beenen, A.S. and F. C. Boogaard. 2007. Lessons from ten years storm water infiltration in the Dutch Delta. *Proceedings NOVATECH 2007 – Sixth International Conference on Sustainable Techniques and Strategies in Urban Water Management*, June 25–28, 2007, Lyon.

Bettencourt, L. M. 2013. The origin of scaling in cities. *Science* 340: 1438–1441.

Bettencourt, L. M. 2015. Cities as complex systems. In: Furtado, B. A. (ed.) *Modeling Complex Systems for Public Policies*. Heidelberg: Springer, 217–238.

Boerma, H. 2016. Overstroming 1916—Van Zuiderzee naar IJsselmeer. https://www.mijnzuiderzee.nl/page/7035/overstroming-1916-van-zuiderzee-naar-ijsselmeer (accessed April 24, 2017).

Brooks, N., Adger, W. N., and P. M. Kelly. 2005. The determinants of vulnerability and adaptive capacity at the national level and the implications for adaptation. *Global Environmental Change* 15(2): 151–163.

Brouwer, J. 2011. *De Eindeloze Trap*. Doetinchem/Enschede: AfdH Uitgevers.

Chen, X., X. Zhang, J. A. Church, C. S. Watson, M. A. King, D. Monselesan, B. Legresy and C. Harig. 2017. The increasing rate of global mean sea-level rise during 1993–2014. *Nature Climate Change Letters*. Advanced Online Publication 26 June 2017, doi: 10.1038/NCLIMATE3325.

Civiele Techniek. 2007. 75 jaar Afsluitdijk. Civiele Techniek 4. http://www.civieletechniek.net/index.php/dossiers?id=9

Climate Central. 2014. 20 Countries Most at Risk from Sea Level Rise. https://weather.com/science/environment/news/20-countries-most-risk-sea-level-rise-20140924 September 25, 2014 (accessed April 24, 2017).

Climate Wire. 2012. "How the Dutch make 'Room for the River' by redesigning cities. Higher and higher dikes will not keep the waters at bay under climate change." *Scientific American* January 20, 2012. Retrieved from: https://www.scientificamerican.com/article/how-the-dutch-make-room-for-the-river/. Accessed: 8 June 2018.

Cohen, J. and I. Stewart. 1994. *The Collapse of Chaos: Discovering Simplicity in a Complex World.* London: Penguin Group Ltd.

Commonwealth of Australia. 2015. *Reef 2050 Long-Term Sustainability Plan.* Canberra: Commonwealth of Australia.

Concevitsch, B. B. 2007. *Counter-Intuitive Selling: Mastering the Art of the Unexpected.* Fort Lauderdale, FL: Kaplan Publishers.

Cressey, D. 2011. "Dubai's artificial islands are affecting marine ecosystems." *Nature* 479(7373): 277.

De Vlieger, B. 2017. *The New Delta; The Rhine-Meuse-Scheldt Delta in Transition.* Heijningen: Japsam Books.

Definedterm.com. undated. Adaptive Planning. https://definedterm.com/adaptive_planning_ap (accessed April 25, 2017).

Dekkers, G. and B. de Jong. 2002. Waterbouwkundige Leen Becu over de Deltawerken. *Historisch Nieuwsblad* 5. https://www.historischnieuwsblad.nl/nl/artikel/5895/waterbouwkundige-leen-becu-over-de-deltawerken.html (accessed April 25, 2017).

Deltacommissie. 2008. *Samen werken met water. Een land dat leeft, bouwt aan zijn toekomst. Bevindingen van de Deltacommissie 2008.* Den Haag: Deltacommissie.

Deltawerken online. undated. http://www.deltawerken.com/Deltaworks/23.html (accessed April 24, 2017).

Engel, H. 1997. The flood events of 1993/1994 and 1995 in the Rhine River basin. *Proceedings Destructive Water: Water-Caused Natural Disasters, their Abatement and Control. Conference*, June 1997, Anaheim, CA. IAHS Publication No 239.

Favis-Mortlock, D. and D. de Boer. 2003. "Simple at heart. The complexity of landscapes." In: Trudgill, S. and A. Roy (eds) *Contemporary Meanings in Physical Geography. From What to Why.* Abingdon: Routledge, 127–171.

Forsman, Z. H., Page, C. A., Toonen, R. J., and D. Vaughan. 2015. "Growing coral larger and faster: Micro-colony-fusion as a strategy for accelerating coral cover." *PeerJ* 3: e1313, doi: 10.7717/peerj.1313.

Frölicher, T. L., Winton, M., and J. L. Sarmiento. 2014. "Continued global warming after CO_2 emissions stoppage." *Nature Climate Change* 4: 40–44, doi: 10.1038/nclimate2060.

Gibling, C. 2013. "Construction process and post-construction impacts of the Palm Jumeirah in Dubai, United Arab Emirates." *Journal of Undergraduate Engineering Research and Scholarship* PT-2013. https://www.google.com.au/url?sa=t&rct=j&q=&esrc=s&source=web&cd=1&ved=0ahUKEwjqvrW51b7TAhXIx7wKHTEhA5QQFggkMAA&url=http%3A%2F%2Fjournals.library.mun.ca%2Fojs%2Findex.php%2Fprototype%2Farticle%2Fdownload%2F458%2F531&usg=AFQjCNHbfrzCvgHIPJi2ZfcpkNbmran2fQ (accessed April 25, 2017).

Goldstein, J. 1999. "Emergence as a construct: History and issues." *Emergence: Complexity and Organization* 1(1): 49–72.

Hansen, J., Sato, M., Hearty, P., Ruedy, R., Kelley, M., Masson-Delmotte, V., Russell, G. et al. 2016. "Ice melt, sea level rise and superstorms: Evidence from paleoclimate data, climate modeling, and modern observations that 2°C global warming could be dangerous." *Atmos Chem Phys* 16: 3761–3812, doi: 10.5194/acp-16-3761-2016.

Huitema, D. 2004. "Harboring water in a crowded European Delta." In: Kuks, S. and H. Bressers (eds). *Integrated Governance and Water Basin Management. Conditions for Regime Change and Sustainability.* Dordrecht, Boston, London: Kluwer Academic Publishers, 59–98.

Kauffman, S. 1995. *At Home in the Universe.* Oxford: Oxford University Press.

Kinder, K. 2011. "Planning by intermediaries: Making cities make nature in Amsterdam." *Environment and Planning A* 43(10): 2435–2451, doi: 10.1068/a4464.

Krugman, P. 1996. *The Self-Organizing Economy*. Oxford: Blackwell.

Lane, J. D. and P. L. Harris. 2014. "Confronting, representing, and believing counterintuitive concepts. Navigating the natural and the supernatural." *Perspectives on Psychological Science* 9(2): 144–160, doi: 10.1177/1745691613518078.

Langton, C. G., Taylor, C., Farmer, J. D., and S. Rasmussen. 1992. *Artificial Life II. Studies in the Sciences of Complexity. Proceedings Vol. 10.* Redwood City, NM: Santa Fe Institute.

Merry, U. and N. Kassavin. 1995. *Coping with Uncertainty: Insights from the New Sciences of Chaos, Self-Organization, and Complexity.* Westport, CT: Praeger.

Ministerie van Verkeer en Waterstaat. 2003. *Procesplan Zwakke Schakels in de Nederlandse Kust.* Den Haag: Ministerie van Verkeer en Waterstaat, Bestuurlijk Overleg Kust.

Ministeries van IenM and EZ. 2016. *Deltaprogramma 2017, Werken aan de Delta. Opgaven verbinden, samen op koers.* Den Haag: Ministeries van IenM and EZ.

Ministeries van V&W, LNV and VROM. 2010. *Deltaprogramma 2011, Werken aan de Delta; Investeren in een veilig en aantrekkelijk Nederland nu en morgen.* Den Haag: Ministeries van V&W, LNV and VROM.

Mitchell Waldrop, M. 1992. *Complexity. The Emerging Science at the Edge of Order and Chaos.* New York, London, Toronto, Sydney: Simon and Schuster Paperbacks.

Nieuw Land. undated. Het plan Lely 1888–1925. www.nieuwlanderfgoed.nl/studiecentrum/canon-van-flevoland/het-plan-lely,-1886-1925 (accessed April 24, 2017).

Nijssen, P. and M. Schouten. 2012. "Dutch national Room for the River project: Integrated approach for river safety and urban development." *Proceedings IS Rivers Conference*, June 26–28, 2012, Lyon.

NPO. 2016. Als de dijken breken. https://www.npo.nl/als-de-dijken-breken/VPWON_1261083?media_type=broadcast (accessed June 21, 2017).

Overpeck, J. T., Otto-Bliesner, B. L., Miller, G. H., Muhs, D. R., Alley, R. B., and J. T. Kiehl. 2006. "Paleoclimatic evidence for future ice-sheet instability and rapid sea-level rise." *Science* 311(5768): 1747–1750.

Poore, R. Z., Williams, R. S., Jr. and C. Tracey. 2011. "Sea level and climate." https://pubs.usgs.gov/fs/fs2-00/ (accessed June 28, 2017).

Portugali, J. 2000. *Self-Organization and the City*. Berlin, Heidelberg, New York: Springer-verlag.

Portugali, J. 2008. "Learning from paradoxes about prediction and planning in self-organising cities." *Planning Theory* 7(3): 248–262.

Provincie Groningen. 2013. *Ontwikkelingsvisie Eemsdelta*. Groningen: Provincie Groningen.

Rittel, H. and M. Webber. 1973. "Dilemmas in a general theory of planning." *Policy Sciences* 4: 155–169. Elsevier Scientific Publishing Company, Inc., Amsterdam. (Reprinted in N. Cross (ed). 1984. Developments in Design Methodology, 135–144. Chichester : J. Wiley & Sons.)

Roggema, R. (ed.) 2013. *Swarming Landscapes: The Art of Designing for Climate Adaptation*. Dordrecht, Heidelberg, London: Springer.

Roggema, R. 2012. "Swarm Planning: the development of a spatial planning methodology to deal with climate adaptation." PhD thesis. Delft University of Technology, Delft.

Roggema, R. 2016. "Research by design: Proposition for a methodological approach." *Urban Science* 1(1): 2–20, doi: 10.3390/urbansci1010002.

Roggema, R. 2017. "The Sydney Barrier Reef: Engineering a natural defense against future storms." *The Conversation* May 16, 2017. https://theconversation.com/the-sydney-barrier-reef-engineering-a-natural-defence-against-future-storms-76862 (accessed June 28, 2017).

Roggema, R. and H. Broess. 2014. "Darwin@home." In: Roggema, R. and G. Keeffe (eds) *Why We Need Small Cows. Ways to Design for Urban Agriculture*. Velp: VHL Press, 17–36.

Savini, F. 2017. "Self-organization and urban development: Disaggregating the city-region, deconstructing urbanity in Amsterdam." *International Journal for Urban and Regional Research* 40(6): 1152–1169, doi: 10.1111/1468-2427.12469.

Schuetze, T. and L. Chelleri. 2011. "Climate adaptive urban planning and design with water in Dutch polders." *Water Science and Technology* 64(3): 722–730, doi: 10.2166/wst.2011.688.

Srivastava, S. 2014. "What is counterintuitive." *The Hardest Science*. https://hardsci.wordpress. com/2014/03/26/what-is-counterintuitive/ (accessed June 26, 2017).

Stive, M. J., de Schipper, M. A., Luijendijk, A. P., Aarninkhof, S. G., van Gelder-Maas, C., van Thiel de Vries, J. S., de Vries, S., Henriquez, M., Marx, S., and R. Ranasinghe. 2013. "A new alternative to saving our beaches from sea-level rise: The sand engine." *Journal of Coastal Research* 29(5): 1001–1008.

Tol, R. S. J. and A. Langen. 2000. "A concise history of Dutch river floods." *Climatic Change* 46(3): 357–369, doi: 10.1023/A:1005655412478.

Verweij, S., van Meerkerk, I., Koppenjan, J., and H. Geerlings. 2014. "Institutional interventions in complex urban systems: Coping with boundary issues in urban planning projects." *Emergence: Complexity & Organization* 16(1): 07–23.

Walsmit, E., Kloosterboer, H., Persson, N., and R. Ostermann. 2009. *Spiegel van de Zuiderzee. Geschiedenis en Cartobibliografie van de Zuiderzee en het Hollands Waddengebied*. Houten: HES & De Graaf.

Waterman, R. E. 2010. Integrated coastal policy via building with nature, PhD thesis. Delft University of Technology, Delft.

Williams, R. S., Jr. and J. G. Ferrigno (eds). 2012. "Satellite image atlas of glaciers of the world." U.S. geological survey professional paper 1386–a–1. Published online: https://pubs.usgs.gov/pp/ p1386a/pdf/pp1386a-1-web.pdf (accessed June 28, 2017).

Williams, R. S., Jr. and D. K. Hall. 1993. "Glaciers." In: *Part VII of Cryosphere* of Gurney, R. J., Foster, J. L., and C. L. Parkinson (eds). *Atlas of Satellite Observations Related to Global Change*. Cambridge, UK: Cambridge University Press, 401–422.

Wirkus, M. 2016. "Adaptive management approach to an infrastructure project." *Procedia—Social and Behavioral Sciences* 226: 414–422, doi: 10.1016/j.sbspro.2016.06.206.

6

The Dutch "Room for the River" Program (2006–2017): Landscape Quality as a Binding Agent

Dirk Sijmons

CONTENTS

Introduction

Throughout the centuries, space for the rivers in the Netherlands has become more and more limited. The closing of the river dikes around the middle of the fifteenth century was the first step. From the fifteenth to the eighteenth century the rivers were regulated by building dams pushing back the rivers to let them erode themselves to navigable depth. In the nineteenth and twentieth centuries, this work was perfected by building smaller groins (breakwaters) to normalize the rivers in their summer beds. Urbanization along the rivers did not limit the space left for them but made the rigid corset that was already tightened irreversible. Nowadays, the rivers of the Netherlands are wedged between high dikes, while the subsidence of the soil behind the dikes is making things worse. If a flood occurred under these conditions, the number of casualties, as well as the economic and emotional damage would be huge. In 1993 and 1995 the river water reached an alarmingly high level, and the dikes just managed to hold their own. In 1995 almost half a million people were evacuated from the river area. Climate change will affect the rivers too. Higher (and lower!) discharges will have to be met now that the Rhine will change character from a melt water/rain water river to only a rain water river. To meet this challenge, the Dutch Cabinet created a package of measures called the Key Spatial Planning Decision "Room for the River" in 2006. The main objectives were flood protection by 2016 and improved overall environmental quality in the river basin region.

Room for the River was a remarkable project in many ways. After the Cabinet acknowledged in 2004 that climate change was real and that the Netherlands had to prepare and adapt itself to it, Room for the River was the first large scale adaptation project—budget 2.3 billion Euro—to make our vulnerable country climate proof and to give better flood

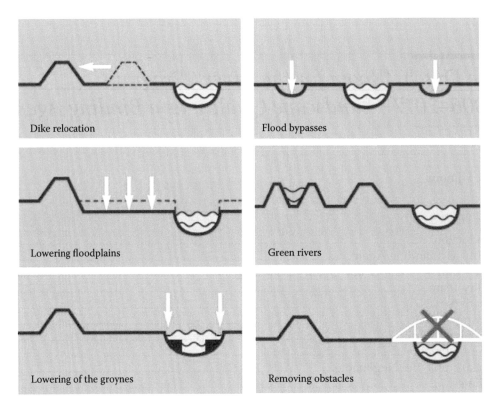

FIGURE 6.1
Six new ways to safety, used by Room for the River.

protection to 2–4 million people. In 2015, the Rhine branches would safely have to be able to cope with a discharge of 16,000 cubic meters per second (m³/s); now 15,000 m³/s. Safety would further be enhanced in an analogue follow-up project after 2015 that would increase the system capacity to discharge up to 18,000 m³/s. In a delta like the Netherlands, river safety goes hand-in-hand with coastal resilience projects, of course. During the Room for the River program many beach and coast nourishment projects were executed. Most of them were traditional, but innovative solutions were also developed. The sand engine near Kijkduin was the first project where working with natural processes was tested. In this pilot, an oversized dune near shore functions as a sand source for the coastal current that deposits the sand on the weak spots of the coast.

The Room for the River project is also remarkable because the Dutch Pavlovian response to answer the new threat with another cycle of raising the dikes was suppressed. Instead the solution is being sought in giving more room to the river. There are six ways to provide more safety (see Figure 6.1): First, dikes could be relocated to allow more room for high discharges. The second is making flood by-passes through the floodplains. Third, shaving off the topsoil from the floodplain is very effective. Forth, lowering the groins decreases the resistance for the water. The fifth way could be to create "Green rivers" between new dams. They can act as spatial reservations that can be used in special cases but are still able to support normal land use. And sixth, removing or mitigating obstacles, such as bridges or old ferry piers are very effective ways to lower the mean high water. The projects in the Room for the River program all use a mix of two or more instruments from this six pack of space makers. And, of course, additionally, the "traditional" strengthening of the levees is employed where urgently needed.

In the twentieth century a project like this would have been organized top-down by our Rijkswaterstaat[1] that once was dubbed by critics "A State within a State." Because times had changed, and we needed all the support that could be rallied—remember Room for the River would have an even more ambitious follow-up—it was decided that the state would facilitate and finance the project as a whole but that the execution would be mostly decentralized. That way, regional governments would be able to make use of the momentum of the project and link local programs to the water planning project. Making local governments responsible for sub-projects would lead to more customized results and, in the end, would be more effective and more swiftly executed than trying things the old-fashioned top-down way. A long list of almost 600 possible measures to give room to the river were identified in the preliminary phase of the project. A design study was executed to see if coherent clusters of measures could lead to more effective and/or more elegant solutions. A shorter list of potential measures was negotiated with local governments, provinces, and district water boards. Those local governments that seemed most eager to take on this responsibility almost all had their own programs that could only materialize by linking up with Room for the River. This new process architecture allowed for a more flexible approach that was able to withstand the very complex context. The state budget was of course restricted to the safety goals, but sometimes the way of reaching these goals can produce synergy with a lot of local spin-offs. Finally, 34 projects, that together would be able to reach the goals set, were selected in the Spatial Planning Key Decision that was approved by Parliament in 2006 (see Figure 6.2). The SPKD Room for the River contained a general description of the types of measures, the locations for most of the measures, and the expected effects. Before a measure could actually be implemented, it had to be worked out further in a planning study in which the exact location and details were determined. An environmental impact assessment was required for many of the measures, giving local residents, authorities, and other stakeholders the chance to have their say.

The motive to address the readers of this book on coastal resilience is that spatial quality was introduced in this project as the second main goal, next to safety.

A quality team was established to advise on all steps in the design processes of the individual Room for the River projects. In this essay, Dirk Sijmons, who put the quality team together and was its first chair, gives an insider's view of how the team functioned and contributed to the ultimate success of the program.

Quality as a Binding Agent

The Room for the River program is yet another project that showcases Dutch expertise, not only because of the engineering and water management ingenuity on display, but above all because of the way flood protection has been combined with landscape quality. The results are remarkable and have attracted considerable interest both at home and abroad.[2] The program clearly and convincingly shows what is possible when engineers, scientists, and designers work together.

Quality is not something that just happens. Projects of quality require exceptional designers and excellence from the commissioning bodies. Strict conditions were imposed in the Room for the River program because of the ambitious landscape quality objectives that had been formulated at the start: the river landscape had to be more attractive after the 34 (the program started out with 39 projects [see also the map]; combining projects,

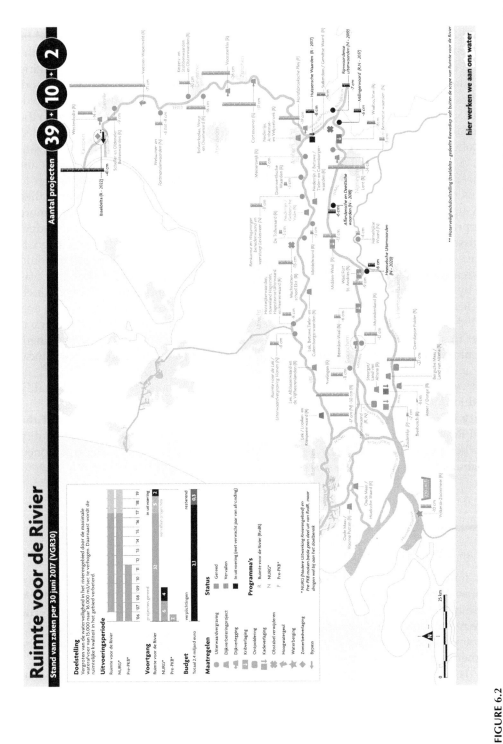

FIGURE 6.2
Map of the 34 Room for the River projects (complemented by 10 + 2 projects in earlier pre-Room for the River projects) and their targets in terms of mean high water reduction. (Courtesy Rijkswaterstaat.)

skipping some ended up in 34 projects) projects had been completed than before the program started.

Design was incorporated into the project set-up, and the Room for the River program office (part of the Rijkswaterstaat organization) handled the tendering and contract process by setting up a separate department to deal with the quality aspects. I believe that the attention given to the design of the projects was not only critical for success, but also a very important contribution to the flexibility of the process and to the admirable budget management.

Domestication of Landscape Quality

Once landscape quality had been firmly embedded as the second objective in the National Spatial Planning Key Decision for the Room for the River program (PKB, 2006), Rijkswaterstaat fully embraced it. In the run-up to the PKB (Planologische Kern Beslissing and would translate into Central or Core Planning Decision), both Rijkswaterstaat and the Office of the Chief Government Architect had already demonstrated their strong support. Deciding on how to tackle this project, though, was problematic. How could it best be incorporated into water management programs, which since 2002 had been subject to a strict procedural framework known as the "SNIP system?"[3]

Landscape quality is an elusive concept. Most people know when quality is absent, but it is much more difficult to define what quality actually is. The more you try to define it, the more the essence of what quality is seems to slip through your fingers. In an era of restless account managers, holding others accountable for reaching targets is a complicating, but perhaps also comforting, state of affairs.

In the water world, legal systems have been established for flood protection, water quality, and shipping, based on cycles of research, assessment, introduction of measures, and then more assessment to see whether the objectives have been achieved. The administrative equivalent of this is the ultra-detailed SNIP system of rules: the precisely defined steps in the process of a water project. SNIP-2a, for example, deals with choosing the preferred alternative in the environmental impact assessment that had to be made for all the Room for the River projects. SNIP-3 governs the crucial moment in the process between planning and implementation when the minister decides whether a project may continue or not. The brand new program office could have taken a rigid approach to this requirement for landscape quality, for example, by drawing up a standard checklist for all project components rather like the *Werkbank Ruimtelijke Kwaliteit*[4] prepared around the same time for quality in area development projects and then just hope that the desired landscape quality would emerge spontaneously.

Nevertheless, it was realized that, unlike the dike-strengthening projects of the past, implementing the measures to create more room for the river would involve a more complex relationship with other stakeholders than just a dialogue with a water authority. The new repertoire of flood risk management measures—dike relocation, removal of obstacles, lowering of floodplains, planning green rivers, excavating new side channels—requires space and thus also a proactive relationship with the land use planning authorities. A modus operandi needed to be created from the strategy of river-expanding measures, one that would be capable of dealing with these new complexities. Landscape quality turned out to be a fruitful way of accessing this public dialogue.

It did not take a great deal of effort for the Office of the Chief Government Architect to persuade the director of the Room for the River program office, Ingwer de Boer, that it was, above all, a "people business." Moreover, he backed our case for establishing a quality team to support the projects. Quality became an active ingredient: projects need care and attention from their inception to implementation. The idea was to have a team that could work step-by-step on the quality of the projects in collaboration with the project developers, the project managers, and the designers. The best way of doing this, however limited, was through peer review. Given his experience with landscape quality, Ingwer de Boer did not take much convincing. He had previously been the Rijkswaterstaat chief engineer for the province of Utrecht, where he had been responsible for route alignment planning for trunk roads.[5] When motorways were redesigned (widened and equipped with new road furniture), it involved ensuring that they blended well into their surroundings;[6] a matter of spatial quality.

Initially, in a similar approach to that taken for route alignment planning, the idea was to create an intermediate design step between the program as a whole and the initial 34 projects.[7] Surely it would be useful to create a vision for each river branch as a kind of landscape framework for the projects containing descriptions of some of the specific characteristics of each branch. This would certainly have been useful but, in the end, it did not happen. There was simply not enough time, as all projects had to be completed in 2016. Instead, it was agreed that these "guidance documents" would be created at the same time as the main body of projects,[8] and it was accepted that this background material would be made available to the designers of the projects at a later stage in the process. Two guidance documents were ready by 2007: one for the IJssel and one for the Rhine, Lek, and Waal rivers.

Quality Team

As Government Advisor on Landscape, I was asked to put together a quality team and then head up that team. The team needed to be a mix of multitalented members, capable of evaluating the special character of all aspects of river projects. It would be a quality team with practical experience and knowledge of the areas, its members capable of working together to make integrated recommendations that would improve and refine the projects. I was given free hand to assemble a group of people who I thought would get on well together. I chose the experienced river expert Dick de Bruin, the urban designer and "city doctor" Maurits de Hoog to bring out the urban potential in the plans, Sjef Jansen, an ecologist and expert on the ins and outs of Dutch nature policy, and Frans Klein, a physical geographer at Deltares.

In the meantime, the program office had its own team led by Regina Havinga that went by the name of "Landscape Quality cluster." The cluster had its own landscape architects, one of whom was Hermine der Nederlanden. The quality team's full-time secretary was also a member of the cluster: Willem de Visser, who came from the National Spatial Planning Agency (RPD), later replaced by Jan van der Grift. Among other activities, the Landscape Quality cluster was responsible for producing the guidance documents for each river branch and a book to provide inspiration for the Room for the River projects.

The quality team, which became known as the "Q-team," was instructed to provide guidance for the 34 projects during the planning phase through to the tender stage. My successor as head of the team, Eric Luiten, managed to convince the program office that the Q-team also had an important role to play in the implementation. The team was independent and could give both solicited and unsolicited advice.

The SNIP (discussed previously) provided the frame of reference for the Q-team. It indicated three points in the process when it would be logical for the team to provide advice. The first visit to the project site was to take place immediately after the administrative agreement had been signed, the moment at which the relevant authority—water authority, municipality, or province—assumed formal responsibility for the project. This first visit was for the team to familiarize itself with the area and the project and give those involved an opportunity to get to know each other. After the first visit the team gave advice on the design process, the design brief, and the development opportunities for the area. The second visit was to take place in the phase during which the project team generated alternatives for the environmental impact assessment. One of the issues to be addressed in this advice was how the preferred alternative would be determined. The third visit was devoted to discussing the preliminary design. At this stage, the Q-team would focus on the design choices to be made in the preferred alternative, how these should be translated into concrete designs and plan specifications, and how quality control would be carried out during implementation. On the basis of this, the team would carry out a final evaluation and advise the minister on how landscape quality was being dealt with in the plan. That was the theory. In practice, about five visits were paid to each project site, and some received as many as seven. Sometimes this was because of complications along the way, but most visits were made because the preliminary design needed to be discussed in several steps with the project team or the designers.

To be perfectly clear: the quality team consulted with the design teams but did not advise them directly. Our advice was directed to the Room for the River program office, who then passed on the advice—sometimes amended—to the project developers. These final positions were officially prepared by the Landscape Quality cluster, which thus played a key role in the quality control process, not only because it was the point of contact for the Q-team's advice, but also because it was responsible for the day-to-day oversight of the designers and project managers on landscape quality matters. Because of its key role, the cluster was indispensable to the managers of the individual projects when it came to the compulsory review of the SNIP procedure for each step of a project.

It was important for the quality team and the Landscape Quality cluster to be alert to the danger of nannying the project teams when advising and consulting with them, which would have smothered all spontaneity. After all, projects can be supervised to death. Fortunately, in the evaluation of the design processes in the Room for the River program it was observed that "both the Q-team and the Landscape Quality cluster succeeded in exercising their influence on the landscape quality without taking over the role of the designer."[9]

Objectives

While the quality team had a large degree of autonomy it obviously did not operate in a political vacuum. The National Spatial Strategy (*Nota Ruimte*)[10] had already outlined four objectives for improving the spatial quality of the river landscape: increasing the landscape diversity between the different river branches; maintaining and strengthening the open character of the area of the major rivers, with their distinctive waterfronts; maintaining and developing the landscape, ecological, geographic, and cultural heritage values, and improving environmental quality; and strengthening the opportunities for the use of the main waterways by commercial shipping and pleasure craft.

Objectives such as these give an indication of the types of solutions that are required but are not necessarily of much practical use in the design process. During the course of their work the quality team developed a definition of landscape quality, and thus plan quality, specific to the river projects. In the talks with the individual project teams this definition also provided clarity on the Q-team's ideas on how the different objectives were related. In a playful reference to the classical Roman architect Vitruvius, who wrote that architectural quality should display *utilitas*, *firmitas*, and *venustas* (utility, solidity, and beauty), we described quality in Room for the River projects as an elegant and convincing marriage between *hydraulic efficiency* (how to achieve the desired effect with minimal means), *ecological robustness* (how the river dynamics can support habitat quality) and *aesthetic meaning* (how the project can make a meaningful cultural contribution to the river landscape, while making use of a design idiom characteristic of the river).

Sparring Partner

As all 34 projects started at about the same time, in 2006, this initial period was particularly labor intensive for the Q-team. Later, the projects ran at different speeds so the work became more spread out. Most visits were to the sites, and during these visits the opportunity was taken to make an extensive tour of the planning area so that the specific landscape qualities and problems could be identified and assessed. Our visits were regular and always had a clear agenda, as they adhered to the SNIP procedure.[11] The "Design Quality Assessment Framework" prepared by the quality team[12] lists the separate components of the design process (ambition, organization of the design process, analysis, conceptualization, synthesis, calculations and drawings, design, and materialization) for the team to consider. The project teams therefore knew what was expected of them at each step.

The Q-team had a more or less preventive function. In nearly all the projects the consultants who were engaged put designers and ecologists in their planning teams to ensure that they were well prepared in the event of a confrontation with the Q-team. Meetings were to the point but also informal. We became a sparring partner for the project managers, but more importantly we assisted the designers, who as members of mainly technical project teams could often do with some support.

The project teams and designers received support in other ways as well. In addition to the guidance documents for the IJssel, Rhine, Lek, and Waal rivers, and the book that included many inspiring examples from the Netherlands and abroad,[13] the Q-team regularly invited external experts to their meetings to share their expertise on specific aspects; from floating dwellings, sustainability in river projects, and shipping and nautical aspects to the complex legislation surrounding Natura 2000 and the EU Water Framework Directive. The independent advisory status of the Q-team also ensured a healthy tension with the program office, which of course was responsible for progress, budget control, and the legal aspects of the program. There were quarterly meetings with the head of the program office to discuss any controversies at an early stage and bring expectations in line with each other.

Painstaking Work

This book is full of attractive illustrations of the completed projects. These of course do not show all the intermediate steps that took place during the planning and implementation phases, nor do they give an indication of the part played by quality control and project architecture. Perhaps it is better not to discuss these at all and just judge the completed projects on their own merits, because describing the work of the Q-team is a bit like showing friends around your recently renovated house. "You've no idea what a mess this all was … there were two walls here that we pulled down … it looks great now, but you can't imagine how much work it was." Your often-reluctant guests would rather not be put through all this and sheepishly make occasional sympathetic noises. Nevertheless, I do want to present a selection of some of the painstaking work that went into the preliminary phases of these designs. The amount of work it took does not mean that the landscape architects were mediocre, the engineers incompetent, or the project managers uninformed. On the contrary, defining a clear design strategy and keeping to it during the course of such complex projects in which so many disciplines had to have their say was no small feat.

In most cases the designers and/or the project managers and/or the competent authorities adopted our advice. All the intermediate steps taken in the projects are recorded in the team's annual reports.[14] There is a clear cartographic overview of the development of the plans in the report on 2009–2011, an extensive report we prepared recording our projects for Parliament. We were under close parliamentary oversight, particularly after the Court of Audit in 2007 made a remark about the measurability of cost-effectiveness. While it is possible to set the designed reduction in floodwater levels against the investments made, how can one do that for landscape quality? For one thing, the way we chose to work made it difficult to identify "extra costs incurred for the sake of quality," and the Court of Audit itself regarded "landscape quality" as a difficult-to-measure concept. Its recommendation was to unpack the concept and draw up a number of evaluation criteria.[15] We were able to reassure the Court of Audit by making landscape quality tangible, albeit not in figures, but by producing a series of maps that describe and illustrate the stepwise planning process.

Fostering Quality

In numerous ways, the Q-team attempted to influence or alter the direction of the design process. This began right at the start, when a salutary situation arose. It transpired that standard design options had been prescribed, which project developers were expected to work up: option one had to show a design that would maximize flood protection; option two was a design that would maximize landscape quality; and option three was a design that would minimize costs. This kind of approach is entirely at odds with the aim of adopting an integrated approach to landscape quality, flood protection, planning and budgeting, and garnering broad support. An integrated approach was chosen because we were convinced that it produces better results *and* is more efficient. "This only frustrates an open-minded search for a commonly shared optimum."[16] After alerting the Room for the River program office, this prescription was rapidly withdrawn.

Our aim was that the plans would impart a true "river idiom": simple, attractive, robust, and without frills. We tried to get the designers to aim for a form of "assertive" modesty that is true to the Dutch tradition of pragmatism, showing reserve rather than making dramatic gestures. A more explicit account of this design vision can be found in an essay in the quality team's annual report. Appropriately, the essay is built around traditional Dutch sayings and proverbs, such as "The simple is the sign of the true," "Why make things difficult if they can be done simply?" "Perfection is the enemy of the good," "You get what you pay for," and "Waste not, want not."[17] In this vision quality is not a luxury, a decoration, or an accessory, but an intrinsic feature. Quality is not the icing on the cake, but the yeast in the bread. Quality is not a drain on the budget, but a slightly different way of spending money.

Sometimes quality costs a little more; on other occasions, it results in savings and speeds things up, sometimes so radically, as happened in Vianen, that the quality team asked for a "clear up option" rather than a "design option," following the saying, "Good riddance to bad rubbish." Could the quality objective not simply be achieved by removing obstacles? Or by stripping down the design in some cases? The new dike in the Munnikenland project was originally designed with broad steps stretching over a great length so that it could serve as a grandstand, which in itself was an attractive idea, as there was a view toward Loevestein Castle. Yet we argued in favor of simplicity. The orientation and above all the gentle slope of the dike made it a natural grandstand as it was; it did not require anything as literal as steps to invite people to sit down. In river landscapes an exaggerated or sophisticated design all too soon comes across as overdone or superfluous. In the end, terraces were dug out of the slopes and edged with sturdy granite blocks. Elsewhere, the team also argued in favor of leaving things out for simplicity, such as favoring "solid low tech" rather than "fragile high tech." But at times the team also insisted on "pushing the boat out" where it would enhance the project. This was the case in Lent, near Nijmegen, where perhaps the most spectacular of the 34 projects of Room for the River were realized (see Figure 6.3a–d). Generally speaking, we were in favor of the projects being easily accessible and designed for everyday informal use, rather than making specific investments for recreational use. For example, no wide access roads were proposed where a narrow path would suffice.

In most projects, the tensions rose around spontaneous habitat development and the need to prevent water discharge being met by extra resistance. Time and again all those involved had to draw on their creativity and inventiveness to ensure that all the important components of the ecology of dynamic river dunes were in place, up to and including rare opportunities for hardwood alluvial forest in slack water. The team was critical where protective measure after protective measure was continually added, apparently in an attempt to buy support, where this implied the use of additional and often inelegant measures. In the case of the channel at Veessen-Wapenveld the team actually presented its advice directly to the minister.

This brings us to the matter of the relationship between the design and the future managing body. The team succeeded in getting the intended managers of the completed projects involved at an early stage of the design process because "management-friendly design" can prevent the need for ugly alterations at a later stage (as at Pannerden). Conversely, we also encountered cases where "design-friendly management" was necessary. In Noordwaard there is an electrical substation on top of an isolated artificial mound in the overflow area. What would happen in the rare event of the substation failing (once in 150 years) during high water when the overflow area is inundated (which happens once a year)? The most obvious design solution would seem to be a bridge, but this would have to be strong enough for a heavy truck to cross it: a very expensive option. In the end, it was decided to construct a soil depot that can be pushed into the waterway should the transformer need replacing during high water: a simpler, more elegant, and much cheaper solution (see Figures 6.4 and 6.5).

(a)

(b)

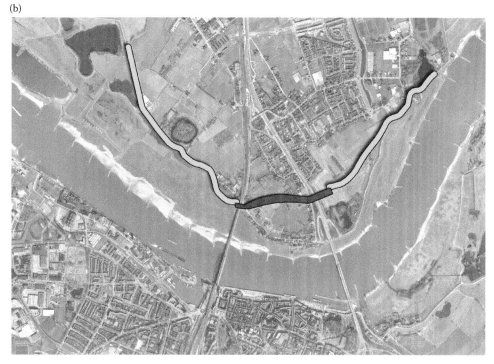

FIGURE 6.3
(a–d) As an example, Lent is not the biggest, but home to perhaps the most spectacular of the 34 projects of Room for the River. (a) Existing dike. (b) Laying the dike back (new alignment makes it possible to make a new urban quay, in red). *(Continued)*

(c)

(d)

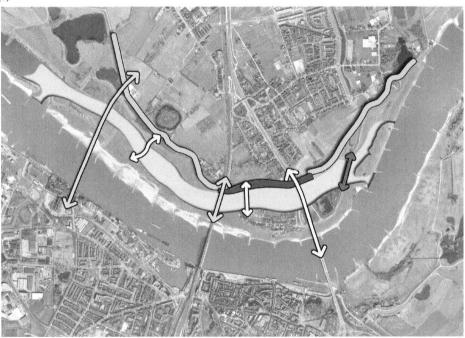

FIGURE 6.3 (Continued)
(a–d) As an example, Lent is not the biggest, but home to perhaps the most spectacular of the 32 projects of Room for the River. (c) Secondary river channel (blue) allows reaching the local target of reducing the mean high water by some 22 cm. (d) New bridges (yellow) and reconstruction of existing bridges (gray) needed to restore the connections.

FIGURE 6.4
The new situation of the secondary channel, the new island, the riverfronts, and the new or renovated bridges seen from the East.

FIGURE 6.5
The new situation of the secondary channel, the new island, the riverfronts, and the new or renovated bridges seen from the West.

All in all, by championing the case for integral quality, respecting and using the river as an instrument of design, and by arguing for a simple design idiom, the Q-team estimated that it saved Dutch taxpayers between 20 and 60 million euros.

Immeasurable But Negotiable

Perhaps one of the best things about working with quality teams is that no matter how complex the process is—conflicting interests, diverging agendas, tight budgets, and other inconveniences—you can always fall back on a simple litmus test. Based on the drawings, the plan document, or the work that has been executed, ask the question: Is this quality? This may be a simplification of the whole process, but the simplicity is appealing because nobody can hide behind other arguments. Working with quality teams forces you to face the harsh realities: you must have a strong sense of purpose and work efficiently to achieve quality. Quality is not measurable, but it *is* negotiable. The municipality of Nijmegen and the Q-team used the Room for the River projects to make the new bridges and quays iconic elements for the city identity, as well as meaningful contributions to spatial quality. Figures 6.6 through 6.9 highlight the design of the underside of the bridges that make them into pleasant and spectacular public spaces.

The integrated approach and quality assurance reduce costs and speed up the process. In their interim evaluation of the PKB Room for the River, the researchers from Erasmus University and Berenschot observe that paying attention to landscape quality "… not only results in a

FIGURE 6.6
Lentloper. The bridge allows pedestrians to pass under it by little balconies that offer a spectacularly different view. (Ney and Paulissen Architects, Photo: Frans Blok.)

FIGURE 6.7
Elongated Waalbridge. The accent of this bridge is its underside. Optimized for water flows, the complex concrete forms offer a "Stonehenge" sensation for the pedestrian. (Zwart and Jansma Architects, Image credit: Frans Blok.)

FIGURE 6.8
De Oversteek. A renovated car bridge crossing the new nature area on the western tip of the new island. (Ney and Paulissen Architects, Image credit: Frans Blok.)

FIGURE 6.9
The new quay on the Waal being inaugurated by the Nijmegen City Run. (Design: Mathieu Schouten, Photo: municipality of Nijmegen.)

substantive contribution, but also influences the process. The Q-team has managed to bring the different parties together and stimulate discussion, which has eased the planning process."[18]

Whether or not we have been successful in preserving the quality of the river landscape or have perhaps even improved it, is up to the readers of this book, visitors to the sites, and local people to judge.

Endnotes

1. Rijkswaterstaat, founded in 1798 as the "Bureau voor den Waterstaat," is part of the Dutch Ministry of Transportation and Water Management that is charged with the task of the practical execution of the so-called waterstaat. This includes the construction of waterways and roads and the maintenance of infrastructure.
2. Talks have been given all over the world and a number of articles have been written about the Room for the River program. The most comprehensive book published so far about the program, its antecedents, strategy, projects, and the architectural process is the book written in Japanese by landscape architect Shiro Takeda and published in 2016. The title translates as: How to teach the rivers of the Netherlands to communicate with nature and improve cities.
3. SNIP translates roughly as "Rules for Wet Infrastructure Projects." This set of ground rules has been applicable to all projects concerning the major waterways since 2002.
4. Peter Dauvellier, et al. 2008. "Werkbank Ruimtelijk Kwaliteit, een praktische hulp bij gebiedsprocessen" ["Landscape Quality Workbench: a practical guide to regional processes"]. *RO Magazine* July/August 2008.

5. The route alignment planning procedure was the most important contribution made by the Ministry of Transport, Public Works and Water Management to the "big projects" in the third policy document on architecture (derde Architectuurnota). Ministries of Education, Culture & Science; Housing, Spatial Planning and the Environment; Health & Welfare; Agriculture, Nature and Food Quality. 2000. Ontwerpen aan Nederland, Derde Nota Architectuurbeleid, 2001–2004 [Designing the Netherlands, Third Policy Document on Architecture]. Den Haag: Sdu.

6. This could be seen as a revival of the long-standing cooperation between civil engineers and landscape designers, who had been collaborating since 1907 on landscaping major roads and canals, a joint undertaking by Rijkswaterstaat and Staatsbosbeheer which ended in the mid-1980s.

7. The process started with 38 projects. Some of these were merged or halted prematurely so that, by 2017, 30 projects had been completed.

8. Bosch and Slabbers. 2007. Handreiking Ruimtelijke kwaliteit IJssel [Guidance on Landscape Quality IJssel]. Terra Incognita, (2009) Handreiking ruimtelijke kwaliteit Rijn, Lek en Waal [Guidance on Spatial Quality Rhine, Lek and Waal].

9. Ecorys. 2011. Walter Hulsker, Manfred Wienhoven, Marlies van Diest, and Steef Buys. Evaluatie ontwerpprocessen Ruimte voor de Rivier [Evaluation of the Room for the River Design Processes]. Rotterdam. 26.

10. Ministerie van VROM. 2009. Nota Ruimte, Kabinetsstandpunt [National Spatial Strategy, Government Decision]. Den Haag: Sdu.

11. Q-team Ruimte voor de Rivier. 2007. Beoordelingskader Ontwerpkwaliteit PDR [Evaluation Framework for Design Quality]. Utrecht: PDR.

12. The annual reports of the quality team show that its workload gradually decreased. About 30 visits and consultations took place in 2006 and 2007, 24 meetings were held in 2008, 22 in 2009, 12 in 2010, and 17 in 2011.

13. Robbert de Koning. 2008. Rivieren en inspiratie [Rivers and Inspiration]. Utrecht: PDR.

14. Q-team (various editions: 2006–2007; 2008; 2009, 2010, 2011). Jaarverslag Kwaliteitsteam Ruimte voor de Rivier [Annual Report Quality Team Room for the River]. Utrecht: PDR.

15. Q-team Annual Report 2006–2007. 4.

16. The Q-team's opinion was later confirmed by the report: Walter Hulsker, Manfred Wienhoven, Marlies van Diest, and Steef Buys. 2011. Evaluatie ontwerpprocessen Ruimte voor de Rivier [Evaluation of the Room for the River Design Processes]. Rotterdam: Ecorys.

17. Dirk Sijmons. 2012. Kwaliteit is geen luxe. Een essay in vijftien oude en nieuwe spreekwoorden [Quality is not a luxury. An essay in fifteen old and new proverbs]. Jaarverslag Kwaliteitsteam Ruimte voor de Rivier 2009/2010/2011 [Annual Report Room for the River Quality Team 2009/2010/2011]. Utrecht: PDR. 4–19.

18. Universiteit/Berenschot. 2011. Tussenevaluatie PKB Ruimte voor de Rivier [Interim Evaluation PKB Room for the River]. Rotterdam: Erasmus. 39–41.

7

Drawing a Line in the Sand: Rebuild by Design, Mathematical Modeling, and Blue Dunes

Kathleen John-Alder

CONTENTS

In the summer of 2013, less than a year after Hurricane Sandy devastated the New York metropolitan region with high winds and a record-setting tidal surge that killed 186 people, damaged more than 600,000 homes, left millions without power, and caused an estimated US$65 billion in damages, the Department of Housing and Urban Development and The Rockefeller Foundation joined forces to sponsor the Rebuild by Design competition. Ten interdisciplinary teams were charged with the ambitious task of devising solutions that would reconnect the social and economic determinants of coastal development with the physical ecology of the shoreline. In light of scientific projections that indicate anthropogenic climate warming will intensify weather-related crises, the reconstruction of pre-storm conditions was deemed unacceptable. The objective was to reverse an entrenched history of decision making that "created or exacerbated [physical] vulnerabilities throughout the region" and ignored "the fragility of life on the coast" (Bisker et al., 2015).

WXY Architects and West 8 responded to the Rebuild by Design challenge with an audacious proposal for a string of barrier islands that stretched from central New Jersey to the eastern tip of Long Island, which they named the Blue Dunes (Figure 7.1). Sophisticated computer modeling indicated the proposed islands would reduce storm surge throughout the region. This would reduce the height of secondary lines of defense, such as the seawalls, making these structures less expensive and less visually intrusive (Figure 7.2). Studies of marine sand and gravel deposits verified there would be enough construction material, and benthic habitat maps indicated the islands would avoid endangered and economically important marine species (WXY/West 8, 2104).

The Blue Dunes scheme began modestly with transect studies of the mid-Atlantic coastline. These cross-sections were to generate site-specific defense strategies that shore communities could implement as the need arose and funding became available. Over time, and at the urging of Adriaan Geuze, the founding principal of West 8 and lead landscape architect for the project, the design grew in ambition and evolved into a regional protection strategy that would reduce storm damage and insurance costs for all waterfront property owners, not just the select few who managed to garner political and financial support. Both the expansion in scope of the project and the promotion of barrier islands to reduce storm surge reflect a storm protection plan Geuze had recently completed for the Netherlands called the Happy Isles (Geuze, 2009). In addition to the use of barrier islands to protect the mainland from storm surge, ideas appropriated from this earlier project, include the mathematical modeling of storm surge to determine the optimum location of

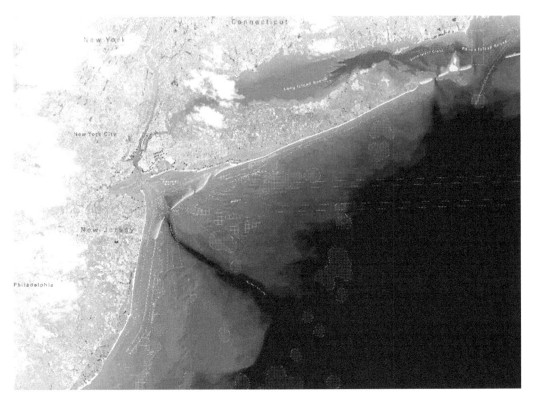

FIGURE 7.1
Blue Dunes plan. (Courtesy of WXY Architects and West 8.)

the proposed barrier islands and an island shape that literally mirrored one proposed for the Happy Isles (WXY/West 8, 2104).

Putting aside for the moment the extensive environmental impact studies that would be required to implement the Blue Dunes, one of the most contentious aspects of the design involved internal team discussions regarding the scale and intent of the project, and the close correlation of these positions to opinions regarding mathematical modeling. As he had previously done for Happy Isles, Geuze fully embraced the conceptual power of mathematical modeling and used it to daringly justify the extensive alteration of coastal topography. In contrast, the physical geologist Orrin Pilkey and his protégée Rob Young, who are skeptical of the predictive power of mathematical modeling, argued for cautious ecological restoration. Given this difference, it is somewhat surprising that Geuze requested that WXY Architects, the lead firm and coordinator of the Blue Dunes team, invite Pilkey and Young to serve as consultants, and that Pilkey and Young accepted the invitation. In hindsight, it is clear that Geuze, Pilkey, and Young envisioned the project as an opportunity to utilize their professional expertise to redefine coastal design. What they failed to understand were the different tactics that they would deploy to address this aim. WXY Architects had to mediate the difference. The task would prove to be impossible.

To fully understand the conceptual divide separating members of the team, it is necessary to step back and briefly examine the relationship of mathematical modeling to weather prediction as exemplified by the work of the mid-twentieth century mathematician Edward Lorenz. Lorenz empirically observed that weather patterns repeat, but never exactly.

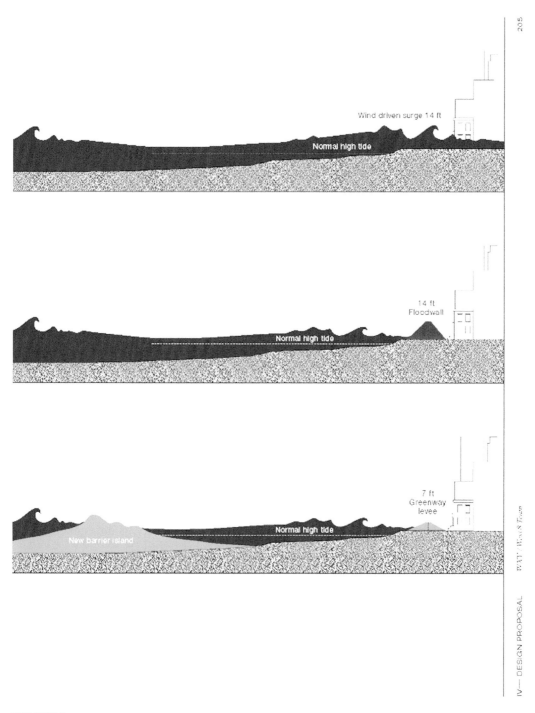

FIGURE 7.2
Storm surge reduction scenario. (Courtesy of WXY Architects and West 8.)

To study this behavior, he devised a 12-variable equation that modeled the rise and fall of air currents under different temperature conditions. This nonlinear, differential equation generated a wave pattern which almost, but never quite, repeated. When Lorenz replicated the computation to verify the result, at first the computer output matched the original. By the end of the run, however, the two patterns varied dramatically. He surmised the alteration was not due to the equation, which had remained the same, but to the incredibly small change in initial conditions that occurred when he mathematically rounded the original six-digit numbers and entered three-digit numbers into the equation—a difference of only one part per thousand. This work, which forms the theoretical foundation of chaos theory, allowed Lorenz to hypothesize that initial conditions determine the behavior of natural systems, and further, that minuscule variations in these conditions generate complex and unpredictable behavior (Gleick, 1987). Lorenz, in other words, demonstrated the ability of mathematical equations to model the behavior of complex natural systems, but his work failed to resolve the issue of predictive accuracy. Proponents of mathematical modeling argue accurate prediction is possible if the initial conditions are carefully specified. Skeptics conversely claim it is impossible to specify initial conditions. The predictive ability of mathematical models has improved over time as data sources have improved and researchers have updated and modified their equations accordingly. Nevertheless, these equations still make assumptions that generalize and simplify the complexity of the real world (Scharnhorst et al., 2012). Lorenz's work was later popularized as the Butterfly Effect, which posits an atmospheric disturbance as small as the flap of a butterfly wing can cause a devastating storm half a world away (Lorenz, 1963).

Today, the applied science of mathematical modeling is routinely used to predict the likely behavior of blood flow, nerve formation, the stock market, a dripping faucet, and wind and water turbulence even though its use continues to raise nagging questions regarding data selection and accuracy. This issue becomes even more problematic when the predictions are intended to mitigate risk and improve health and welfare. Should we, in other words, allow our fate to be decided by inherently uncertain mathematical equations that can only be validated in hindsight?

In spite of these unresolved issues, Geuze found justification for the Blue Dunes barrier islands in the sophisticated set of two- and three-dimensional hydrodynamic equations developed by the oceanographer Alan Blumberg to model a storm surge (WXY/West 8, 2104; Gleick, 1987; Kellert, 1993; Dizikes, 2011). This model, which transforms data generated by an extensive grid of ocean and weather monitoring sites into bright, color-coded predictions of water movement, wave height, and flood level, had accurately predicated the devastating storm surge of Hurricane Sandy. This work led to early evacuation notices and it is credited with saving many lives. During an initial conversation with Blumberg, Geuze soon realized he could adapt the mathematical modeling procedure that he had used to develop the Happy Isles design to coastline conditions in North America.

For the Blue Dunes project, Blumberg used data from three storms that produced significant flooding—Hurricane Donna in 1960, a Nor'easter in 1992, and Hurricane Sandy in 2012. To minimize the need for expensive computation runs—each computation used 256 computers and required approximately 20 hours to complete—proposed island locations were carefully considered. During these discussions, Geuze would draw an island configuration, and Bloomberg, based upon his understanding of seafloor morphology and his in-depth knowledge of the mathematical equations in his model, would suggest alterations to the layout. Twelve scenarios were eventually tested, and this required over 200,000 hours of computer time. The resulting maps documented the height of the storm surge with and without the islands for an area that extended from central Connecticut

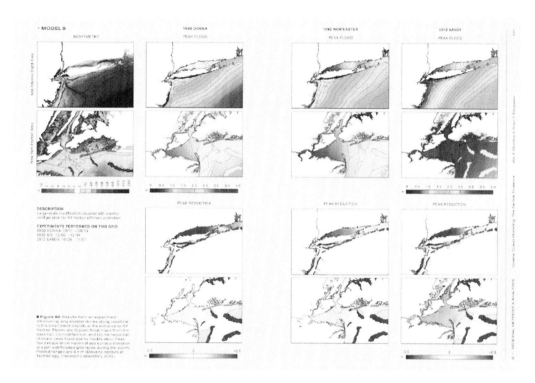

FIGURE 7.3
Storm surge mathematical modeling diagrams. (Courtesy of WXY Architects and Dr. Alan Blumberg.)

to central New Jersey (Figure 7.3). The optimum island configuration, which was the one presented to the Rebuild by Design competition jury, provided the second greatest reduction in storm surge over the greatest geographic area. Although this string of islands did not provide the most protection, it was selected because it was generally considered easier and less costly to construct (Keenan and Weisz, 2016).

An equally important factor in the development of the Blue Dunes design and its barrier island scenario, is the fact that Geuze is captivated by grand feats of engineering—the visionary kind that transform whole regions of the landscape—a predilection, no doubt, related to the fact that he is Dutch. His fascination with transformative infrastructure is readily seen in the essay "Flatness," which he wrote for his firm's monograph *Mosaics West 8* (Geuze, 2007). This essay traces the history of the Dutch landscape as the inhabitants of this precarious Netherlands region of Europe built the physical foundation of a nation using an innovative combination of drainage canals, dikes, and windmills. The result of this seabed colonization, as Geuze duly notes, was a paradisiacal landscape of ethereal light, clouds, and water in which there was such beauty that "the landscape itself became a subject of painting" (Geuze, 2007). Although Geuze did admit this reclamation process had unforeseen failures, such as the 1953 storm whose devastating high tides and tidal surge caused 1,853 deaths, he also observed the tragedy provided the impetus for a heroic response epitomized by increasingly sophisticated environmental knowledge and technological acumen that is now firmly entrenched within the physical structure of the landscape and the historical memory of its inhabitants.

Intriguingly, Geuze's deep appreciation of the polder landscape of his homeland, and its co-evolution of people and place, suggests a less explicit agenda in the Blue Dunes proposal, which is the belief that the new barrier islands would insert a carefully choreographed genius loci, and thus cultural authenticity, back into the chaotically suburbanized New Jersey and Long Island landscapes. To reinforce this point, idyllic presentation images prepared by his office in Amsterdam employ a visual iconography and palette of colors that romantically recall the golden era of Dutch landscape painting (Figure 7.4a and b and Figure 7.5). These images tie the Blue Dunes to the seventeenth century founding of New Amsterdam and the Dutch colonization of Manhattan, suggesting, in turn, that it

(a)

(b)

FIGURE 7.4
(a) and (b) West 8 perspective studies. (Courtesy of WXY Architects and West 8.)

FIGURE 7.5
West 8 perspective study. (Courtesy of WXY Architects and West 8.)

is only natural for the Dutch to once again apply their particular brand of technological acumen to reconfigure American culture and its coexistence with nature.

Moreover, Geuze knew from the empirical observation of built work post-construction that landscape design could produce serendipitous ecological surprise. This was notably seen, for example, in the unexpected animal behavior that emerged in response to the bold black and white ground cover pattern he designed for the Scheldt storm surge barrier in the Netherlands—white birds gravitated to white shells and black birds gravitated to black shells (Skjonsberg, 2015; West 8, 2017). As an exemplary model of how to "build with nature," this project, in spite of (or perhaps because of) its drastic alteration of the coastal landscape, illustrated the innate potential for landscape design to transcend the utilitarian and even outperform nature. In a later summary of the project he wrote, "Narratives will evolve, surprises will happen, and beauty is already always there" (Geuze and Skjonsberg, 2012). Thus, for Geuze, landscape design is not only an art, it is also an act of faith that sets in motion a complex chain of events in anticipation that these alterations will ultimately create environments that exceed expectation, surpass the ordinary, and intensify the experience of life.

To help sell the Blue Dunes proposal, Geuze created a narrative that situated the project's coastal engineering within the historical legacy of infrastructure projects in the United States completed during the Great Depression. His list of projects included the Triborough Bridge in New York City (Keenan and Weisz, 2016). This reference was intended as a provocative reminder to the Rebuild by Design jury, the general public, and his team that people in the United States once thought big and acted big, and, further, that the grand scale of this thinking fostered an economic expansion that improved the everyday lives of millions. However, praising the Triborough Bridge, and by extension the planning genius of Robert Moses, was a risky move given Moses' dictatorial reputation as a domineering powerbroker and conversely Rebuild by Design's desire for collaborative community-based design. Nevertheless, the tribute is fitting considering the way Blue Dunes sought to promote the common good through drastic alteration of the region's marine and coastal landscape. To Moses' credit, the Triborough Bridge did provide a long overdue and much needed linkage of Manhattan to the Bronx and Queens. The toll revenue generated by the bridge underwrote construction bonds, which financed new highways, parks, and community beaches. Moses even had his own island—Randall's Island—that served as his base of operation. Depending upon one's point of view, Moses was either a hero or a villain, and his projects illustrate both the favorable and unfavorable outcomes inherent to all large-scale planning and design projects (Caro, 1974). By acknowledging Moses, Geuze indicated that he was not averse to heroic action, big ideas, and provocation. Geuze knew how to build this type of project. He had already constructed several mountains on Governors Island using construction techniques applicable to Blue Dunes (Ulam, 2014). Yet, he failed to fully appreciate the impact his bold plans would have on Pilkey and Young.

Pilkey and Young study shorelines and barrier islands around the globe. Their work, which is based upon extensive empirical evidence gathered in the field, illustrates the unique qualities of these landscapes and why and how they react differently to sea level rise, coastal storms, and development (Pilkey and Field, 1973). They frequently use this expertise to help shore communities determine the best course of action in situations where buildings and infrastructure are threatened by beach erosion or damaged by storms (Pilkey and Yooung, 2009). In fact, Pilkey, who began his career as a deep seafloor morphologist in the early 1960s, around the same time that Lorenz began to mathematically model the weather, switched his research to barrier islands and coastal shorelines following the destruction of his parent's home by Hurricane Camille in 1969.

A key principle underlying their consulting work, which they characterize as unbiased and flexible, is ecological resiliency. "Any attempt at restoration after a storm," Young observes, "should return the shoreline back to its natural trajectory" (Pilkey, 2015). This, however, does not mean that once natural processes are reinstated, the beach and its dunes will grow in size, and community life will continue as it has in the past. It means that the natural processes inherent to that site, whether accretive or erosional, will resume. It is, therefore, not surprising that both men are critical of costly, and temporary, sand stabilization regimes dedicated to maintaining existing beach conformations, or that they advocate managed retreat, which Young defines as "a gradual change in the footprint of communities based upon the reality of coastal shoreline dynamics and the reality of sea level rise" (Young, 2013).

To represent his vision of ecological resiliency, Pilkey often collaborates with the batik artist Mary Edna Fraser who portrays barrier islands as bright splashes of color unburdened by political boundaries, storm protection, or people. They began working together in 1993 when Fraser asked Pilkey to supply explanatory text for an exhibit of her work at the Smithsonian in Washington, D.C., entitled "Aerial Inspirations." Pilkey agreed, as the request provided an excellent opportunity to educate the public. He used the platform to argue that barrier islands may protect the mainland from the ravages of storms, but nature did not design them for this purpose. Instead, he argued, they exist because their innate ability to accrete and erode makes them a naturally efficient way for the continental edge to meet the sea. Fraser's imagery not only captures the fluid geomorphology of this edge condition, but it also implicitly states the survival of the islands, and by extension the coastal communities they protect, depends upon this fluidity (Pilkey, 2003).

In addition to their stanch defense of ecological resiliency, another unwavering constant in the Pilkey and Young approach, as mentioned previously, is a deep-seated skepticism of the mathematical models used to justify groins, seawalls, and beach sand replenishment. Their skepticism increases, and rightly so, when human choice and political objectives become variables in these equations. As presented by Pilkey in the text *Useless Arithmetic* (Pilkey and Pilkey-Jarvis, 2007), this is a nuanced argument about the inherent uncertainty of mathematical modeling that is related to the assumptions and generalizations it makes concerning real-life conditions. To explain his position, Pilkey divides mathematical modeling into two interrelated but distinct applications. The first application is qualitative. This type of model asks why, how, and what if, and predicts the direction and magnitude of a process over time. For example, it may indicate that an eroding shoreline will continue to retreat, and the rate will probably accelerate. According to Pilkey, these models are useful when applied correctly and verified by field observations. The second application is quantitative. This type of model asks where, when, and how much, and provides a precise number rather than a trend. For example, an eroding shoreline will retreat almost 52 meters in the next 10 years. Pilkey dismisses quantitative models, and this is due to the fact, as earlier demonstrated by Lorenz, that minuscule variation in the initial conditions of complex natural processes makes this type of accuracy impossible. To illustrate his point, he calls attention to the ever-changing angles of waves as they crash against the shore. This behavior, he notes, reflects a sequence of over 29 events in which both the initial conditions and their order of occurrence is never the same. As a result, every wave hitting the beach is unique and represents an all but impossible to unravel modeling scenario that Pilkey calls "a convoluted bird's nest of interrelationships" (Pilkey and Pilkey-Jarvis, 2007).

One of the biggest abuses of mathematical modeling in Pilkey's estimation is the Bruun Rule. This quantitative equation models the dynamics of wave energy and sand supply and is used to predict the severity of beach erosion. As Pilkey explains, the Bruun Rule accurately assumes shorelines erode as sea level rises, but inaccurately assumes the profile of the shore will remain the same during this process. Even more egregious for Pilkey, however, is that the proponents of this model ignore the evidence from scientific studies of shoreline erosion that illustrate the equation's inaccuracy. Pilkey contends the continuing popularity of this model, despite its limitations, reflects the fact that it is the product of a world "dominated by engineers rather than scientists." In this landscape, he argues, it is imperative to find an answer, rather than "admit defeat and walk away," as he and Young sometimes suggest beach communities do when confronted with the inevitability of sea level rise and shoreline erosion. He concludes this discussion with a reprimand to a Dutch modeler who claims the "Bruun model isn't wrong; it is just incomplete and needs to be more comprehensive" (Pilkey and Pilkey-Jarvis, 2007). In this sense, Pilkey's critique of the Blue Dunes proposal was not about the use of the Blumberg model to qualitatively predict the direction and magnitude of storm surge. The model had, after all, accurately predicted wave height and flood levels for Hurricane Sandy. Rather, it involved the incorrect application of this model to justify an island construction scenario concerned with the quantitative specifics of where, when, and how much sand would be required to construct the islands and stabilize their shorelines. Indeed, for Pilkey the whole project was nothing more than a maintenance intensive and extremely expensive exercise in beach sand replenishment that had the potential to destroy, or greatly diminish, the existing barrier islands. For both Pilkey and Young, this was the point in the project when hard to substantiate design preferences, competition politics, and engineering bravado trumped science, and the work took an exponential leap into environmental fallacy. They subsequently removed themselves from the team (Pilkey, 2016).

A classic landscape architecture example of the Pilkey/Young approach to coastal protection is the conceptual case study "Sea and Survival" done by the landscape architect Ian McHarg at the University of Pennsylvania (McHarg, 1968; McHarg, 1969). Similar to Blue Dunes, Sea and Survival originated in response to a devastating storm—in that case, the Ash Wednesday Nor'easter that pounded the Atlantic coastline of the United States in March of 1962. For three days, gale force winds and ten-meter-high waves buffeted the barrier islands of New Jersey. Dunes were breached. Homes, roads, and utilities were destroyed. Several people were killed and many more were injured (Salvini, 2012). These events prompted McHarg and his students to suspend work on a housing studio and spend the rest of the semester studying littoral currents, sand flow, and barrier island formation. Their research called attention to the dynamic nature of the shifting sand; the deleterious impact of hardened shore stabilization structures; and the relative stability provided by the native plants that colonize the sand dunes and the coastal wetlands. To protect the dunes and wetlands, and by extension the people living in shore communities, they restricted development to the central spine of barrier islands and the back slope of the dunes that faced the bay. A central road, elevated to function in a manner analogous to the flood control dikes in Holland, ran down the central spine of the barrier island. It was argued the economic penalty imposed by this action would be offset by reductions in future storm damage. This type of design, McHarg observed, was "the stuff of kindergarten classes" in the Netherlands (McHarg, 1968).

Unfortunately, New Jersey shore communities ignored the common-sense recommendations presented in Sea and Survival and decided, for the most part, that their continued existence depended upon the revenues generated by small businesses and residential property taxes rather than the preservation of dune fields and wetlands, and they rebuilt their communities to match pre-storm conditions. Nevertheless, Sea and Survival is noteworthy for the way it limits human action in order to protect the wind and waves—a reversal of normative shoreline development procedures, such as seawalls and groins, which attempt to limit the action of natural processes.

When asked to participate in Rebuild by Design, Geuze believed, in a response that echoed McHarg's reaction to the Ash Wednesday storm in 1962 and Pilkey's reaction to Hurricane Camille in 1969, that this post-disaster call-to-arms was a serendipitous, albeit tragic opportunity to generate new dialogue and fresh ideas about development in shore communities along the eastern seaboard of the United States (Geuze, 2015). Many of these shore communities, as just noted, had destroyed naturally occurring wetlands and coastal dunes and replaced them with groins, seawalls, and beach replenishment programs. Sandy had devastatingly proved these artificial defense systems were inadequate to meet the harsh reality of coastal weather. Yet as Geuze reasoned, this was a form of ecological reckoning that could be overcome through a defense strategy, which like the work of McHarg, Pilkey, and Young, looked to natural processes and forms as a means to honor both human life and the dynamic energy of the wind and waves. But unlike these individuals, Geuze did not consider abandonment an option. Culturally conditioned to look upon shore defense as a survival mandate, he deployed barrier islands to draw a proverbial line in the sand.

But Geuze was not operating in Holland. The Dutch approach that underlies the design of the Blue Dunes was antithetical to the small-scale, stakeholder facilitated designs championed by Rebuild by Design. Furthermore, the time required to construct the project did not align with the expediency demanded by politicians and shore communities alike, who instead favored more traditional proposals with less expensive and easier to install, stylistically updated groins and seawalls. Consequently, the jury shied away from the project questioning its cost and complexity, and it was not selected as one of the 6 competition finalists (Gleick, 1987). Indeed, unlike other projects, the team had to explain and justify the project cost, which it did with an innovative property risk assessment (Keenan and Weisz, 2016).

Claire Weisz and Mark Yoes, the founding partners of WXY Architects, observed in their defense of the Blue Dunes project that "the idea of proposing climate adaptation through land creation is neither simple to advocate, nor is it simple to defend" (Keenan and Weisz, 2016). This comment highlights the complexity of the Rebuild by Design challenge and its problematic request that no design reconstruct pre-storm conditions. Strictly speaking, none of the proposals put forward by the ten multidisciplinary teams in the competition reconstructed what was there before. But they didn't really propose anything new. Even Blue Dunes reflected precedents explored previously by Geuze in the Netherlands. Another glaring omission is that none of the proposed coastal defense schemes fully acknowledge the impact of sea level rise and that existing shore communities, if not whole regions of the country, will inevitably be abandoned as a consequence. Unlike the Netherlands where shoreline protection is an existential act and the commitment to extraordinary defensive measures is deeply embedded in the national psyche, in America there is more than enough room for abandonment, planned or otherwise, and therefore no need to draw a proverbial line in the sand. Nevertheless, prudent retreat remained unmentioned. What all the team submissions emphasized instead, as clearly seen in their illustrative perspectives, were happy people, safely protected from damaging storms, enjoying their close relationship with the water. But as Pilkey observes, the coexistence of people and the beach is never peaceful or placid, and the richer the country the less placid the coexistence (Pilkey, 2003). And to deny this reality, as both Pilkey and Young made clear in their objection to Blue Dunes and its expensive but temporary fix, was the competition's greatest fault.

Yet, it may just be that the expensive and temporary fix detailed in the Blue Dunes proposal is the very thing that makes it commendable. What the proposal does, and does admirably, is to force us to radically alter the assumptions that govern our vision of shoreline development. More than any of the other projects, it captures through sheer size and audacity the immense consequences that centuries of small everyday activities have had upon the world's climate and the physical conformation of its coastlines. The construction of several hundred miles of new barrier islands makes it difficult to ignore the dire physical and social repercussions of global climate change and its economic burden. Whether or not one accepts such massive intervention, the vast scope of the project and its contentious modeling debate, calls into question the historical relationship of coastal development to progress, control, environmental knowledge, technological acumen, politics, economics, and landscape aesthetics. Put simply, if we decide that abandonment is not an option and we want life in coastal communities to continue as it has in the past, then it will be necessary, as Geuze rightly argued, to engineer solutions at scales the Dutch have been doing for centuries.* But then again, is the idyllic heroism of Dutch-inspired terraforming any more effective in the long run than an expedient series of small-scale, protection measures that are combined with planned retreat?

As this essay has demonstrated, it is difficult to claim with certainty that one type of coastal protection is better than another, even for members of the same competition team. Too many factors involving hydrodynamics, geomorphology, professional training, visions of nature, and concepts of control are involved. The choice, as Robert Young argues, comes down to values (Young, 2013). What environments do we value? How should we spend our money to protect those environments? Is the proposed solution temporary? In the case of the Blue Dunes, Geuze valued constructed landscapes, permanent solutions, and the art of landscape design, while Pilkey and Young predicated temporality and the fluid beauty of natural processes. And each of these individuals had valid reasons for doing so. In this scenario, there are no heroes or villains, just different sets of initial conditions. There is, however, one irrefutable fact. When designers, scientists, and engineers are asked to combine their expertise, embrace complexity, and build with nature, no matter what margin of error they idealistically factor into their imaginative equations, the fluid dynamics of the coastal terrain, particularly when it becomes entwined with politics and the dynamics of human choice, makes it all but impossible to accurately predict what will emerge from the chain of events set into motion.

---

* Geuze applied the same barrier island strategy to protect the shoreline of the Mississippi Delta, and in this case, perhaps due to the history of intensive engineering in this region of the country, his barrier island scheme was declared the winner of the Changing Course competition. See: Moffatt & Nichol/West 8/LSU-CSS Team Declared Winner in Changing Course Competition. West 8 Press Release. http://www.west8.nl/press_releases/20_aug_2015/ (accessed July 17, 2017).

References

Bisker, J., Chester, A., and T. Eisenberg. 2015. *ReBuild by Design*. New York: American Printing Company, pp. 11–19.

Caro, R. A. 1974. *The Power Broker: Robert Moses and the Fall of New York*. New York: Knopf, p. 20.

Dizikes, P. 2011. When the Butterfly Effect Took Flight. *MIT Technology Review*. http://www.technologyreview.com/article/422809/when-the-butterfly-effect-took-flight/ (accessed August 7, 2015).

Gleick, J. 1987. *Chaos: Making a New Science*. New York: Viking, pp. 11–31.

Geuze, A. 2007. *Flatness, in Mosaics West 8*, Smelik et al. (Eds.). Basel: Burkhäuser, pp. 6–20.

Geuze, A. 2009. Blue Isles Plan. *Topos*, 22: 36–38.

Geuze, A. 2015. Unpublished Blue Dunes project statement, courtesy WXY Architects.

Geuze, A., and Skjonsberg, M. 2012. Dancing with Entropy. *Architectural Design*, 82(5): 124–129.

Kellert, S. 1993. *In the Wake of Chaos: Unpredictable Order in Dynamical Systems*. Chicago: University of Chicago Press, pp. x–xiii.

Keenan, J. M. and Weisz, C. (Eds.). *Blue Dunes: CLimate Change by Design*. New York: Columbia University Press, p. 45.

Lorenz, E. N. 1963. Deterministic Nonperiodic Flow. *Journal of Atmospheric Science*, 20: 130–141.

McHarg, I. L. 1968. Ecology for the Evolution of Design, in *Via 1 Ecology in Design*, Rolf Sauer et al. (Eds.). Philadelphia: Via, The Student Publication of the Graduate School of Fine Arts, University of Pennsylvania, pp. 47–49.

McHarg, I. L. 1969. *Design with Nature*. New York: Wiley & Sons Inc., pp. 7–17.

Pilkey, O. H. 2003. *A Celebration of the World's Barrier Islands*. New York: Columbia University Press.

Pilkey, O. H. 2015. Orin Pilkey's Research Site http://www.sites.nicholas.duke.edu/orrinpilkey/ (accessed July 17, 2017).

Pilkey, O. H. 2016. Personal email to the author dated March 17, 2016.

Pilkey, O. H., and Field, M. E. 1973. Onshore Transportation of Continental Shelf Sediment: Atlantic Southeastern United States, in *Shelf Sediment Transport: Process and Patterns*, Swift, D. (ed.). Stroudsburg, PA: Dowden, Hutchinson & Ross, Inc., pp. 429–446.

Pilkey, O., and Pilkey-Jarvis, L. 2007. *Useless Arithmetic: Why Environmental Scientists Can't Predict the Future*. New York: Columbia University Press, pp. 23–36.

Pilkey, O. H., and Young, R. 2009. *The Rising Sea*. Washington: Island Press/Shearwater Books.

Salvini, E. R. 2012. The Great Atlantic Storm of http://www.1962.njtvonline.org/news/uncategorized/the-great-atlantic-storm-of-1962/ (accessed July 17, 2017).

Scharnhorst et al. 2012. An Introduction to Modeling Science: Basic Model Types, Key Definitions, and a General Framework for Comparison of Process Models, in Models of Science Dynamics: Encounters Between Complexity Theory and Information Sciences, Scharnhorst et al. (Eds.). Berlin: Springer-Verlag, pp. 3–22.

Skjonsberg, M. 2015. Counterpoint: The Musical Analogy, Periodicity, and Rural Urban Dynamics, in *Revising GreenInfrastructure*, Daniel Czechowski et al. (Eds.). Boca Raton, Florida: Taylor & Francis Group, pp. 225–243.

Ulam, A. 2014. Opening a Forbidden Land: Governors Island, New York. *Topos: The International Review of Landscape Architecture and Urban Design*, 89: 23–29.

West 8. *Moffatt & Nichol/West 8/LSU-CSS Team Declared Winner in Changing Course Competition*. West 8 Press Release. http://www.west8.nl/press_releases/20_aug_2015/ (accessed July 17, 2017).

West 8. http://www.west8.nl/projects/all/landscape_design_eastern_scheldt_storm_surge_barrier/ (accessed July 17, 2017).

WXY/West 8. 2014. *Blue Dunes: The Future of Coastal Protection*. Competition Report (accessed July 17, 2017).

Young, R. 2013. A Year After Sandy, The Wrong Policy on Rebuilding the Coast. Yale Environment 360, 31 Oct. 2013. http://www.e360.yale.edu/feature/a_year_after_sandy_the_wrong_policy_on_rebuilding_the_coast/2705/ (accessed July 17, 2017).

8

Designing Resiliency: Interdisciplinary Geovisualization for Complex Coastal Environments

Forbes Lipschitz

CONTENTS

In the age of climate change and planetary urbanization, coastal communities around the world are facing unprecedented social and environmental challenges. As populations increase along the land's edge, anthropogenic disturbances and ongoing resource depletion strain coastal ecosystems. Meanwhile, the incremental rise in the world's oceans is increasingly accompanied by stronger more frequent tropical storms. These changes will continue to threaten human habitation by forcing population displacement and altering economic, ecological, and agricultural systems in unforeseeable ways. The complexity and dynamic nature of climate-stressed coastal environments defy simple, linear design solutions. No single discipline or expertise is capable of solving such a multi-scalar, multi-faceted problem. Responding to climate change in coastal landscapes will require coordinated interventions across multiple sectors at multiple scales.

Although the importance of interdisciplinarity in formulating a coordinated response to coastal threats is generally recognized, there remains much work to be done in improving the process in which it is carried out. Effective interdisciplinary interventions require much more than giving representatives from different disciplines a voice in the discussion. Successful interdisciplinary teams must break through disciplinary silos in order to meaningfully exchange knowledge across disciplinary boundaries. They must engage politicians and diverse stakeholder groups in order to develop solutions that consider ecological, social, and economic outcomes. How can collaboration be facilitated when experts, politicians, and stakeholders often have different values and interests, operate at different scales, and lack a common technical language?

The design process employed by landscape architects is uniquely positioned to facilitate interdisciplinary problem solving because of our capacity to think and communicate visually. We embrace the power of visual representation to communicate ideas across disciplines, cultures, and even language barriers. By revealing underlying spatial patterns and relationships—and presenting them in a way that non-experts can observe and understand—we can enhance the interdisciplinary process in ways that produce better problem solving and decision making outcomes. One key tool in brokering meaningful interdisciplinary collaboration is geovisualization—the creation and use of interactive maps to facilitate thinking and problem solving.

From the discipline's beginnings, landscape architecture and landscape architects have employed an arsenal of visual tools ranging from maps and diagrams to models and renderings. Data visualization is foundational to the discipline. Yet coastal resiliency presents unique information management challenges to visual representation. The complexity of coastal systems requires the integration and analysis of large, and at times, disparate datasets, while capturing and synthesizing results in ways that allow for iteration and collaboration. The emerging practice of geovisualization shows great potential in addressing these issues. Using Geographic Information Systems as the primary platform, geovisualization can convey complex datasets and spatial relationships to subject experts, while three-dimensional models and landscape simulations can be easily understood by non-experts. As a design tool, these representations can help to generate new hypotheses, develop problem solutions, and facilitate non-linear thinking (Kraak, 2003). The suitability of specific visualization techniques varies by problem scale, type, and intended audience. In the initial stages of a collaborative design process, geovisualizations can reveal previously unseen information in the data. This involves a high degree of human-map interaction. In the diagram of map use created by geographers Menno-Jann Kraak and Alan MacEachren (see Figure 8.1), geovisualization objectives are organized in a three-dimensional space by their level of interaction, their audience, and the addressed tasks. The map can therefore

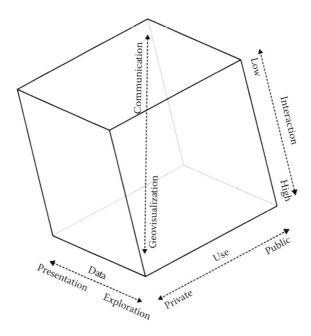

FIGURE 8.1
Geovisualization and communication. (Adapted from Kraak, M. J. and A. M. MacEachren. 1994. "Visualization of spatial data's temporal component." *Proceedings, Spatial Data Handling, Advances in GIS Research*, Edinburgh, Scotland, September 5–9, 1994. 391–409.)

go beyond its traditional role as a presentational device, becoming an interactive interface to explore and analyze geospatial data (Kraak and MacEachren, 1994).

Focusing on contemporary coastal resiliency projects in Louisiana, this paper explores the ways in which interdisciplinary design teams use geovisualizations in research, design, and planning projects. Case study projects completed at Louisiana State University demonstrate the role that geovisualization can play in promoting dialogue and knowledge production through the process of interdisciplinary design. From social science research projects to regional master planning activities, the projects reveal the ways in which geovisualization maps can uncover patterns in complex data, articulate the scale of the coast, and facilitate collaborative resiliency planning. As sea levels rise and the frequency and intensity of tropical storms increase around the world, the findings produced in Louisiana and the interdisciplinary design process that facilitated their production will be increasingly relevant to a variety of different actors and contexts.

Representing Louisiana

Louisiana is an ideal laboratory for understanding the advantages and limitations of geovisualization as a tool for coastal resiliency. Home to approximately two million residents, coastal Louisiana faces a variety of acute and persistent environmental risks, including sea level rise, subsidence, and wetland loss. Increasingly strong and frequent storms greatly affect residents, with four of the worst hurricanes in history having made landfall there in the past 10 years. Human disturbance has also drastically altered Louisiana ecosystems. Indeed, the Mississippi River Delta is one of the most anthropogenically disturbed deltaic systems in the world. Dams restrict flows on its major tributaries while constructed levees and bank revetments abut its straightened river banks. Moreover, the Delta sits at the outflow of the Mississippi River Basin, the third largest drainage basin in the world, draining 41% of the contiguous United States. The basin is also the most productive agricultural region in the country, with highly intensive production systems discharging over a billion tons of nitrogen into the Mississippi River each year, subsequently depleting oxygen and negatively impacting aquatic ecosystems in the Gulf of Mexico.

The complexity of Louisiana ecosystems is matched only by the scale and pace of environmental change. Coastal Louisiana wetlands comprise the seventh largest delta in the world, containing nearly 40% of the estuarine marshland in the contiguous United States, which currently account for 90% of the coastal wetland loss in the country. The pace of land loss is staggering. Between 1932 and 2012, coastal Louisiana lost 25% of its total land area and continues to lose more by the day (Couvillion, 2011). Though the area of land lost comprises more than 487,694 hectares, the rates of topographic change come down to a few centimeters, a blurry wet line between water and land. Seemingly micro-scale phenomena have enormous consequences when applied across the landscape.

A variety of geospatial datasets are available to document and analyze this rapid social and environmental change. Recent advances in remote sensing and aerial imaging technologies are expanding access to geospatial information on the coastal environment. This data is produced and managed by different agencies—often at different resolutions, datums, and extents. To name just a few examples, a designer analyzing hurricane vulnerability in Louisiana would need to obtain topography and bathymetry from the National Oceanic and Atmospheric Administration, shapefiles of Louisiana levees from the Army Corps of Engineers, designated flood zones from the Federal Emergency Management Agency, and evacuation routes from the Louisiana Department of Transportation. Geovisualizations that examine coastal resiliency must therefore simultaneously employ multiple disparate datasets and communicate multiple scales.

It is clear that the problems facing coastal Louisiana are complex, entrenched, and enmeshed and cannot be confined to the study of ecology, engineering, or design. By synthesizing disparate datasets and revealing hidden patterns and processes, geovisualization can help to break through disciplinary silos and facilitate transdisciplinary knowledge exchange. As a mapping tool, geovisualization allows the designer to organize, synthesize, and visualize data from disconnected datasets.

A Brief History of Geovisualization

Mapping and geospatial data visualization has a long history in the design, planning, and engineering professions. Throughout the twentieth century, manual mapping methods were the primary means by which researchers and designers could study and communicate geospatial data. Analog overlay techniques, popularized by Ian McHarg, assigned value to thematic datasets as a measure of the suitability of that condition to a given use. Multiple data layers are then overlaid and their values calculated. The resulting map indicates the most suitable locations for the land use in question. The McHargian overlay technique laid the groundwork for the computer-based Geographic Information Systems (GIS) that emerged in the 1980s and became widespread throughout professional and design and planning practices in the mid- to late 1990s. In this context, maps served as analytical tools to inform design, as well as presentation devices that could appear dispassionately objective.

Mapping has gained renewed importance to design professions in the twenty-first century. The silkscreen prints and maps of Mathur and da Cunha's *Mississippi Floods* (2001) reveal how different representations of the river have shaped its design, engineering, and control. In *Petrochemical America* (2013), Richard Misrach and Kate Orff couple a photographic record of Louisiana's industrial landscape with speculative maps that create an "Ecological Atlas" of the region. Most recently, Jill Desimini and Charles Waldheim's book and accompanying exhibit *Cartographic Grounds: Projecting the Landscape Imaginary* (2016) explores the trajectory of cartographic techniques through historic and contemporary examples of map and plan drawings. Such projects highlight maps as projective and subjective, rather than objective representations.

What remains to be explored in the design professions is the capacity of maps to facilitate interdisciplinary collaboration. When properly designed, maps can produce knowledge. Cartographic problem-solving dates back to the mid-nineteenth century, when a serious outbreak of cholera in London's Soho district defied explanation. By mapping specific cases, Doctor John Snow identified that cholera was transmitted by water, not polluted air, as previously thought. In the seminal essay, "The Agency of Mapping," landscape architect and theoretician James Corner states that the power of maps "lies in neither reproduction nor imposition but rather in uncovering realities previously unseen or unimagined" (Corner, 1999). Similarly, in the scientific community, visualization is described as a range of graphic techniques that "transform the symbolic into the geometric, enabling researchers to observe their simulations and computations. Visualization offers a method for seeing the unseen. It enriches the process of scientific discovery and fosters profound and unexpected insights" (McCormick et al. 1987). By synthesizing Corner's "Agency of Mapping" with the scientific definition of visualization, we arrive at geographic visualization, or geovisualization.

Contemporary geovisualization provides a method for visual exploration, analysis, and synthesis of geospatial data. While traditional static maps have a limited exploratory capability, geovisualization allows for heightened user interaction. Denis Cosgrove notes that "the representational spaces of the map are theoretically unlimited, constrained only by the imagination of the map-maker and the practicalities of legibility and comprehension" (Cosgrove, 1999). The two-dimensionality and static nature of conventional maps makes exploring complex and time variant data difficult. An important facet of geovisualization is the capacity to view geospatial datasets in a variety of alternative ways, unconstrained by traditional cartographic rules and conventions. The ease of iteration and high level of interactivity that geovisualization affords can reveal previously unseen spatial patterns. For example, if a designer makes a single-color gradient map of post-hurricane vacancy, the user's visual interpretation of that map might be a function of the color scheme. In an interactive data exploration environment, one can change the color scheme or symbolization regime instantaneously, and thus learn something about the pattern of vacancy that was latent in the single-map solution.

Interactivity is also an increasingly important facet of geovisualization, exemplified by a range of interactive sea level rise maps currently available on the web. The North American Oceanographic and Atmospheric (NOAA) Office for Coastal Management, has an interactive digital sea-level viewer that renders community-level impacts from coastal flooding and sea level rise. A sliding scale allows users to map sea level rise in 304 millimeter (one foot) increments and overlay transparent blue water levels with flood frequency, socio-economic vulnerability, and wetland loss datasets. The inundation scenarios do not, however, include coastal storm surge, riverine flooding, erosion, or other coastal processes. The sea level rise slider also limits scenario testing to a single variable. In the creation of such interactive platforms, designers must not only design the maps themselves, but also how those maps are meant to be manipulated. When combined with the existing expertise of the map user, the capacity to manipulate the maps can facilitate creative thinking and knowledge construction. To this end, modes of interaction need to be designed as carefully as the visual symbols on the map display.

Visual Thinking and Discovery: Louisiana Case Studies

Designers appreciate the power of visualization as a research tool and acknowledge that the ways of representing data can be as important as the data itself. This is particularly true in an interdisciplinary venue, in which technical and nontechnical participants can often struggle to relate. When rendered as a visual graphic, complex and abstract data can be seen and understood. Geovisualization should therefore focus not on generating new imagery, but on revealing hidden patterns and relationships. In so doing, new images can generate new ideas. In this section, we will explore three different case studies that explore the applicability of geovisualization to coastal resiliency design, planning, and research.

Case Study 1: Revealing Patterns

A recent interdisciplinary research project completed at Louisiana State University demonstrates the ability of maps to reveal significant patterns and scientific hypotheses. With researchers from the Coastal Sustainability Studio, the Robert Reich School of Landscape Architecture, the Department of Sociology, and the Public Policy Research Lab, the study examined the effects of geographically differentiated support resources on perceptions of household preparedness for different persistent and acute environmental threats (Cope et al. 2018). The study area consisted of two coastal Louisiana parishes (Plaquemines and Lafourche), which have been struck by multiple hurricanes and heavily impacted by the Deepwater Horizon oil spill—the worst oil spill in American history. The area also faces a range of slow, but persistent environmental threats, including sea level rise, saltwater intrusion, and coastal erosion. A household level survey of approximately 1,000 respondents was conducted to understand how local residents perceive both vulnerability and preparedness to these wide-ranging threats. The researchers asked how maps might be employed to reveal additional dimensions of data represented on a regional scale. As each layer interacts with others, could patterns emerge to suggest trends in behavior?

Survey respondents were asked to identify their community, which was then assigned a spatial coordinate. Individual results and community aggregates were imported in ESRI ArcMap as point data and mapped in relation to the ecological, infrastructural, and urban systems that characterize coastal Louisiana. Geospatial datasets were obtained from a variety of publicly available sources, including USGS, NOAA, and FEMA. The maps were developed using a common template, symbology, and visual language, to allow for comparison across the survey results. Researchers employed a broad range of data exploration methods, including varying the symbolization, highlighting portions of the datasets, and employing multiple views. In this way, maps were generated as both visual interfaces to access and explore geospatial data and presentation devices. Preliminary maps thus allowed for the visual exploration of the survey results in order to reveal unanticipated spatial patterns, while finalized maps communicated project findings.

Exploratory mapping demonstrated that perceptions of threats and household preparedness were geographically differentiated. Base maps indicated the location of major highways and interstates, federal flood protection levees, areas of historic land loss, and FEMA-designated high-risk flood zones. Survey respondents identified threats of environmental change (including sea level rise, coastal erosion, saltwater intrusion, and seasonal flooding), hurricanes (including wind damage and flooding), as well as threats to fishing harvests and environmental pollution. Many of these communities are located in high risk flood zones and experience high rates of land loss and coastal erosion. One observable spatial trend is that the communities further inland are more concerned with storms while those more exposed along the coast are more concerned with environmental change.

Mapping also revealed patterns in risk preparedness (see Figure 8.2). Respondents in communities with low levels of flood risk felt more prepared to deal with hurricanes than those in high flood risk areas. Within high risk areas, those communities that were closer to the coast had fewer regional resources and exhibited lower levels of preparedness. However, mapping also indicated that respondents who identified themselves as underprepared had no plans to leave their community. A high percentage of respondents in underprepared communities stated that they "will not" leave or "do not know" if they will leave their community.

Finally, maps were used to examine survey respondents that would not move away after what they have identified as a "most serious threat" happens or continues to happen (see

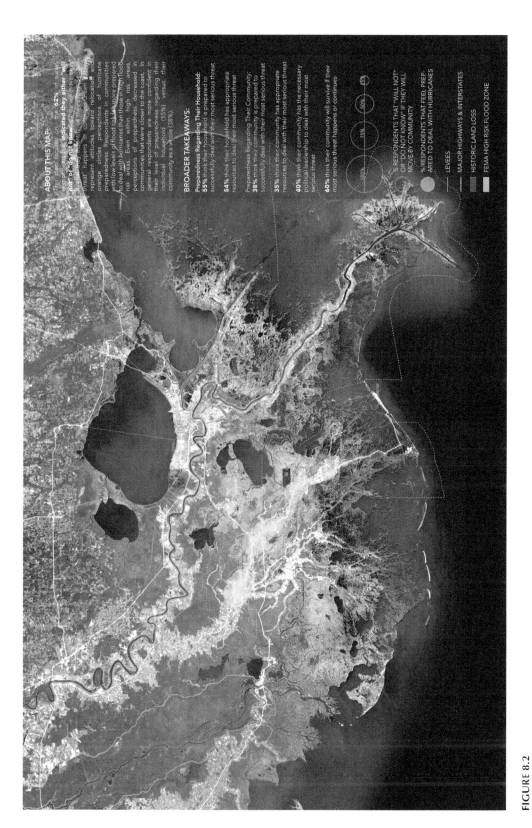

FIGURE 8.2
Patterns of household preparedness perception. (Adapted from Lee, M. R. et al. 2014. Geographically Distant Support Resources Elevate Perceived Preparedness for Environmental Threats in Disaster Prone Coastal Communities 2014.)

Figure 8.3). Green lines visualized responses of those who listed the place they would move to. The graph at the bottom of the page charts driving distance from Lafourche Parish and plots circles scaled to represent the percentage of people who named each location. The map revealed that many respondents would move above or along the I-10 corridor, to communities in Baton Rouge, New Orleans, and Mississippi. Overlaying the survey results with topographic data revealed that the interstate corridor is located along the Pleistocene ridge that runs just north of the Mississippi River Delta. This ridge makes the corridor a natural refuge above the rising sea levels and high inundation rates in the shifting landscape of the delta. The mapping processes therefore revealed spatially relationships between household perceptions of vulnerability and preparedness and the surrounding ecological and geographic systems that would not be evident through statistical analysis alone.

Case Study 2: Articulating Scale

While geovisualizations are primarily constrained to the digital environment, designers and planners have been increasingly employing digital fabrication techniques to bring them into three dimensions. The 3-D map has a long history in cartography and design, with topographic relief and contour models being the most prevalent. Abstract data that is translated into physical models can be more understandable and engaging than two dimensional maps and plans. The 3-D map allows users to project their own spatial knowledge onto the physical model. In this way, the 3-D map can become a powerful research tool that provides users with a common platform to think creatively about the project at hand.

Modeling is a particularly useful tool in Louisiana, where the flat expansiveness of the coast makes representing and revealing topographic patterns in two dimensions difficult. The processes that shape the delta extend across thousands of acres, yet the topography only varies by a few critical centimeters. The horizontal compression of the wetland cannot be understood through traditional surface mapping methods. A recent seminar at Louisiana State University developed novel three-dimensional geovisualization techniques for representing a particularly dynamic portion of the Louisiana Coast, the Wax Lake Delta.

In 1942, the U.S. Army Corps of Engineers erected a river diversion to protect Morgan City from the floodwaters of the Atchafalaya River. This diversion directed water and sediment to the Wax Lake, resulting in unexpected and measurable land building. Geologists first noticed sediment deposition in Atchafalaya Bay during the 1950s and following a severe flood in 1973 when new land rose above the water line for the first time. Since then, the Atchafalaya and Wax Lake deltas have continued to grow by 260 hectares (one square mile) per year (Rosen and Xu, 2013). The area provides valuable insight into the ecological dynamics of coastal restoration.

Shelby Doyle, then Visiting Assistant Professor in the School of Architecture, taught the seminar "Fabricating the Wax Lake Delta" with support from the LSU Coastal Sustainability Studio in collaboration with the Coastal Protection and Restoration Authority River Model Exhibition Project. Drs. Robert Twilley and Leanna Heffner provided research from the LSU College of Coast & Environment to support the geovisualizations. The course explored a range of digital modeling and fabrication techniques for producing dynamic and interactive models of Louisiana's riparian and deltaic landscapes.

Students began the seminar by mapping quadrants of topographic and bathymetric terrain in a digital modeling environment, each exploring different methods of vertical exaggeration. The resulting digital surface was 3-D printed and then aligned in a grid to

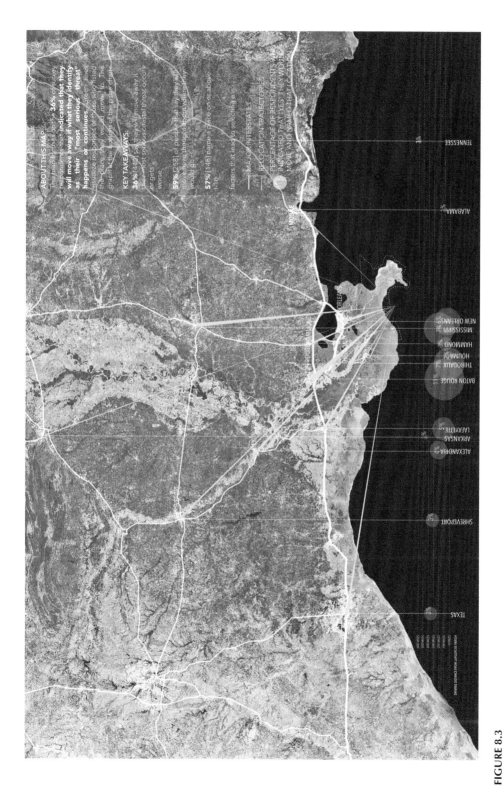

FIGURE 8.3
Regional relocation patterns. (Adapted from Lee, M. R. et al. 2014. Geographically Distant Support Resources Elevate Perceived Preparedness for Environmental Threats in Disaster Prone Coastal Communities.)

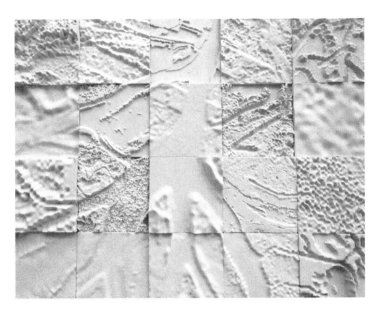

FIGURE 8.4
3-D-printed topography bathymetry models of the Wax Lake Delta. (Image credit: Shelby Doyle.)

represent the Wax Lake Delta region (see Figure 8.4). Additional data was then animated, mapped, and projected onto the 3-D-printed surface (see Figure 8.5). The vertical exaggeration of the terrain revealed patterns of erosion and sediment deposition. The projection mapping showed the dynamic linkages of terrain to hydrologic, vegetative, and climatic conditions across the deltaic ecosystem. This geomodeling method can reveal

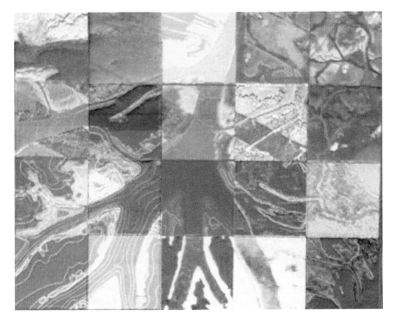

FIGURE 8.5
Systems projection on 3-D printed models. (Image credit: Shelby Doyle.)

hidden interactions between micro-topographic shifts and landscape systems, enabling a productive dialogue about land building in Louisiana.

Case Study 3: Facilitating Collaboration

As the use of geovisualization for research and design expands, a growing interest in the effectiveness of said mapping techniques for interdisciplinary planning has emerged. This demand warrants an enhanced understanding of how geovisualizations are prepared, implemented, and perceived in the design process. Historically, analog sketching has been used to collaborate and document ideas in group-based design activities. Freehand drawing provides non-design participants with an entry into the creative process, promoting dialogue and providing base drawings for further development. While such analog methods are accessible and user friendly, they cannot provide the comprehensive contextual information necessary when designing and planning in complex coastal landscapes.

Geovisualization can be a powerful tool to stimulate collaborative design in interdisciplinary design and research projects. Complex ecological, social, and infrastructural data can be presented in a format that designers, scientists, decision makers, and stakeholders can understand and debate. A recent interdisciplinary design competition in Louisiana revealed the ability of geovisualizations to facilitate interdisciplinary collaboration. The 2015 Changing Course Design Competition challenged teams to create innovative solutions for the Lower Mississippi River Delta that enhance the Louisiana State Master Plan. The design proposals needed to maximize the river's sediment budget for land-building without impacting navigation. The design also needed to achieve optimal outcomes for ecological restoration and flood mitigation, while improving local socio-economic conditions.

The Giving Delta proposal (led by the Moffat & Nichol team which included the LSU Coastal Sustainability Studio, Deltacrest, Maritime Institute of Technology and Graduate Studies, RAND Corporation, UNO Pontchartrain Institute of Environmental Sciences, and West 8) employed geovisualization to facilitate collaboration across disciplinary boundaries. Following data collection and synthesis, the team mapped the coast in four distinct systems: ecological networks, river control infrastructure, transportation infrastructure, and human settlements. Geovisualizations represented these systems as layered networks without discrete boundaries. The resulting maps provided varied yet comparable layers that could be overlaid and reworked as part of interdisciplinary design workshops.

With the systems maps as the foundation, the team conducted scenario planning charrettes that envisioned alternative futures across the various sectors of the delta system. Such alternative scenarios included sea level rise, oil and gas exploration, urban displacement, and novel sediment diversions. Break-out sessions explored each system, producing analog sketches that showed major design concepts and their system-wide effects. Rough sketches were then incorporated into the digital drawings, which were later reviewed and refined by technical experts (see Figure 8.6). By overlaying systems, synergies and opportunities were revealed, allowing design interventions to yield multiple benefits for cultural, infrastructural, economic, and ecological systems. Take, for example, the strategic siting of "flood-pulse" structures. The structures needed to be sited in a way that would maintain wetlands, reduce nutrients, and enhance delta productivity, while minimizing construction costs and population displacement. Maps revealed the location of uninhabited farmland

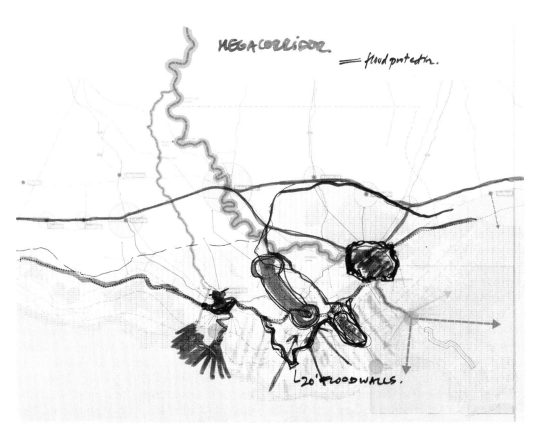

FIGURE 8.6
Participatory mapping of human settlement patterns. (Image Credit: Moffat & Nichol, LSU Coastal Sustainability Studio, Deltares, Maritime Institute of Technology and Graduate Studies, RAND Corporation, UNO Pontchartrain Institute of Environmental Sciences, and West 8.)

tracts, existing hydraulic conveyances, and relic geologic crevasses that could accommodate five controlled floodways into Barataria and Maurepas Basins.

The Changing Course Competition demonstrates that geovisualization can support collaborative use and become, in the words of James Corner, a "game-board." Corner conceives of such mapping platforms as shared surfaces upon which multiple stakeholders and users can articulate their differences. He states:

> As a representation of contested territory, the map assumes an enabling or facilitating status for otherwise adversarial groups to try and find common ground while 'playing out' various scenarios. Ideas of drift and layering are developed here, as the former allows for personal engagement between mapper and constituents, while the latter permits the analytical separation of multiple issues and agendas.
>
> **Corner, 240**

In this context, geovisualizations enable a range of alternative scenarios. The map becomes an open-ended architecture, in a constant state of iteration.

Though the Giving Delta scheme employed digital geovisualizations as base drawings for collaboration, the design charrettes still relied on analog overlay techniques to facilitate collaborative design. Designers were then responsible for translating the design proposals and iterations into the digital environment. The primary benefit of this method is that the hands-on approach actively engages the participants. Due to the limitations of printed paper base maps, however, the next step toward truly interdisciplinary geovisualization will be to develop an interactive and collaborative visualization interface, that allows users to quickly iterate the map in real time. Such an interface could allow for data exchange between Geographic Information Systems and parametric modeling systems, resulting in flexible workflows that could be modified to suit a wide variety of interdisciplinary design scenarios. Though workflows can incorporate geospatial data into automated and parametric modeling environments, they are not tightly integrated.

The Limits of Geovisualization

Geovisualization can be a powerful tool to stimulate collaborative design in interdisciplinary design and research projects. Dynamic, interactive, and participatory mapping will undoubtedly be critical to the processes of coastal resiliency design. Though designers, cartographers, and researchers have historically focused on the map as a communication tool, information is not only useful in public or collaborative settings but also for individual exploration. The process is a way for the designer or researcher to gain a more detailed understanding of the processes and issues across the study area. Creative discoveries in both design and science often emerge from the moments when one is forced to think unconventionally. In this way, alternative mapping methods and geovisualization that defy cartographic conventions can become an internal tool for discovery and design, or in the words of Jacques Bertin, "a storage mechanism and a research instrument" (Bertin, 1983). This method of visual thinking produces insights on patterns and relationships, exposes anomalies in data, and enables transdisciplinary dialogue.

Geovisualization can be hindered, however, by the very complexity that it is designed to embrace. The case studies discussed herein have focused primarily on nuanced applications in which the end user is a knowledgeable designer and/or researcher. Such complexity is often not easily understood by outsiders. Detailed maps intended for use by those immersed in the subject may be confusing to broader audiences. Interdisciplinary design teams have addressed this dilemma by simplifying the representations intended for a public audience. For example, perhaps the clearest design diagram of the Changing Course Giving Delta submission was sampled from the iconic 1958 Project Design Flood diagram of Mississippi River flow capacity (see Figure 8.7). The diagrammatic reinterpretation presents an alternative controlled flood framework that employs diversions to support wetland building. Meandering rivers are simplified and straightened as in a simple plumbing diagram. Though an additional color is used to distinguish wetlands from water and dry land, the indeterminate flux of the deltaic hydrology is rendered fixed and rigid. While the diagram does not communicate the complexity of the coastal landscape, its engineered simplicity is actionable, which is precisely what makes the Army Corps of Engineers diagram so effective in the first place.

In addition to abstracted diagrams, interdisciplinary design teams are increasingly relying on photorealistic mapping. This is in part because satellite imagery, reinforced by the widespread availability of web viewing platforms like Google Earth, has also surpassed planimetric maps as the most commonplace mode of representing and understanding large scale landscapes. The Giving Delta team employed this approach to communicate potential futures for the Louisiana coast if no decisive action is taken (see Figures 8.8 and 8.9). Two maps are presented side by side: one depicting the familiar satellite image of the "Louisiana Boot,"

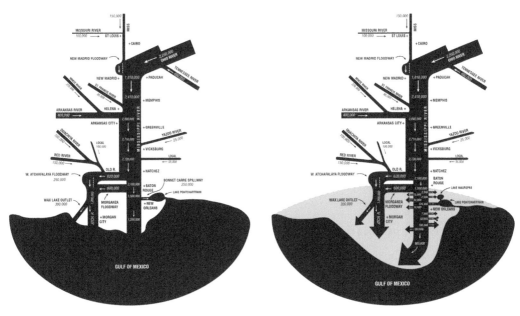

Army Corps of Engineers 1958 Project Design Flood   The Giving Delta Framework

FIGURE 8.7
The Giving Delta framework. (Image Credit: Moffat & Nichol, LSU Coastal Sustainability Studio, Deltares, Maritime Institute of Technology and Graduate Studies, RAND Corporation, UNO Pontchartrain Institute of Environmental Sciences, and West 8.)

FIGURE 8.8
The coast in 2015. (Image Credit: Moffat & Nichol, LSU Coastal Sustainability Studio, Deltares, Maritime Institute of Technology and Graduate Studies, RAND Corporation, UNO Pontchartrain Institute of Environmental Sciences, and West 8.)

FIGURE 8.9
Predicted outcomes of a status quo river approach. (Image Credit: Moffat & Nichol, LSU Coastal Sustainability Studio, Deltares, Maritime Institute of Technology and Graduate Studies, RAND Corporation, UNO Pontchartrain Institute of Environmental Sciences, and West 8.)

the other a stark image representing the dissolution of coastal wetlands. The verisimilitude of the rendering is compelling and easily understood by lay participants.

The challenge to designers and researchers working on interdisciplinary coastal projects is how to synthesize the large amounts of geospatial data and communicate landscape complexity and dynamics, while creating a graphically clear and legible final product. The aforementioned examples reveal that while geovisualization is a powerful tool for the design process, it cannot capture all of the ranges of meaning and information necessary to effectively communicate a design proposal. Just as no single discipline is capable of solving the problem of coastal resiliency, no single visualization method can meet the needs of every phase of the interdisciplinary design process.

Conclusion

Meaningful interdisciplinarity is critical to the design and planning of resilient coastal landscapes. Yet the integration of different disciplines, ontologies, and methods into a single project presents unique and difficult challenges. Different disciplines often have different technical languages and different core values. Because geospatial datasets are often produced by multiple agencies at inconsistent resolutions, conflicting datums and extents further complicate the problem. As a tool for visual data exploration and synthesis, geovisualization is valuable for harnessing transdisciplinary synergies and promoting meaningful collaboration. From rapid manipulation of symbology, 3-D fabrication techniques, and participatory mapping, novel visualization strategies can reveal patterns that are not necessarily evident when traditional mapping methods are employed. The use of alternative and unconventional cartographic methods stimulates visual thinking and creativity. Indeed, there is often something to be "gained from translation" when representing information outside the established framework of a particular discipline can result in new insights.

The case studies explored herein demonstrate the value of geovisualization to reveal patterns, articulate the landscape, and facilitate a collaborative design process. Louisiana case studies were selected because the pace and scale of climate-induced environmental change across the coast make the state an ideal testing ground for interdisciplinary design processes. They also uncover the challenge in developing graphic strategies that can engage both subject experts and lay people. These strategies are far from comprehensive and show room for improvement. Though platforms for animating, parametric modeling, and 3-D printing geospatial data exist, known workflows are not user friendly. In order for interactive geovisualizations to more effectively support interdisciplinary collaboration, parametric plugins within geospatial platforms could be designed to support quick and iterative data exploration and design generation. Animation and rapid 3-D printing technologies might allow users to explore spatio-temporal datasets across multiple dimensions. Animations could also allow multiple datasets to be broken down and sequenced in order to reduce visual complexity. This would improve the potential of geovisualizations to support both high and low levels of user interaction.

While the geovisualization techniques employed on a given project will vary based on the problem at hand, the interdisciplinary design toolkit is growing. Advancements in geovisualization technologies will undoubtedly continue, resulting in platforms that are more dynamic, interactive, and open ended. This will likely improve the efficacy of geovisualization as a framework for interdisciplinary design in climate stressed coastal landscapes.

References

Bertin, J. 1983. *Semiology of Graphics*. Madison, WI: University of Wisconsin Press.

Cope, M., M. R. Lee, T. Slack, T. C. Blanchard, J. A. Carney, F. E. Lipschitz and L. Gykas. 2018. "Geographically distant social networks elevate perceived preparedness for coastal environmental threats." Population and Environment 39.3: 277–296.

Corner, J. 1999. "The agency of mapping: Speculation, critique, and invention." *Mappings*, ed. D. Cosgrove. 231–252. London: Reaktion Books.

Cosgrove, D. 1999. "Introduction: Mapping meaning." *Mappings*, ed. D. Cosgrove. 11. London: Reaktion Books.

Couvillion, B. R. 2011. *Land Area Change in Coastal Louisiana (1932–2010)*. Reston, VA: U.S. Dept. of the Interior, U.S. Geological Survey.

Kraak, M. J. 2003. "Geovisualization illustrated." *Journal of Photogrammetry and Remote Sensing* 57: 390–399.

Kraak, M. J. and A. M. MacEachren. 1994. "Visualization of spatial data's temporal component." *Proceedings, Spatial Data Handling, Advances in GIS Research*, Edinburgh, Scotland, September 5–9, 1994. 391–409.

McCormick, B. H., T. A. DeFanti and M. D. Brown. 1987. "Visualization in scientific computing." *Computer Graphics* 21.6: 3.

Rosen, T. and Y. J. Xu. 2013. "Recent decadal growth of the Atchafalaya River Delta complex: Effects of variable riverine sediment input and vegetation succession." *Geomorphology* 194: 108–120.

9

Best Practices for Interdisciplinary Collaboration for Coastal Resiliency

Jill Allen Dixon

CONTENTS

The Need to Look beyond Traditional Collaborations

> All too often, experts forget that "problems of society do not come in discipline-shaped blocks."
>
> **Moti Nissani, quoting R. Roy (1997: 209)**

Historically, coastal resilience strategies were considered the domain of engineering and science; the goal of most strategies was to keep water out of existing communities. Today, as broader resiliency ideas are taking hold—with more diverse goals that include ecological health and social resilience—we need to look to new kinds of collaborations to grow more resilient coastal communities. Design thinking, in particular, plays a special role in successful teams. Because interdisciplinary collaboration already comes naturally to many designers, it can be a powerful integrator and core component of these projects (Figure 9.1).

Drawing on scholarly literature, as well as experience in coastal design collaborations, this chapter discusses the potential for interdisciplinary collaborations to develop innovative ideas for coastal resiliency. Given the diverse challenges coastal communities face, interdisciplinary collaboration should be a fundamental component of any process focused on coastal resilience. "Interdisciplinary" in this context does not just mean landscape architects working with planners, architects, and/or urban designers. Coastal resilience demands an even broader strain of interdisciplinary teamwork—with ecologists, social scientists, engineers, and beyond.

And yet—importantly—interdisciplinary teamwork is not just a matter of a broader team. It is also about different ways of solving problems and a deeper need for creativity *across* disciplines. It requires fundamentally different skills, different attitudes, and new ways of communicating. Unsurprisingly, it can be extremely difficult to do well.

FIGURE 9.1
The focus of the Changing Course Design competition, coastal Louisiana's rapidly disappearing wetlands are an example of a coastal resilience challenge that requires expertise from many disciplines to address. (Luke Nairn.)

This chapter focuses on applying interdisciplinary thinking in practice by discussing the following questions:

- How does interdisciplinary collaboration enrich ideas for coastal resilience?
- How are great teams built? What factors lead to successful collaboration?
- What methods or practices help overcome common challenges with interdisciplinary teamwork?
- What is the value of combining science with design thinking for coastal resilience?

This chapter begins by defining interdisciplinary collaboration and what it means when applied to coastal resilience challenges. The next section draws on scholarly literature to summarize recognized benefits and challenges; the many advantages of interdisciplinary teamwork are well-recognized, but effective collaboration does not always come easily. The third section proposes ten best practices for collaborating as an interdisciplinary team. This section draws from experience with several coastal resilience projects, including the Changing Course Design Competition in the lower Mississippi Delta and the Rebuild by Design Competition, an initiative of the Hurricane Sandy Rebuilding Task Force (Changing Course, http://changingcourse.us/; Rebuild by Design, http://rebuildbydesign. org/). The final two sections discuss the rewards of successful collaboration and point to opportunities to promote interdisciplinary teamwork beyond individual projects.

Defining Interdisciplinary Collaboration

With many variations of collaborative research being discussed today, what exactly is "interdisciplinary collaboration," and how does it compare to "multidisciplinary" or "transdisciplinary" practices? The most common distinctions relate to the degree of collaboration. In "interdisciplinary" teamwork, individuals representing two or more disciplines work in an integrated team and combine knowledge from different perspectives to solve a common problem (Nissani, 1997; Brewer, 1999; Szostak, 2002; Blackwell et al., 2009; Stock and Burton, 2011). In contrast, "multidisciplinary" and "transdisciplinary" represent opposite ends of the collaboration spectrum. "Multidisciplinary" refers to situations where different disciplines work in parallel without significant integration, while "transdisciplinary" describes groups that work in such an integrated fashion that they transcend their individual disciplines (Stock and Burton, 2011:1098). With transdisciplinarity, groups can include stakeholders or the general public and may require crossing boundaries at the levels of institutions or the forming of new disciplines (Blackwell et al., 2009; Stock and Burton, 2011). Of these three types of collaboration, interdisciplinary collaboration holds the most promise for coastal resilience. It offers a balance of allowing different disciplines to contribute in their own way while also promoting more integrated collaboration. In other words, it combines disciplinary perspectives with more integrated, hybrid methods of working. This combination can be the spark of novel ideas.

In the context of coastal resiliency, interdisciplinary means deep collaboration that includes designers, as well as a diverse range of engineers and natural and social scientists—for example, hydrologists, geomorphologists, ecologists, sociologists, and beyond. Traditional design approaches need the contributions of other disciplines to craft more comprehensive, systems-based solutions for coastal resiliency.

Benefits of Interdisciplinary Collaboration

Interdisciplinary thinking has been on the rise for the past few decades, and today it could be at an all-time high with more than one-third of the references in scientific papers now pointing to other disciplines (Ledford, 2015; Van Noorden, 2015). Part of the success of interdisciplinary research may be due to a growing appreciation of its relevance for today's toughest issues. Here are four of the benefits that are especially relevant for coastal resiliency collaborations:

Novel Combinations of Ideas

A single disciplinary perspective is insufficient to understand complex topics like coastal resiliency; teams need many viewpoints to more fully comprehend all relevant factors (Nissani, 1997). By bringing together ideas from many backgrounds and perspectives, interdisciplinary collaboration facilitates creative breakthroughs and yields richer ideas capable of addressing complex problems (Nissani, 1997; Paulus, 2000; West and Sacramento, 2006; Sovacool, 2014). In addition to generating innovative ideas, interdisciplinary collaboration can also help facilitate implementation. For example, combining technical expertise in stormwater technologies with social science perspectives on urban water governance has helped increase implementation of sustainable water practices in Australia, Singapore, China, and Israel (Brown et al., 2015).

Opening Up Traditional Assumptions

Interdisciplinary collaboration also helps avoid common traps or errors by promoting the questioning of ideas that are accepted as standard practice in other disciplines. One expert, for example, observes that individuals outside of a given field are more able to identify errors that go unnoticed by practitioners within that discipline.[1] Discussing ideas from one discipline with individuals from other disciplines forces the articulation of underlying assumptions, raising the likelihood of innovation.

Reframing of the Questions to Open Up Broader Possibilities

One of the benefits of interdisciplinary collaboration comes at the beginning of the process. Interdisciplinary groups will often realize that the given question is only part of the real challenge. By reframing the question, groups give themselves a "broader canvas" which helps enable them to see more options; in turn, this can lead to "radical successes" (Blackwell et al., 2009:82) and better solutions which can more comprehensively address the underlying challenges (Szostak, 2002:118). As one researcher summarizes, "Successful interdisciplinary outcomes … involve not only new answers, but also new questions. … All too often interdisciplinary teams are brought together 'to fix the plumbing'—only to tell the client—'We can fix the plumbing sir, the problem is you're living in the wrong house'" (Blackwell et al., 2009:82).

  In design projects, this benefit can be especially helpful. For example, the Changing Course Design Competition was focused on addressing rapid wetland loss in the Mississippi Delta. Through discussions, our team realized that wetland loss was only one part of a much broader set of social, ecological, geological, political, and economic challenges.

These individual challenges had been recognized by others but had not been considered together. Our team's early innovation was combining all existing information to obtain a comprehensive view of the many challenges facing the region. By understanding the true breadth of the problems, we were able to craft integrated solutions appropriate for the scale of the challenge.

Systems-Based Solutions Appropriate to the Scale of the Challenge

Some aspects of coastal resilience demand regional approaches; other challenges may be able to be solved with creative, localized solutions. Teams need diverse disciplines to develop strategies appropriate for nested scales of challenges. Having all disciplines collaborating ensures solutions are at the most effective scale and that solutions work together across scales to promote greater synergies.

Challenges of Interdisciplinary Collaboration

> We witnessed biophysical researchers accusing social scientists of poor rigour and of spending too much time conceptualizing problems without exploring and offering solutions. Conversely, social scientists were often frustrated that biophysical researchers were too focused on solutions, reductively overlooking the wider societal implications of their proposed solutions.
>
> **Brown et al., 2015: 316**

Interdisciplinary collaboration brings together diverse ideas to spark innovation, yet this is easier said than done. In fact, many of the aspects that contribute to the promise of innovation are also what make it exceptionally difficult. Communicating across diverse fields is challenging, and disciplinary biases can be difficult to overcome (think of how designers and engineers traditionally might view one another!). An individual's ways of thinking, worldview, and modes of interaction are all shaped by his or her formative education and experiences in practice which tend to be discipline-specific. Three of the most significant challenges are teamwork and communication, team management, and institutional structures.

Teamwork and Communication Can Be Difficult—Especially When It Involves a Diverse Group

With the potential for teammates to jockey for position during discussions or, on the other hand, slack off while others think, teamwork in general can be difficult (Paulus, 2000; Chatenier et al., 2009). Add to this challenge individuals with very different perspectives and backgrounds, and the difficulties can be even greater (Nissani, 1997; Brewer, 1999).

These communication challenges are often a result of team members speaking different disciplinary languages (Davis and Masten, 1996; Brewer, 1999; Paulus, 2000; Oskam, 2009). Overcoming these challenges and developing constructive working relationships is possible, but requires time, perseverance, and effort.[2] While different perspectives can complement one another, they can also make communicating difficult. In a team brainstorm, an engineer may feel concerned that bold ideas sketched by a landscape architect are not fully considering all the constraints, and both may have a hard time understanding a sociologist's focus on the dynamics of people and other non-spatial considerations.

Team Management

Team management, unsurprisingly, is another challenge. Getting a diverse group to work well together can be a demanding, grueling charge in itself. Furthermore, the unpredictable nature of collaboration can add to the challenges of management. Because one of the benefits of interdisciplinary teamwork is reframing of the problem, leaders cannot easily predict the final outcomes or necessary path to get there at the beginning of a process. One researcher demonstrates these challenges by imagining a hypothetical interdisciplinary sales pitch: "You might not get what you expect but trust us, it will be valuable, but it may be difficult to measure in a comprehensive way, and oh yes, it might take longer than we initially said" (Blackwell et al., 2009:81). Leaders must be great communicators, good listeners, and advocates for the process.

Institutional Structures

Institutional structures and practices that have been organized along disciplinary lines can deter interdisciplinary research. Promotion decisions, journals and publishing, and funding tend to prioritize specialization over interdisciplinary development (Brewer, 1999; Blackwell et al., 2009; Brown et al., 2015). Promoting new structures that reward diverse collaboration is essential for promoting coastal resiliency but will require major cultural change in academia and beyond.

Ten Best Practices for Better Collaboration in Coastal Resiliency Projects

> It is not sufficient to merely cut up the assignment into small sections, allow each discipline to deal with its own bit of the project and then glue the pieces back together again (multidisciplinary cooperation). Interdisciplinary teamwork is needed through which the design problem is investigated jointly, solutions are elaborated, and models are made and tested.
>
> **Oskam, 2009: 7**

Interdisciplinary collaboration can lead to richer ideas, but as the previous section shows, it is inherently difficult. What makes interdisciplinary collaboration work well? How can teams maximize benefits while overcoming challenges? This section combines best practices from research about building productive teams with experience in the Changing Course Design and Rebuild by Design competitions.

Assemble a Team with "The Biggest Box" Possible

When you are beginning to form a new team, how should you decide who to include? One pair of researchers finds that "assembling a group of people only capable of thinking 'outside the box' clearly won't be sufficient for success, as the generation of the idea is only one part of the process"; they posit instead that the foundation of a good team is starting with "the biggest box possible" (Fudge and Roca, 2012:3).

Which disciplines to include is a key question, and it can be more complicated than you might expect. Diversity of knowledge is needed, but is more always better? Experts disagree about the desired level of diversity on a team. Some researchers find that more diversity is a good thing (West and Sacramento, 2006), but another believes that too disparate knowledge foundations can hinder communication, leading to lower levels of innovation (Paulus, 2000). Experts find demographic diversity to have mixed results as well (Paulus, 2000; West and Sacramento, 2006). Too little demographic diversity can mean less interesting ideas, but too much diversity can inhibit communication. When forming a team, leaders should be conscious of the many types of diversity that could improve the team dynamic. Research suggests that teams including women (West and Sacramento, 2006; Sovacool, 2014) minorities (Sovacool, 2014), social scientists in addition to engineers and other scientists (Sovacool, 2014; Brown et al., 2015), and diverse personality types, leadership attitudes, and educational backgrounds (Sovacool, 2014; Brown et al., 2015) can produce better ideas if communication works well.

Compared to a typical planning and design competition team, interdisciplinary teams focusing on coastal resilience will likely require a much broader range of disciplines (Figure 9.2). Including designers, planners, engineers, scientists, and social scientists helps ensure ideas will be scientifically grounded, technically sound, socially considerate, and clearly communicated. Depending on the problem at hand, groups may also benefit from including even more diverse perspectives—philosophers, artists, anthropologists, historians, or economists, for example. Knowing which disciplines to include may not be

FIGURE 9.2
For coastal resilience challenges, teams must go beyond traditional design or engineering disciplines. For example, in the Changing Course Design competition, the Baird Team included a broad range of local and national experts. (Courtesy of Sasaki.)

obvious at first. In fact, you may want to build your team in several stages. The first set of recruits may be better positioned than the convener to identify needs from neighboring disciplines.

Build Your Team from Disciplinary Experts

Many find that the best teams, especially for complex challenges, are composed of individuals who are experts in their own disciplines. In an aptly titled paper "Superman or the Fantastic Four: Knowledge Combination and Experience in Innovative Teams," one pair of researchers explores whether individuals or teams are more effective at innovation. Their findings suggest:

> Individuals are capable of more creative integration of diverse experiences than teams are. Combining knowledge requires a deep understanding of knowledge, rather than information scanning or exposure. When seeking innovation in knowledge-based industries, it is best to find one "super" individual. If no individual with the necessary combination of diverse knowledge is available, one should form a "fantastic" team, with each team member having deep knowledge and experience working with the other team members.
>
> **Taylor and Greve, 2006: 735–737**

Given the challenges of the most complex problems today, the possibility of finding a single individual with the right combination of deep, yet diverse, experience across all relevant fields for a particular challenge is slim at best. Therefore, team members need to be experts in their own discipline (Brewer, 1999:333). Furthermore, a particular strain of expertise—a "T-shaped" expert—can make an individual an even better collaborator (Oskam, 2009; Brown et al., 2015). "T-shaped" describes an individual who has deep expertise in one discipline (represented by the tall part of a "T"), combined with knowledge of related disciplines (the top horizontal line of a "T"). This related knowledge helps "T-shaped" individuals more easily bridge gaps between disciplines while also contributing their deep perspectives to the issue at hand.

Ensure Team Members Are Curious, Open Minded, and Excellent Communicators

Disciplines bring more than just expertise about different topics; they also offer fundamentally different ways to approach a problem. For example, different disciplines operate at different scales—across both time and space. Planners may think regionally and across large areas; geomorphologists bring perspectives that consider natural processes across much broader time spectrums. These different ways of thinking led to the core ideas of the Sasaki team's Rebuild by Design proposal to create a more resilient Jersey Shore. The strong collaboration between designers, ecological scientists, and community leadership contributed to building a new type of resiliency—one that not only protects the beach, but also enhances social capital and connectivity (Figure 9.3).

While integrating these diverse ways of thinking is the very thing that generates new ideas, it is also precisely what can make communication difficult. Creating integrated solutions requires overcoming language barriers caused by different disciplinary perspectives, backgrounds, and training. A way of thinking that is commonly accepted by one discipline can feel like a foreign language to another. Vocabulary about SWAN and ADCIRC models may feel second-nature to a flood risk modeler, but entirely foreign to an urban planner, who may speak using a completely different type of alphabet soup—FARs,

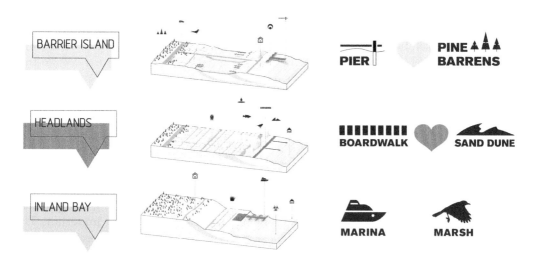

FIGURE 9.3
The Sasaki-Rutgers-Arup team's Rebuild by Design concepts were based upon three distinct coastal typologies that combined ecology with icons of the Jersey Shore, reflecting team expertise that blended ecology, planning and design, and social science. (Courtesy of Sasaki.)

TIFs, and PPPs. Accordingly, when forming a team, do not overlook the importance of communication. Seek out team members who have excellent communication skills and are able to articulate complex ideas in a way that is understandable to someone from another discipline (Davis and Masten, 1996).

The best team members are intensely curious, open to new ideas, and have strong synthetic and analytic abilities (West and Sacramento, 2006; Blackwell et al., 2009). In our Changing Course Design team, brainstorms, ensuring all individuals had time to share their perspectives while team members listened openly, helped break down disciplinary barriers over time as a shared language and knowledge base was developed. A breakthrough moment came when the team's port and navigation expert exclaimed after listening to an oyster scientist, "We're talking about an economic delta!" This opened up possibilities to see synergies between oysters, fisheries, navigation, land building, and economies.

In summary, when building a team, look for team members who are expert, collaborative, and curious. Each core team member should be an expert in his or her discipline to bring deep knowledge to the team and should also contribute a curious and collaborative personality.

Find the Leader with the Right Personality and Background to Direct an Interdisciplinary Team and Process

The leader plays a particularly important role on an interdisciplinary team, helping to integrate knowledge while also building the team. He or she must carefully facilitate the process and cultivate team dynamics toward a shared mission (Paulus, 2000; West and Sacramento, 2006; Chatenier et al., 2009). A special type of leader is needed—one with passion and humility (Blackwell et al., 2009:65–66). Leaders need curiosity and drive to keep the process moving forward despite setbacks; an open attitude is key to recognize contributions from other disciplines. One set of researchers found that the leader's technical background was especially important for facilitating innovation, playing a key role in problem solving by recognizing potential hurdles and opportunities (West and Sacramento, 2006).

Work Together, Not in Parallel

With multiple disciplines, it may be tempting for a team to operate with each discipline working in parallel, rather than collaborating in a more integrated manner. While this approach may feel more comfortable for teammates, it hinders the potential of innovation because it does not enable the cross-pollination of ideas across disciplines. By tackling complex problems using disparate and contained approaches, bigger opportunities and synergies are lost. Breaking silos is critical. Early in the process, all disciplines should be present for team brainstorming sessions; later, once ideas are beginning to emerge, smaller working groups can continue to refine ideas. These smaller groups can be topic-based and should include multiple relevant disciplines.

In-person meetings are especially valuable for the team-building process. When teams are built for a particular complex challenge, team members are likely to be geographically dispersed, but finding opportunities for in-person meetings periodically is important. The Changing Course Design and Rebuild by Design teams both found conference calls and emails to be less conducive to collaborative discussions; in-person meetings were critical to developing high-functioning teams.

For example, at the heart of our Changing Course Design team's interdisciplinary collaboration were team workshops at the University of New Orleans. These were full-day gatherings, conducted on donated time over weekends. Navigation experts learned about the Mississippi river's rich geologic history; sociologists and ecologists discussed governance considerations with the team's proposed managed distributaries concept; landscape architects suggested adapting the idea of oysters and salinity gradients to a layered approach of flood risk tolerance for community planning and land use considerations. The days were intense, but overall quite enjoyable for our team. Many ideas—even seemingly crazy ones—were suggested and considered. Through these debates and discussions, our team's core ideas took shape.

Cultivate a Collegial Atmosphere of Trust and Mutual Respect That Fosters Constructive Dialogue

With communication a critical factor for team success, cultivating good team dynamics is essential. Researchers recommend teams "forge a shared mission" and "nurture constructive dialogue" (Brown et al., 2015:316 and 317). Building personal relationships among team members and an atmosphere of mutual trust is important to ensure that everyone is comfortable sharing their ideas and ideas are considered with an open mind (West and Sacramento, 2006; Chatenier et al., 2009; Ledford, 2015). Collaborative dialogue is key to success and innovative thinking.

In coastal resiliency projects, our teams have found it helpful to focus on team bonding and discussion early to jump-start collaboration. Our teams often include members who have prior experience working together. Developing an effective interdisciplinary team is not an immediate, easy task; forming a cohesive team takes time. While it is unlikely that an entire pre-existing team would have the exact combination of knowledge and skills for a particular coastal challenge, it can be helpful to build a team comprising at least some team members that have previous experience working with one another.

Let Everyone Speak and Listen Openly

Value all teammates and their contributions and share openly with one another. What seems simple to one discipline is ground-breaking for another; this is how the deepest benefits of interdisciplinary collaboration are unlocked.

One tool you may find helpful is structured brainstorming. Structured brainstorming ensures all team members have a chance to contribute in an organized manner. The benefits of this approach are that it helps team members develop a common language and helps decrease the likelihood of unsupportive judgment that can result when different disciplines approach an idea from different perspectives. Our Changing Course Design team used de Bono's "Six Thinking Hats"[3] in facilitating group brainstorming. We found that this tool helped promote collaboration, ensuring that when "green hats" were on (representing a time for creativity and new ideas), other disciplines did not prematurely judge and instead helped elaborate and consider unusual ideas.

Leave Time and Space for New Ideas: You Cannot Rush Innovation

As literature shows, one of the largest benefits of interdisciplinary collaboration is its potential to reframe issues, opening up new possibilities and more effectively solving root problems. For this reason, allowing sufficient time for early brainstorming is critical to ensure the real challenges are fully understood. You will often find that the given challenge is usually only part of the problem.

Value Storytelling

Complex solutions that effectively address complex problems can be tough to explain, yet sharing the potential of the ideas is critical to obtain buy-in and funding for implementation. Explaining complex ideas in a simple, compelling, and understandable manner can help others appreciate key ideas. The value of designers and design-thinking is especially helpful here. Visualizations can be powerful tools to increase understanding, especially when paired with technical statistics and scientific findings. Together, numbers and images build a strong case that can be understood by many different kinds of learners (Figures 9.4 and 9.5).

FIGURE 9.4
This rendering helps the viewer imagine the potential of expanding coastal tourism further inland, promoting a more resilient, year-round economic driver. (Courtesy of Sasaki.)

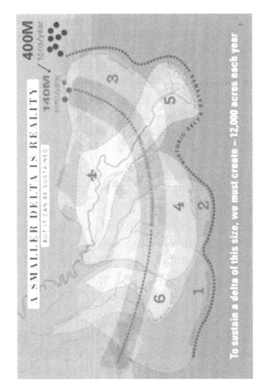

FIGURE 9.5
Designers play an important role in translating technical ideas. For example, the two graphics from the Changing Course Design Competition show the original engineering calculations for a sustainable delta and how the findings were adapted for a broader audience. (Left image: Courtesy of Baird; Right Image: Courtesy of Sasaki.)

Be Bold

Do not underestimate the power of an interdisciplinary team (especially your own!); the outcomes might surprise you. Tough challenges can seem impossible, but with the right group of people, a path forward may emerge. Set targets that are such a reach they might seem unattainable. The goals can help push the team to stretch, promoting innovative ideas.

For example, the idea of "A Delta for All" inherently reflected our Changing Course Design team's interdisciplinary spirit. Our solution for creating a sustainable delta needed to benefit the delta's full range of stakeholders; unlike previous planning efforts, it would not focus on land creation at the expense of others (estuarine fisheries, especially oysters) or without regard to possible synergies (navigation). The innovation at the heart of our team's work was an optimism and a desire to seek solutions that would work together to address the complex challenges of the region, building more land and at the same time improving navigation capacity, reducing flood risk and uncertainty for delta communities, growing regional economies, and creating a diversity of delta ecosystems to ensure the long-term health of fisheries. Creating a sustainable delta is not possible by "tweaking" the existing system; the scale of the solution must align with the scale of the challenge. Therefore, our team's ideas reinvent the system in a revolutionary—not evolutionary—way (Figure 9.6).

The Rewards of Fruitful Collaboration

The promise of collaboration promoted in literature is confirmed by experiences in interdisciplinary design teams tackling coastal resilience challenges. Interdisciplinary collaboration can be challenging, especially at the beginning, but when pursued persistently, it can be a powerful tool.

Fundamentally, interdisciplinary collaboration holds the potential to generate feasible ideas for problems previously considered impossible. The types of ideas generated through interdisciplinary collaboration are likely to look different than strategies created by a single discipline working in isolation. Interdisciplinary teamwork can lead to systems-based solutions, which are appropriate to the scale of the challenge. In contrast, individualized solutions focused on a narrow goal can come with high opportunity costs; while these kinds of singular solutions may demonstrate measurable benefits, they risk solving one aspect of a problem but—in doing so—ignoring a bigger issue.

Working with others from different backgrounds helps planners and designers see challenges differently and ultimately develop richer ideas than any single discipline could have working in isolation. The balance of designers, social scientists, and others with more technical backgrounds helps solutions combine qualitative and quantitative approaches at multiple scales, as well as clearly visualize and communicate ideas to a broad range of audiences. This combination holds promise for implementation as well: by approaching a challenge from multiple outlooks, solutions can be technically feasible, scientifically sound, and politically savvy. With this sort of broad appeal and diverse strengths, interdisciplinary ideas for coastal resiliency can prove more resilient, themselves, over time, with an inherent greater degree of flexibility to adapt to changing conditions.

Finally, a significant reward of interdisciplinary collaboration lies in the teamwork itself. Through collaboration, designers can meet new colleagues and make friends—and find opportunities to learn new ways of thinking.

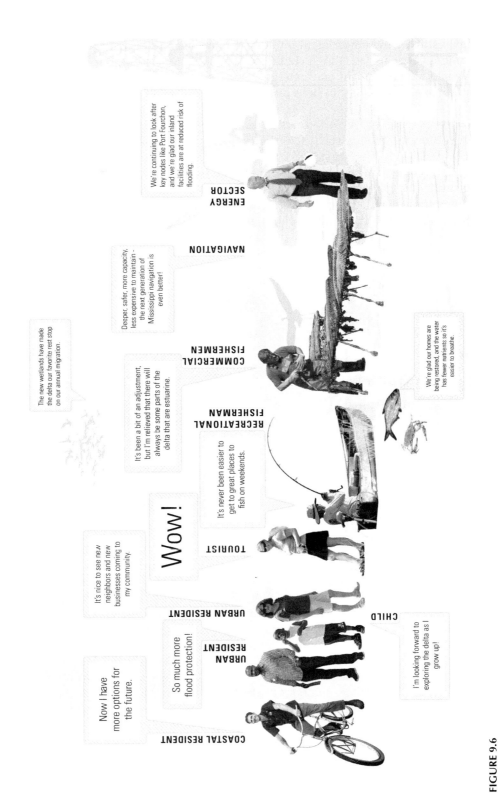

FIGURE 9.6
In the Changing Course Design Competition, our Baird Team set an early ambition of a "Delta for All," an ambitious goal to ensure ideas would lead to improved outcomes for all stakeholders. This goal set a high standard for ideas and pushed the team to rise to the challenge. (Courtesy of Sasaki.)

Paths Forward

> By now, most of us no longer think it possible to become a Renaissance Scholar à la Leonardo da Vinci. Gradually during the nineteenth century, the ideal of the unity of knowledge—that a genuine scholar ought to be familiar with the sum total of humanity's intellectual and artistic output—gave way to specialization.
>
> **Nissani, 1997: 202**

With the complexity of today's issues, it is increasingly difficult for a single person to have the knowledge necessary to tackle the toughest problems. While the era of Renaissance individuals may be past, perhaps this century can be the era of "Renaissance Teams"—where diverse skills, knowledge, and personalities combined through interdisciplinary collaboration craft the next generations of innovative ideas.

Project-specific collaborations are demonstrating the potential benefits of interdisciplinary teamwork, and broader changes could further cultivate the next generation of Renaissance Teams. New methods of education and different structures within and around design firms could help break down disciplinary silos before a project begins. For example, if design studios began to connect with ecology or social science departments, students would get an early taste of interdisciplinary collaboration—and perhaps common disciplinary biases might form a little less deeply. Over time, interdisciplinary collaboration could become a more natural way of working, and design firms may even begin to become more diverse, themselves, or to more commonly partner with academics or more non-traditional collaborators on projects.

Endnotes

1. "The natives live and breathe their customs; the perceptive foreigner doesn't ... Disciplinarians often commit errors which can be best detected by people familiar with two or more disciplines" (Nissani, 1997:205).
2. The length of time typically required for successful interdisciplinary collaboration can be another challenge (Blackwell et al., 2009; Brown et al., 2015).
3. Edward de Bono's Six Thinking Hats is a brainstorming tool. Six colored hats each represent a different type of information/thinking and are used sequentially to allow for a thorough consideration of ideas. During a brainstorming session, team members all wear the same hat (i.e., think in the same way) at the same time. Hats include new ideas/creativity (green hat), information/facts (white hat), overall procedure/process (blue hat), caution/critical thinking (black hat), optimism/benefits (yellow hat), and feelings, intuition, and emotion (red hat). For more see de Bono, E. (1999). *Six Thinking Hats* (2nd ed.). Boston, MA: Back Bay Books.

References

Blackwell, A. F., Wilson, L., Street, A., Boulton, C., and J. Knell. 2009. *"Radical innovation: crossing knowledge boundaries with interdisciplinary teams."* Technical Report Number 760. Cambridge, UK: University of Cambridge Computer Laboratory. https://www.cl.cam.ac.uk/techreports/UCAM-CL-TR-760.pdf.

Brewer, G. D. 1999. "The challenges of interdisciplinarity." *Policy Sciences*, 32, 327–337.

Brown, R. R., Deletic, A., and T. H. F. Wong. 2015. "How to catalyse collaboration." *Nature*, 525, 315–317.

Chatenier, E. D., Verstegen, J. A. A. M., Biemans, H. J. A., Mulder, M., and O. Omta. 2009. "The challenges of collaborative knowledge creation in open innovation teams." *Human Resource Development Review*, 8 (3), 350–381.

Changing Course. Competition website. http://changingcourse.us/.

Coastal Protection and Restoration Authority of Louisiana, Louisiana's Comprehensive Master Plan for a Sustainable Coast. 2012. http://coastal.la.gov/a-common-vision/2012-coastal-master-plan/.

Davis, M. L. and S. J. Masten. 1996. "Design competitions: Does 'multidisciplinary' contribute to the team building experience?" *Proceedings of the 1996 26th Annual Conference on Frontiers in Education*. 276–279. Salt Lake City, UT.

de Bono, E. 1999. *Six Thinking Hats* (2nd ed). Boston, MA: Back Bay Books.

Fudge, C. and J. Roca. 2012. "Ten tips for successful innovation teams." http://innovationmanagement. se/wp-content/uploads/2012/04/Ten_Tips_for_Successful_Innovation_Teams_Fudge_Roca_ im.se_.pdf.

HUD. April 3, 2014. "HUD's 'Rebuild by Design" teams unveil final proposals in regional resiliency design competition." HUD No. 14-031. [Press Release]. Retrieved from http://portal.hud.gov/ hudportal/HUD?src=/press/press_releases_media_advisories/2014/HUDNo.14-031.

Ledford, H. 2015. "How to solve the world's biggest problems." *Nature*, 525, 308–311.

Nissani, M. 1997. "Ten cheers for interdisciplinarity: The case for interdisciplinary knowledge and research." *The Social Science Journal*, 34 (2), 201–216.

Oskam, I. F. 2009. "T-shaped engineers for interdisciplinary innovation: an attractive perspective for young people as well as a must for innovative organisations." *37th Annual Conference–Attracting students in Engineering*, Rotterdam, The Netherlands (Conference Paper). http://www.sefi.be/ wp-content/abstracts2009/Oskam.pdf.

Paulus, P. B. 2000. "Groups, teams, and creativity: The creative potential of idea-generating groups." *Applied Psychology*, 49 (2), 237–262.

Rebuild by Design. Competition website. http://rebuildbydesign.org/.

Sovacool, B. K. 2014. "Energy studies need social science." *Nature*, 511, 529–530.

Stock, P. and R. J. F. Burton. 2011. "Defining terms for integrated (multi-inter-trans-disciplinary) sustainability research." *Sustainability*, 3, 1090–1113.

Stokols, D. 2011. "Transdisciplinary research in landscape architecture: Prospects and challenges." *Landscape Journal: Design, Planning, and Management of the Land*, 30 (1), 1–5.

Szostak, R. 2002. "How to do interdisciplinarity: Integrating the debate." *Issues in Integrative Studies*, 20, 103–122.

Taylor, A. and H. Greve. 2006. "Superman or the fantastic four? Knowledge combination and experience in innovative teams." *Academy of Management Journal*, 49 (4), 723–740.

Van Noorden, R. 2015. "Interdisciplinary research by the numbers." *Nature*, 525, 306–307.

West, M. A. and C. A. Sacramento. 2006. "Chapter 3: Flourishing in teams: Developing creativity and innovation." In: J. Henry (Ed.), *Creative Management and Development* (3rd ed.) 25–44. London, UK: Sage. http://www.corwin.com/upm-data/11445_03_Henry_Ch03.pdf.

10

Engaging the Community to Envision the Coastal Climate Future

Karen M. O'Neill and Heather Fenyk

CONTENTS

Introduction

Adaptation to climate change cannot be left to experts or officials. It requires a transformation in how residents, owners, and other community members envision and use shore areas. Spatial planners, landscape architects, and other design professionals can work collaboratively with community members and leaders through three processes: *learning* about varied local values and how the shore is changing, *convening* to deliberate about the future, and *inspiring* participants by helping them to develop appealing options for adaptation. Communities will have to weigh their tolerance for risk in light of varied and uncertain local projections for sea level rise (Kopp et al., 2014). They will have to continually collect and share information and as conditions change, players will have to reconvene, and inspiration will have to once again be sparked.

Collaborating to manage change is what planners and designers do, and climate change presents a peak challenge. Coastlines naturally change due to sediment or flooding from rivers, tectonics, waves, wind, and storm surges. Human uses and structures, such as seawalls and groins, interact with these natural dynamics in complex ways to expand or erode beaches (Lazarus et al., 2016). But risks along coasts are increasing overall, because more people are living in hazardous zones than before, and because sea level rise and other effects of climate change threaten coastal settlements in new ways (World Bank, 2010). There is particular trouble in predicting the extent and timing of climate change, particularly because scientific models cannot predict sea levels and storm behavior at fine, local scales. Local knowledge of tides and storms and scientific data about past storms will also become less useful, as the underlying dynamics of the climate and ocean physics and chemistry change (French et al., 2016). If planners and designers assist community members and organizations to observe and report local changes, join in deliberations, and weigh options, adaptation projects are more likely to be adjustable, to cope with uncertainty, and to be socially useful and functional.

A good deal is known about what works in public engagement and about how people perceive risk. Planners and designers can apply these ideas when collaborating with coastal communities to reduce coastal hazards. This chapter focuses on spatial planning, design, and landscape architecture at the local and regional levels. We describe the skills and experiences that suit planners and designers to this work. We present findings from research and practice that are most relevant to planning for coastal climate change. Finally, we consider the three strategies for coastal adaptation—protect, accommodate, and move—and analyze how public engagement tasks differ when communities weigh one or more of these strategies.

Collaborating, Analysis, Visioning, Visuals: The Skills of Planners and Design Professionals

Community engagement in coastal adaptation presses planners and design professionals into roles that other professions or institutions do not typically address or are unlikely or unprepared to take on. Activists and politicians may be motivated to seek changes but need help to create a unified vision and set of policies that they can make into a specific campaign for action. Government lawyers and other staff members can review existing laws and other constraints and help build the adaptation program if the campaign succeeds. But planners and design professionals have distinctive skills and tools that can enable these efforts to cohere. They are skilled in collaboration with institutions and individuals, they are experienced in planning under conditions of uncertainty, and they use case-based methods to provide site-specific solutions. One way of appreciating the value of these skills and tools is by considering how they enhance the three processes of learning, convening, and inspiring.

Learning about local climate change and possible adaptive strategies can be enhanced by planners' skills in gathering and analyzing technical information, such as market assessments, measures of coastal elevation, and engineering studies. Perhaps even more important is planners' and designers' ability to learn about community needs, perceptions, and values. This is information that cannot be captured well without concerted and direct involvement with those most affected by a project. A planning process for adaptation that involves genuine participation of community members in all aspects of the project is most likely to generate valid and useful information about values (Klain and Chan, 2012).

Convening is the centerpiece activity for participatory decision making that planners and designers can facilitate in official or unofficial venues. Working with elected officials, planners in government can convene projects to meet regulatory mandates (e.g., updates to a municipality's master plan) or to respond to informal opportunities. Working with community groups, firms, or other organizations, planners and designers can help frame local concerns in ways that invite participation in projects outside government. Asserting the legal power or moral authority to convene discussions about adaptation can affirm community members' concerns about the future and empower them to weigh their values and options before emergency conditions force them to act (Koslov, 2016).

Inspiring the public to participate in decision making taps into the value of creativity at the heart of the planning and design professions. By listening and showing examples of approaches that might appeal to community members, planners and designers can encourage participants to collaborate in creating their own forms of inspiration. Professionals can then write preliminary concepts, design logos and installations, and suggest policy and economic strategies that community members can revise and debate (Manzini and Rizzo, 2011).

Throughout these processes, planning offices and planning and design consultants act as generalists who bridge divisions between government and outside professionals and between officials and members of the public. Climate change presents new types of uncertainty that elude expertise based on past data and local knowledge based on experience of past conditions. The generalist's role will therefore become ever-more important in planning and design work.

These techniques come together in the case-based approach that planners and designers use. They do not plan for a community in the abstract or use a set of fixed principles. Instead, they engage the community at a particular place with particular needs at a particular time. The case-based approach recognizes that we live in the world through its realities, not through abstractions. It is therefore useful for learning and for practice because it treats conditions and potential solutions in context, while providing tools for recognizing similarities across cases and in different scenarios. The case approach highlights the conditions under which problems emerge and under which solutions succeed or fail (Ragin and Becker, 1992; Byrne, 2009). The processes of defining what the case is and of understanding the context of the case, are keystones of the case approach (Yin, 2008) that can help consultants and other participants in planning as they learn, convene for discussion, and devise inspiring responses.

Engaging Community: Best Practices

Decisions about coastal climate adaptation have special characteristics that make broad community involvement vital but that may also make it more difficult. We discuss these special characteristics as they relate to basic findings about public participation and perceptions of risk.

Guidelines for engaging the community can be informed by understanding the ways people perceive risks and seek to regulate those risks. We note that much of this research focuses on wealthier countries. Individuals may be more motivated to take personal action or support regulations to avoid outcomes that evoke a sense of dread (e.g., radioactive waste), even if scientists estimate that the dreadful outcomes are much less likely to occur than other similarly harmful outcomes (Pidgeon et al., 2008; Norgaad, 2011). Demographics such as gender or ethnicity may mediate perceptions but are less important than personal experience with a natural hazard or trust (or mistrust) in institutions that are communicating about the hazard. Individual and community memory of disasters fades quickly, however. And many individuals who perceive that they are at high risk do not make preparations for hazards or take steps to reduce hazards they are exposed to. This is the "risk paradox" (Wachinger et al., 2013). Processes for participating in assessing risks, such as participatory vulnerability assessments, can enhance trust in the institutions sponsoring the assessments and may appear less threatening to implement than proposals made from the top down

(Kates et al., 2012; Buckecker et al., 2013). Summing up an admittedly large and varied body of research, perceptions across the public do not necessarily match with assessments of risk made using scientific methods; perceptions are influenced by culture, experience, and social position; and some people are more unrealistically optimistic about their exposure to risk than others.

Perceptions about climate change and its potential harms have additional special characteristics. Climate change is not directly perceivable and so is difficult to contemplate (Milkoreit, 2016). Furthermore, the human role in climate change can seem abstract and be difficult to grasp. Acceptance of climate science findings is generally associated with education level, although ideology plays a key role as well. Some climate skeptics believe that humans are blameless or that efforts to reduce climate change are futile (Lee et al., 2015). Fossil fuel interests and conservative ideological groups have deliberately fostered doubt about the overwhelming scientific consensus that climate change is occurring and is caused by humans. This doubt has become tied to conservative political beliefs in the United States and in other countries that have high levels of climate skepticism. For these individuals, climate change proposals may contribute to their suspicions about experts and may become a proxy for the broader question of whether governments should regulate economic behavior. Yet doubt can be reduced by repeated and simple messages about the scientific consensus by trusted sources (Maibach et al., 2014), such as local emergency managers or officials.

Focusing on the local effects of climate change is a winning approach. Framing climate change as a threat to polar bears and arctic ice or to Pacific island nations is directly relevant to some people, but for others, this framing encourages a psychological distance that dampens the urgency to act (Slocum, 2004). Thinking about extreme outcomes that would remake one's community, such as permanent inundation, might seem likely to provoke dread, but it can provide the chance to ponder what really matters to members (Koslov, 2016). Possible effects on local culture, such as changes in species that are fished recreationally, have likely been little discussed in the community. These conditions affect community members' interest in planning, the ability to recruit them to participate, the amount and kind of information that they would like to have, how they will define the tasks for planning, and the range of interested and affected parties who should be recruited to join discussions. Being aware of these problems can help organizers to see that engagement needs to be integrated throughout the planning and implementation processes.

The idea about community engagement most often emphasized by researchers and practitioners is that participation should go beyond mere talk. Tacking on community outreach at the end of a project invites cynicism and fails to capture knowledge that could improve basic project decisions. Creating public planning and design processes that encourage members of the community to collaborate to gather information and to make decisions (Arnstein, 1969; Cohen and Uphoff, 1980) may help push a project toward implementation and result in a project that will be well used and appreciated. Having community members directly involved in decision making is especially important in defining what the problem is (agenda setting) and in setting out the ranges of options that will be on the table, because it is these framing processes that leave some matters aside. Direct involvement of members of the public may not be needed or desirable for all decisions. Decision making should itself be a topic for deliberation in the early stages of planning. Avoiding participation fatigue will be important over the long run in dealing with climate change. When deliberations drag on, the people who still show up tend to be those with the most resources who have the greatest direct interest in the outcomes, in other words, people who may not fairly represent others in the community. Any participatory approach should therefore consider participation by whom, for what, and in what? (Cohen and Uphoff, 1980).

These concerns about involvement are important because social values are fundamental to debates about climate adaptation and cannot be elicited without genuine community engagement. Questions about allowing the sea to erode some parcels or changing how people can use the shore area are questions about tax money and about what people want. Many policy deliberations have not acknowledged this and have treated environmental problems as highly technical matters of science, technology, and law instead. This discourages non-specialists from feeling they can contribute to discussions, and it often obscures what is at stake. Identifying value conflicts, as well as areas of agreement across various interest groups, should be a priority for public engagement (Layzer, 2002).

Broad engagement is important not only for reasons of equity and fairness but also for the basic functioning of a project. It helps each planning participant to learn about varied local cultures and how other community members understand climate change and its local effects. Knowledge about local patterns of risk perception that could affect project success cannot be predicted adequately by general findings from research. For instance, in New Zealand, people closest to the shore were more likely to be concerned about climate change than others who are less exposed (Milfont et al., 2014), a finding that may not hold true elsewhere.

Broad engagement means recruitment across a diverse set of community members and other interested parties. Some groups are not likely to attend meetings. Depending on local circumstances, these may include children, members of minority ethnic or cultural groups, adults working more than one job, families with low incomes, new residents, the elderly, and those with mobility problems (Roberts, 2004). Targeted recruitment may be needed if particular threats pose harm to particular groups that are under-represented in deliberations. Planners and designers can gain an integrated view of these issues from the environmental justice movement, which has been the most active in insisting on procedural fairness and equity (e.g., having the community well represented in deliberations), examining fairness and equity in outcomes (e.g., whether environmental harm and benefits are equitably distributed), empowering under-represented groups, building organizational capacity, and seeking respect for the diversity of values across the community (Smith and Pangsapa, 2008).

Efforts to spread new ideas generally work best by using pre-existing forms of social organization and trusted institutions and using a variety of approaches and venues to reach those who are otherwise under-represented. This requires people who are active in the planning process to learn about and get access to the venues where people most affected by hazards congregate, and the individuals or groups whom they trust, online and in person (Roberts, 2004). Having a wide range of people engaged through various forums can help identify emerging problems, such as localized flooding, and suggest solutions that are locally appropriate, welcome, and will encourage safe living near the shore. Recognizing engagement processes beyond the purview of one's particular planning process will also be important. Processes, such as regional transit planning, will become more relevant to local planning, because climate change will have such broad effects. Community engagement in these other venues may provide opportunities for gathering information about local preferences and about the possible emergence of initiatives by other organizations that may conflict with local plans.

The challenge of learning about local climate changes that are actually underway can become an opportunity to attract active participants. Citizen Science is a set of practices that recognizes the value of involvement. Community members identify current problems, generate information, and apply it (Mayan and Daum, 2016). Citizen Science typically gathers data that may be highly localized, widely dispersed, or otherwise not feasible for scientists to get funding to collect (Dickson et al., 2010). Examples include participatory mapping of ecological and human resources and environmental problems, and ongoing monitoring of species and habitat quality (Dickson et al., 2010; Klain and Chan, 2012). As they interact with professional scientists, participants in community-based research projects may also gain competence in assessing what science can and cannot contribute to decision making. Community members who take on these tasks can increase their sense of ownership of local adaptation projects (Kallio et al., 2015).

These are just some of the ideas available from research studies and practice guidelines that can inform planning participants about improving community engagement. The challenges in creating best practices for community engagement will be amplified when planning for the long process of climate change adaptation.

Pathways for Engaging Community

In countries with active formal environmental regulatory processes, public participation has often been narrowly defined as a matter of holding public hearings, or possibly design charrettes, which have shortcomings. Planners and designers can encourage other forms of participation as well.

Because climate change calls for all-hands responses, any forum that is affected by coastal land use and coastal hazards is a potential venue. Some forms of engagement are almost universally suitable, such as outreach through mass media. Other forms are locally specific, such as tribal governance, immigrant cultural organizations, or religious organizations. Social media networks built by local governments, non-governmental organizations, or local activists can become the basis for crowd-sourcing information about problems and solutions (Brabham, 2013). Other venues including regional planning processes concerning topics like watershed conditions, regional business forums, and organizations of emergency managers, are examples of pre-existing networks that can attract a wide variety of participants and that are relevant to climate adaptation.

Some planners and designers are also creating novel forms of engagement through their practice, often in response to environmental issues. These include online game-like exercises (also called serious games), in-person or online exercises for expressing aesthetic preferences (e.g., Visual Preference Surveys) or preferences for policy aims (see Figure 10.1), and pop-up installations to test how members of the public use redesigned public spaces (e.g., "pop-up parks" and temporary bike ways). Scientifically sampled opinion surveys of residents can capture a wide range of residents, users, and potential users of shore areas who do not participate in other forums (Swinomish Indian Tribal Community, 2010). Other experimental forms of engagement might be adopted within a community, such as citizen panels or citizen juries (convening community members selected through lotteries or other selection processes to deliberate policy options) (Roberts, 2004).

The public sphere is where collective decisions about adaptation will be shaped, but it is also important to consider how planning and design processes are affected by the cumulative actions of individuals in real estate markets, in private groups, and in actual uses of the shore. Actions by individuals, households, firms, and others that are not coordinated may still result in broader patterns. For instance, a local "disaster culture" (Bankoff, 2003; Kulatunga, 2010) may encourage the voluntary spread of techniques that make private residences better adapted to wind or storms. These decisions are made by

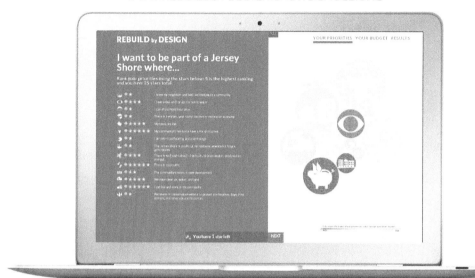

FIGURE 10.1
Online tool for assessing resident preferences. (Image courtesy of Sasaki Associates, http://crowdgauge.org/.)

individuals but may be influenced by institutions such as government incentives, nonprofit educational campaigns, and shifts in public opinion, and so they constitute a form of public engagement. Ultimately, attitudes about what is a livable space will determine the demand for shore area land. Changes in attitudes about hazards could reduce the market value of coastal properties or cause migrations out of informal settlements. Through regulations, planning, learning, and outreach, adaptation planning could therefore aim to develop a disaster culture that bolsters the usefulness of local land by reducing hazards.

Collaborating to Reduce Coastal Hazards and Envision a Community's Future

Considering these lessons about community engagement together with research on coastal adaptation projects leads to several suggestions about planning for coastal adaptation. These reflections reinforce our sense that planning and the design professions can take special roles in facilitating coastal climate adaptation.

Three Strategies for Adaptation

Coastal scientists and coastal managers have been the most active in creating and experimenting with coastal adaptation. They identify three strategies: protection,

FIGURE 10.2
Three adaptation strategies at one site (Union Beach/Keansburg, New Jersey, USA): (1) *Move* from areas along streams and coastline that become inundated or subject to repeated storm surges and redirect human uses to places that are higher and drier; (2) *accommodate* rising waters by reconnecting a man-made lake to the ocean and introducing salt-tolerant species; (3) *protect* some areas with engineered barriers (super-levee) and artificial dunes and beaches. (Image courtesy of Sasaki/Rutgers/Arup.)

accommodation, and moving (Harman et al., 2015). These strategies may be used in combination along a stretch of coast (e.g., building barriers around a critical piece of infrastructure while allowing the nearby shoreline to be exposed to the sea; see Figure 10.2) or by being built in succession at a particular site as conditions change (e.g., protecting a section of cottages with barriers until sea level rise overwhelms the barriers).

Protection is the most familiar and visually obvious strategy along many coastlines around the world. Most are hard engineering projects that wall off the ocean from the land (e.g., seawalls, revetments, floodgates, and gabions). Residents, owners, and officials often perceive barriers as providing certainty by fixing the shoreline to suit human uses (Jones and Clark, 2014). Coastal scientists in some places are succeeding in getting coastal management practitioners and members of the public to recognize the natural dynamism of coastlines and the protective roles of coastal wetlands, mangroves, dunes, and natural beaches. Protection may therefore include preserving natural shorelines that remain, restoring ecological functions through regrading and planting, building engineered protections, or creating hybrid systems of natural and engineered barriers (Sutton-Grier et al., 2015).

Accommodations are less visually obvious and less familiar in most places. Accommodations tend to be developed to meet specific regional or local conditions and needs, so that some technologies may not be easily transferrable. The most recognizable built accommodation is the raising of structures on stilts or piers, a practice that occurs in places as varied as the Philippines and the Outer Banks of North Carolina (Bankoff, 2003). Other broadly applicable accommodations promoted by engineers include floodable streets and structures, salt-tolerant infrastructure, and check valves to prevent high tide flow from entering storm drains (Butler et al., 2016). However, accommodations may function only within a narrow range of sea level and storm conditions before becoming overwhelmed by sea levels or frequent storm surges. Overall, protection and accommodation can keep functions in place, at least for a time, and so they are viewed as less disruptive in the short-run than the moving strategy would be.

The moving strategy, or retreat, is the extreme case among adaptation strategies because it requires that owners remove structures and that users change how they use and think of these places. In most places, the moving strategy is the least familiar strategy. Surveys in these places usually find that retreat is the least preferred option (Apine, 2011; Jones and Clark, 2014). Moving may be difficult to fund, legally complicated, and emotionally and culturally wrenching. Moving from built areas would be the most radical form of retreat. A related strategy is preventing new building in a vulnerable area (avoidance), an approach that may be legally and politically less threatening than removing existing buildings but that would likely depress the market value of existing buildings in the targeted area. Because retreat is the strategy that marks the most dramatic break from current uses, and because residents use a range of rational calculations, values, and emotions to weigh it as an option (Alexander et al., 2012), retreat is also the strategy that can most benefit from having planners and designers at work engaging the community.

Distinctive Skills for Engaging Community on Adaptation

How can planners and designers respond to these lessons and challenges? We derive suggestions based on the skills of planners and designers, best practices for community engagement, and the three adaptation strategies. This exercise affirms both the central value of the *analytically based vision* that planners and designers can provide in response to a problem, and planners' and designers' embeddedness in *institutions for managing collective concerns*. Analysis can incorporate community values by providing not only facts about past and current conditions but also projections of future desired conditions. Visions based on these values can be expressed through words, sounds, images, structures, and landscapes. The three processes of learning, convening, and inspiring benefit from analytically based visioning that planners and designers can facilitate.

The process of *learning* about coastal adaptation emphasizes analytic skills but takes guidance from community values and visions that unfold from deliberations about adaptation. Because climate change is an emerging phenomenon, there is a greater burden than usual in learning about local changes and in articulating the reasons why new decisions are needed. Climate change was detected and framed as a problem by climate scientists, and so information about climate change has tended to be pushed from experts to the public, using concepts like probability that may be difficult to grasp. Organizers of adaptation projects may also face community members who reject climate science, but experienced planners and designers are equipped to handle these conditions. They know how to communicate technical matters as part of a broader vision for the community, and they know that condescending messaging can harm a planning project. This is why we label the process "learning" rather than "communicating," which has too often been taken to be a one-way process in policy making. Learning should be used to change the goals and methods for the project as participants gather new data and clarify the values they want expressed in the adaptation project. Methods for learning will vary across cultures and locales, as discussed by Jane Wolff in Chapter 4 of this volume. These processes are familiar to designers and planners, who are accustomed to doing research and using iterative steps and participation in producing designs and plans.

As practitioners of case-based approaches, designers and planners are also suited to facilitate learning about the feasibility of various protection, accommodation, and moving options for the target site. For the protection strategy, learning may be focused on exploring drawbacks that may be unfamiliar, such as the potential for structures to erode nearby beaches or to cause ecological degradation. For the accommodation strategy, learning may require a basic review of potential technologies and studies of whether community members would accept them. For the moving strategy, learning would require broad and deep inquiries about the current uses of the site, values, and the potential for moving uses elsewhere. All three strategies could be assessed for costs and benefits, aesthetic and cultural values, and flexibility in the face of uncertainties about change.

The process of *convening* would typically derive from an official planner's involvement in government regulation, but it may also originate in other institutions for managing collective concerns, such as watershed planning agencies. If the learning process finds evidence that the effects of climate change are likely to be suffered unequally within the community or with a neighboring community, a likely finding, it becomes ever-more important to ensure equity in the processes of participating and in setting the rules that will lead to decision making. Philosophers and others have developed the concept of environmental democracy to characterize the vibrant array of participatory processes that have emerged in many venues, many of which have been taken up by designers and planners (Hester, 2006). Extending concepts of equity, fairness, and justice, philosophers highlight how seemingly neutral elements of process, such as formal meetings, may systematically exclude potential participants, such as members of indigenous groups who may see debate as disruptive or disrespectful (Eckersley, 2004). The venues used by under-represented groups that have been identified through the local learning process (e.g., places of worship or social clubs) may also be appropriate for hosting deliberations. Planners and designers are familiar with the need to identify trusted members of the community to facilitate this outreach.

Convening tasks would differ for the three adaptation strategies. Convening to consider protection strategies may not require an elaborate planning process if the site for protection is small or if this strategy would extend existing protections that are functioning well at reasonable maintenance costs. For more extensive protection strategies and for accommodation strategies, convening topics would include the costs of replacing structures, encouraging adoption by owners of private property, analyzing whether the accommodations would function without universal participation by owners, and assessing whether community members would tolerate inconveniences such as floodable streets. For moving strategies, the potential trauma of reducing or abandoning the use of some sites would in many cases be the largest challenge. Planners' considerable experience in redevelopment points to the need to plan not only for retreating from particular sites, but also for relocating the human activities that had been in those sites. We would hope that the fraught history of urban redevelopment and displacement would bring planners to all of these tasks with humility.

Having designers and planners convene discussions about coastal adaptation would be especially valuable because existing coastal management institutions in some places have done a poor job of integrating the public in decision making. The United Kingdom created some of the earliest projects for coastal retreat through national and regional coastal science and engineering agencies. Initial projects failed to consult residents at an early stage or presented projects as benefiting ecosystems or as reducing maintenance costs, not as providing benefits for humans (Myatt-Bell et al., 2002). Some coastal managers still see community members' resistance as resulting from their lack of knowledge about coastal dynamics and their failure to take the long view (French, 2004), rather than as the agency's problem of engagement and establishing trust. This is a critical misunderstanding because social science research finds that trust in institutions, social networks, and other forms of social capital shape how community members perceive climate risk and approach decisions about coastal management (Jones et al., 2013; Jones and Clark, 2014). More recent projects have integrated spatial planners and designers from local authorities along with coastal and other agencies into the earliest convening efforts, with more positive results (Milligan et al., 2009).

The process of *inspiring* draws from planners' skills in apportioning space and its uses, and from designers' skills in envisioning space and translating ideas and emotions into visual forms. As they consult with experts and local opinion leaders, and absorb information about community members' values and preferences, planners and designers can offer provisional options that can be used to further stimulate public discussion about a community's future. They can present principles from ecology and geology, and concepts such as sediment transport and watersheds as potential frameworks for community members' discussions of goals (Dramstad et al., 1996; Burger et al., 2017). Few other professions that are likely to become involved in climate adaptation are capable, or interested in, encouraging inspiration or in working to integrate values from science, culture, and community.

Inspiration and creativity will have large roles in discussions about the three adaptation strategies because adaptation to ongoing but unpredictable change is not a task that has been routinely included in local spatial planning. Design can show how various approaches would look and function under a variety of future scenarios, helping to extend participants' imagination. This can become an exercise in making the future seem less distant and in creating empathy for the lives of future residents and users. Visions for the protection strategy may have to respond to local perceptions favoring existing approaches. Residents of a town that has relied on beach replenishment or on high seawalls may have difficulty envisioning and accepting different technologies that engineers may be recommending (Jones and Clark, 2014). Visions for the accommodation strategy could be made more attractive by providing designs that highlight the flow of water and the changing of seasons by showing how ecologies can adapt to change and by illustrating public benefits (Handel, 2015, see Figure 10.3). Visioning for the moving strategy shares elements of visioning for new communities. This tends to be positive, but is also associated with redevelopment, which has displaced millions and broken community ties, and so any visioning of moving strategies must be driven by community needs and values (Koslov, 2016). Attracting activities and development to places that are relatively higher and drier may be more effective than centering discussion on the idea that some places will become unsafe (Burger et al., 2017). For all three adaptation strategies, providing an array of possible solutions, rather than emphasizing fear messages, is likely best for envisioning the future. Creating visual representations about current and future conditions in the locality, and an array of policy options, are distinctive contributions that design professionals and planners can provide. Rather than showing images of polar bears trapped at sea on small bits of ice, showing future visions for local sites can bring climate adaptation home.

Evolution: Habitat Engine

Existing conditions – limited habitat

Limited palette of habitats, no space for habitat migration in the case of sea level rise.

20 years – create additional habitat

Carve habitat terraces adjacent to existing habitat, creating space for water to take over salt marsh/wetland and salt marsh/wetland to take over meadows as sea level rises.

50 years – sea level rises, habitats migrate

Sea level rises, flooding salt marshes and some freshwater wetland.
Terraces provide new salt marsh and wetland habitat for species to move into.

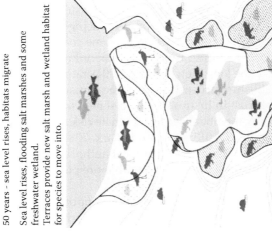
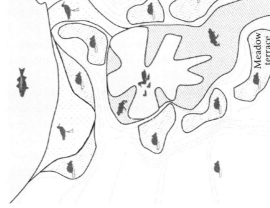
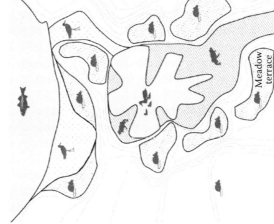
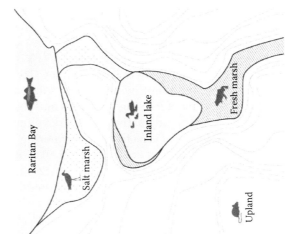

Raritan Bay

Salt marsh

Inland lake

Fresh marsh

Upland

Meadow terrace

Spotify

FIGURE 10.3

A site designed to adapt to changing sea levels. (Union Beach/Keansburg, New Jersey, USA; Sasaki/Rutgers/Arup.)

Conclusion

Planners and design professionals can take on important roles in coastal climate adaptation by using case-based approaches to engage the community. When planning is coupled with design, it can facilitate learning, encourage communities to convene discussions, and inspire them to adapt. For planners and designers to take on these roles, officials and community leaders have to understand and value their skills and tools and view them as complementing their own roles. Chapter 8 by Forbes Lipschitz explains the challenges and benefits of communicating coastal management across disciplines and professions, and Chapter 9 by Jill Allen Dixon provides an example of an interdisciplinary project for the Mississippi Delta area at the Gulf Coast. Projects for specific places will still be the basis for the work of planners and design professionals, but ideas from these cases can encourage broader discussion of adaptation and inspire projects elsewhere. Following the long tradition of case-based work in design and planning, case studies can become stories that officials and community leaders share about how to engage the community and how to adapt. There are many examples of cities borrowing project ideas (policy mobility). This process of borrowing and remaking ideas may be driven by politicians seeking renown, designers and planners exchanging information with fellow professionals, or activists seeking new tactics (Gonzalez, 2011; Soderstrom and Geertman, 2013).

We have outlined why we feel that planners and design professionals are well positioned to facilitate these collaborations. Planning can highlight the benefits of protective efforts that extend beyond hazard reduction, resulting in new visions for a community in the face of change. Design can offer a new visual and experiential vocabulary for coastal adaptation that links engineering concerns to livability and community values. These are ideals that will not often be met in practice, but setting these ideals is helpful for clarifying what design and planning could achieve, and what would be missed in its absence.

References

Alexander, K. S., Ryan, A. and T. G. Measham. 2012. "Managed retreat of coastal communities: Understanding responses to projected sea level rise." *Journal of Environmental Planning and Management* 55:409–433.

Apine, L. 2011. "Residents' attitude towards possible adaptation measures to the sea coast erosion in Latvia." *International Journal of Climate Change Strategies and Management* 3:238–249.

Arnstein, S. R. 1969. "A ladder of citizen participation." *Journal of the American Planning Association* 35:216–224.

Bankoff, G. 2003. *Cultures of Disaster: Society and Natural Hazard in the Philippines*. London: Routledge Curzon.

Brabham, D. C. 2013. "The four urban governance problem types suitable for crowdsourcing citizen participation." In: Carlos Nunes, S. (ed.) *Citizen E-Participation in Urban Governance: Crowdsourcing and Collaborative Creativity*. Hershey, PA, USA: IGI Global, pp. 50–68.

Buckecker, M., Salvini, G., Di Baldassarre, G., Semenzin, E., Maidl, E. and A. Marcomini. 2013. "The role of risk perception in making flood risk management more effective." *Nat. Hazards Earth Syst. Sci.*, 13:3013–3030.

Burger, J., O'Neill, K. M., Handel, S. N., Hensold, B. and G. Ford. 2017. "The shore is wider than the beach: Ecological planning solutions to sea level rise for the Jersey Shore, USA." *Landscape and Urban Planning* 157:512–522.

Butler, W. H., Deyle, R. E. and C. Mutnansky. 2016. "Low-regrets incrementalism: Land use planning adaptation to accelerating sea level rise in Florida's coastal communities." *Journal of Planning Education and Research* 36:319–332.

Byrne, D. 2009. "Introduction: Case-based methods, why we need them; what they are; how to do them." In: Byrne, D. and Ragin, C. C. (eds.) *The SAGE Handbook of Case-Based Methods*. London: SAGE Publications Ltd., pp. 1–10.

Cohen, J. M. and N. T. Uphoff. 1980. "Participation's place in rural development: Seeking clarity through specificity." *World Development*, 8:213–235.

Dickson, J. L., Zuckerberg, B. and D. N. Bonter. 2010. "Citizen Science as an ecological research tool: Challenges and benefits." *Annual Review of Ecology, Evolution, and Systematics* 41:149–172.

Dramstad, W. E., Olson, J. D. and R. T. T. Forman. 1996. *Landscape Ecology Principles in Landscape Architecture and Land-use Planning*. Washington, DC: Harvard Univ. Graduate School of Design, Island Press, and the American Society of Landscape Architects.

Eckersley, R. 2004. *The Green State: Rethinking Democracy and Sovereignty*. Cambridge, MA: MIT Press.

French, J., Payo, A., Murray, B., Orford, J., Eliot, M. and P. Cowell. 2016. "Appropriate complexity for the prediction of coastal and estuarine geomorphic behaviour at decadal to centennial scales." *Geomorphology* 256:3–16.

French, P. W. 2004. "The changing nature of, and approaches to, UK coastal management at the start of the twenty-first century." *The Geographical Journal* 170:116–125.

Gonzalez, S. 2011. "Bilbao and Barcelona 'in motion.' How urban regeneration 'models' travel and mutate in the global flows of policy tourism." *Urban Studies* 48:1397–1418.

Handel, S. N. 2015. "Jump-starting the habitat engine." *Ecological Restoration* 33:1–2.

Harman, B. P., Heyenga, S., Taylor, B. M. and C. S. Fletcher. 2015. "Global lessons for adapting coastal communities to protect against storm surge inundation." *Journal of Coastal Research* 31:790–801.

Hester, R. 2006. *Design for Ecological Democracy*, Cambridge, MA, MIT Press.

Jones, N. and J. R. Clark. 2014. "Social capital and the public acceptability of climate change adaptation policies: A case study in Romney Marsh, UK." *Climatic Change* 123:133–145.

Jones, N., Koukoulas, S., Clark, J. R. A., Evangelinos, K. I., Dimitrakopoulos, P. G., Eftihidou, M. O., Koliou, A. et al. 2013. "Social capital and citizen perceptions of coastal management for tackling climate change impacts in Greece." *Regional Environmental Change* 14:1083–1093.

Kallio, K. P., Hakli, J. and P. Backlund. 2015. "Lived citizenship as the locus of political agency in participatory policy." *Citizenship Studies* 19:101–119.

Kates, R. W., Travis, W. R. and T. J. Wilbanks. 2012. "Transformational adaptation when incremental adaptations to climate change are insufficient." *Proceedings of the National Academy of Sciences* 109:7156–7161.

Klain, S. C. and K. M. A. Chan. 2012. "Navigating coastal values: Participatory mapping of ecosystem services for spatial planning." *Ecological Economics* 82:104–113.

Kopp, R. E., Horton, R. M., Little, C. M., Mitrovica, J. X., Oppenheimer, M., Rasmussen, D. J., Strauss, B. H. and C. Tebaldi. 2014. "Probabilistic 21st and 22nd century sea-level projections at a global network of tide-gauge sites." *Earth's Future* 2:383–406.

Koslov, L. 2016. "The case for retreat." *Public Culture* 28:359–387.

Kulatunga, U. 2010. "Impact of culture towards disaster risk reduction." *International Journal of Strategic Property Management* 14:304–313.

Layzer, J. A. 2002. "Citizen participation and government choice in local environmental controversies." *Policy Studies Journal* 30:193–207.

Lazarus, E. D., Ellis, M. A., Brad Murray, A. and D. M. Hall. 2016. "An evolving research agenda for human–coastal systems." *Geomorphology* 256:81–90.

Lee, T. M., Markowitz, E. M., Howe, P. D., Ko, C. Y. and A. A. Leiserowitz. 2015. "Predictors of public climate change awareness and risk perception around the world." *Nature Climate Change* 5:1014–1020.

Maibach, E., Myers, T. and A. Leiserowitz. 2014. "Climate scientists need to set the record straight: There is a scientific consensus that human-caused climate change is happening." *Earth's Future* 2:295–298.

Manzini, E. and F. Rizzo. 2011. "Small projects/large changes: Participatory design as an open participated process." *CoDesign* 7:199–215.

Mayan, M. J. and C. Daum. 2016. "Beyond dissemination: Generating and applying qualitative evidence through community-based participatory research." In: Olson, K., Young, R. A. and I. Z. Schulz (eds.) *Handbook of Qualitative Health Research for Evidence-Based Practice*. New York, NY: Springer New York, pp. 441–452.

Milfont, T. L., Evans, L., Sibley, C. G., Ries, J. and A. Cunningham. 2014. "Proximity to coast is linked to climate change belief." *PLOS ONE* 9, e103180.

Milkoreit, M. 2016. "The promise of climate fiction: Imagination, storytelling, and the politics of the future." In: Wapner, P. and H. Elver. (eds.) *Reimagining Climate Change*. Abingdon, UK: Routledge, pp. 171–190.

Milligan, J., O'Riordan, T., Nicholson-Cole, S. A. and A. R. Watkinson. 2009. "Nature conservation for future sustainable shorelines: Lessons from seeking to involve the public." *Land Use Policy* 26:203–213.

Myatt-Bell, L. B., Scrimshaw, M. D., Lester, J. N. and J. S. Potts. 2002. "Public perception of managed realignment: Brancaster West Marsh, North Norfolk, UK." *Marine Policy* 26:45–57.

Norgaad, K. M. 2011. *Living in Denial: Climate Change, Emotions, and Everyday Life*. Cambridge, MA: MIT Press.

Pidgeon, N. F., Lorenzoni, I. and W. Poortinga. 2008. "Climate change or nuclear power—no thanks! A quantitative study of public perceptions and risk framing in Britain." *Global Environmental Change* 18:69–85.

Ragin, C. C. and H. S. Becker. 1992. *What Is a Case? Exploring the Foundations of Social Inquiry*. Cambridge, MA: Cambridge University Press.

Roberts, N. 2004. "Public deliberation in an age of direct citizen participation." *The American Review of Public Administration* 34:315–353.

Slocum, R. 2004. "Polar bears and energy-efficient lightbulbs: Strategies to bring climate change home." *Environment and Planning D: Society and Space* 22:413–438.

Smith, M. J. and P. Pangsapa. 2008. *Environment and Citizenship: Integrating Justice, Responsibility and Civic Engagement*. London: Zed Books.

Soderstrom, O. and S. Geertman. 2013. "Loose threads: The translocal making of public space policy in Hanoi." *Singapore Journal of Tropical Geography* 34:244–260.

Sutton-Grier, A. E., Wowk, K. and H. Bamford. 2015. "Future of our coasts: The potential for natural and hybrid infrastructure to enhance the resilience of our coastal communities, economies and ecosystems." *Environmental Science & Policy* 51:137–148.

Swinomish Indian Tribal Community. 2010. *Swinomish Climate Change Initiative: Climate Adaptation Action Plan*. La Conner, WA: Swinomish Indian Tribal Community, Office of Planning and Community Development.

Wachinger, G., Renn, O., Begg, C. and C. Kuhlicke. 2013. "The risk perception paradox—implications for governance and communication of natural hazards." *Risk Analysis: An International Journal* 33:1049–1065.

World Bank. 2010. *Cities and Climate Change: An Urgent Agenda*. Urban Development Series, Knowledge Paper. Washington, DC: The International Bank for Reconstruction and Development/The World Bank.

Yin, R. K. 2008. *Case study research: Design and Methods*. Sage Publications, Inc, CA, USA.

PART 2

NEW STRATEGIES

Section III

Deltas, Bays, and Estuaries

Across the globe, delta landscapes epitomize some of the most acute issues for human settlements under threat. Broad reengineering strategies offer improved resilience for deltas, bays, and estuaries.

11

Structures of Coastal Resilience: Adaptive Design for Jamaica Bay, New York

Catherine Seavitt Nordenson

CONTENTS

Since Superstorm Sandy's landfall along the Atlantic seaboard of the United States in October 2012, and given the very active hurricane season of 2017, with hurricanes Harvey, Irma, Maria, and Nate striking the Gulf Coast, Virgin Islands, and Puerto Rico, the buzzword "resilience" has loomed large as coastal communities and government agencies assess their preparedness for the potential of future storm events and the certainties of climate change and sea level rise (Schultz et al., 2012). Post-Sandy, former New York City Mayor Michael Bloomberg was quick to organize the Special Initiative for Risk and Reconstruction (SIRR), with a dedicated team producing a 445-page report and recommendations that were released in the summer of 2013 (Figure 11.1). At a national level, the United States Army Corps of Engineers (USACE), defenders of the nation's navigable waterways and coastlines, was tasked by Congress in January 2013 to conduct a comprehensive study of the eastern seaboard affected by Sandy, a close scrutiny of the possible future of the post-Sandy coastal landscape from Virginia to Maine. The report resulting from this $19.5 million study, led by the North Atlantic Division of USACE and titled the *North Atlantic Coast Comprehensive Study* (NACCS), was delivered to Congress in January 2015 (USACE, 2015).

Structures of Coastal Resilience (SCR) was conceived in 2013 as a progressive multi-university academic research paradigm, working in parallel with USACE's two-year *Comprehensive Study* schedule. This 18-month multi-university collaborative study, fully funded by the Rockefeller Foundation in New York City, provides a focused look at designing for resilience at four coastal embayment sites along the East Coast, all within the greater study area of USACE. The four sites and four university teams, each led by a landscape architect or an architect, are Narragansett Bay, Rhode Island—Harvard University; Jamaica Bay, New York—City College of New York; Atlantic City, New Jersey—Princeton University; and Norfolk, Virginia—University of Pennsylvania. Though the work of SCR was not officially

FIGURE 11.1
Jamaica Bay, drone imagery of Little Egg Marsh, Big Egg Marsh, and Broad Channel, Queens, New York. (Image courtesy of Catherine Seavitt Nordenson/Structures of Coastal Resilience, Jamaica Bay.)

part of USACE's study, these projects were reviewed, advised, and informed by continuing discussions with both members of the North Atlantic Division and those in the regional district offices, and references to the projects were included in the final *Comprehensive Study* document. This direct contact with members of USACE reframed the designers' resiliency thinking, pushing the limits of comfortable and known niches. The ambition is that these academic studies, and the design thinking they engendered, might indeed have actionable impact, transforming the more traditional practices of the United States Army Corps of Engineers' storm risk management practices.

The *Comprehensive Study* addresses storm risk reduction strategies and promotes resilient communities along the coast, while considering future climate change and sea level rise scenarios. Perhaps most interestingly, given USACE's traditional reliance on hard engineered structures for coastal engineering, the study examines the integration of what USACE has termed natural and nature-based features (NNBFs) into their lexicon of risk reduction measures—a radical paradigm shift for the agency (Bridges, 2013; USACE, 2013). For the Army Corps, nature-based features include such elements as dunes, oyster and coral reefs, barrier islands, maritime forests, and wetlands (The Nature Conservancy, 2013). Often the benefits and performance of these features increase with scale. And scale is arguably the most invaluable quality of Jamaica Bay, New York, the research site of the City College of New York's *Structures of Coastal Resilience* design team.

The State of the Bay

The Rockaway peninsula and the back bay communities of Jamaica Bay were massively impacted by Hurricane Sandy's surge and extensive flooding in 2012 (Figures 11.2 and 11.3).

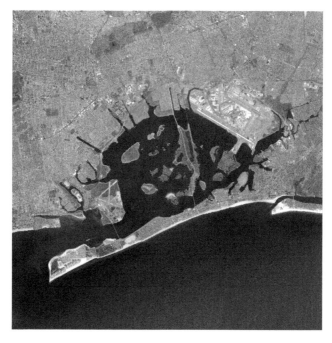

FIGURE 11.2
Jamaica Bay high-resolution orthoimagery. (Image courtesy of Catherine Seavitt Nordenson/Structures of Coastal Resilience, Jamaica Bay.)

The bay has been environmentally challenged since its early days as New York City's dumping ground (Moses, 1938; NYCDEP, 2007, 2012). Since the 1970s, the reduction of the marsh island footprints within the bay has rapidly accelerated, likely due to the combined anthropogenic impacts of poor water quality, nutrification, erosion, and sediment starvation (Hartig, 2002). Yet, given its vast scale, Jamaica Bay has the potential to be recast as an impactful ecological, infrastructural, and community asset, becoming an anchor of the region's resiliency. And for the Army Corps of Engineers, the bay presents an incredible design opportunity for exploring the performance of nature-based features, particularly salt marshes, maritime forests, and dunes. Jamaica Bay is a dynamic ecological entity, an estuarine embayment in a post-glacial outwash plain. This sandy and naturally shifting terrain is geologically capable of functioning resiliently both during and after disturbance events. The City College of New York's adaptive design proposals for Jamaica Bay seek to enhance that capability.

Experimental Research Studio

The *Structures of Coastal Resilience* Jamaica Bay research team was not structured as an academic studio, but rather as an ongoing experimental research studio within the academic setting at City College of New York's Spitzer School of Architecture (Figures 11.4 and 11.5). This grant-funded research studio reflects a distinct shift in the landscape architecture discipline. In recent years, a significant number of landscape architecture educators and

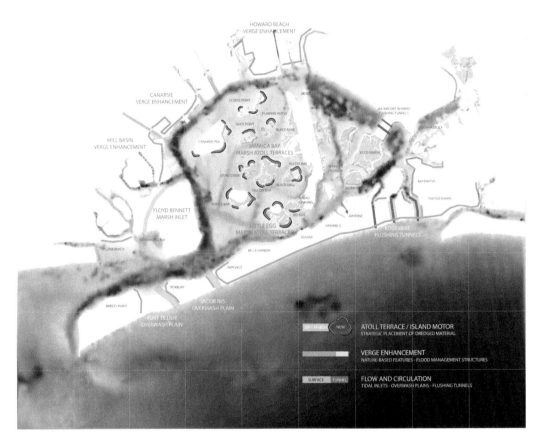

FIGURE 11.3
Resiliency Strategies Key Plan, Jamaica Bay. (Image courtesy of Catherine Seavitt Nordenson/Structures of Coastal Resilience, Jamaica Bay.)

practitioners have argued that the range of the discipline is both expanding and evolving. And several expect that landscape architecture's range of inquiry in the earth, ocean, and atmospheric sciences is approaching, or even becoming a STEM (the acronym for the academic disciplines of science, technology, engineering, and mathematics) or a STEAM (the acronym for STEM disciplines as influenced by art and design) field of study. In some ways, the Jamaica Bay experimental research studio, and the SCR project in general, may represent another phase shift toward a new model of a research-driven, laboratory-based academy, with research inquiries led by principal investigators in the discipline of landscape architecture.

Inspired by interaction with the United States Army Corps of Engineers and investigation into their historic methodologies during a visit to Vicksburg, Mississippi, the Jamaica Bay

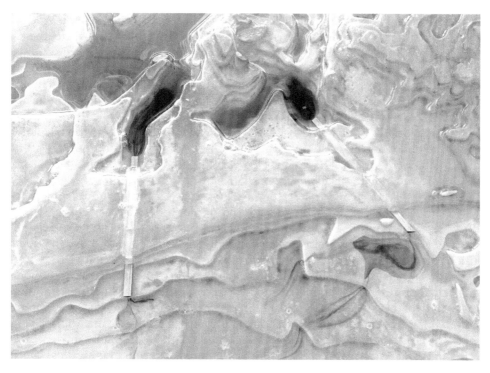

FIGURE 11.4
Water tank model of Far Rockaways flushing tunnels. (Image courtesy of Catherine Seavitt Nordenson/ Structures of Coastal Resilience, Jamaica Bay.)

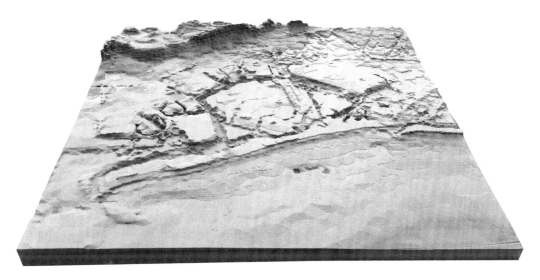

FIGURE 11.5
CNC-milled baywide physical model. (Image courtesy of Catherine Seavitt Nordenson/Structures of Coastal Resilience, Jamaica Bay.)

experimental research studio is modeled as a contemporary version of the Waterways Experiment Station (WES), established by USACE in 1930 in response to the Great Mississippi Flood of 1927. A 272-hectare complex in Vicksburg, Mississippi, WES is the Army Corps' hydraulics research laboratory facility on the site of the Engineer Research and Development Center (ERDC) headquarters. WES built large-scale physical hydraulic models of the Mississippi River, Chesapeake Bay, and New York Harbor in order to test flood control structures and the effects of coastal longshore drift. While ERDC still works with physical hydraulic models, their research has substantially shifted to digital modeling. Yet the think-tank atmosphere of experimentation is still present, and ERDC's affiliated engineers and scientists are actively involved with the study of the nature-based features that figure prominently in the *Comprehensive Study*.

City College's Jamaica Bay laboratory has produced a large number of physical models, from the scale of the full watershed to that of a detailed salt marsh platform, resulting in a deep understanding of the physical morphology of the bay. Five of the models were designed as hydraulic water tank models, constructed to test the dynamic conditions of water flow, residence time, overwash, sediment transfer, and surge. Though the hydrodynamic effects produced are somewhat coarse, the precision of the merged Jamaica Bay topographic/bathymetric model through which the water flows allows the iterative study of design scenarios via subtle physical transformations. These water tank models are topographic contour models cast in glycerin, a malleable material that allows for quick design transformations and manipulations to the topography during the flow experiments. The empirical results of the water tank experimental studies, captured in video, have provided integral feedback to the development of the proposed design strategies.

In the digital realm as well, a highly refined digital elevation model (DEM) of the topobathy of Jamaica Bay has been built, seamlessly merging topographic (land) and bathymetric (water depth) data. This continuous surface is integral to the conceptual understanding of terrain as a gradient surface, a transformable basin containing the dynamic medium of water. The DEM is used in both its existing condition and with proposed design modifications to create comparative analyses of the efficacy of the design proposals using digital hydrodynamic modeling tools such as the SLOSH (Sea, Lake, and Overland Surges from Hurricanes) model, SLAMM (Sea Level Affecting Marshes Model), and the wind fetch raster generator WAVES2012.

But getting out of the lab in order to ground-truth in the field proved critical (Figure 11.6). Broad Channel, Queens, is an unexpected community poised on a marsh island in the middle of the bay. Like Jorge Luis Borges' fabled one-to-one map of the kingdom, Jamaica Bay itself serves as a full-scale research lab. On-site research enabled friendships with local activists and environmentalists living in Broad Channel. The EcoWatchers (Dan Mundy, Jr. and Dan Mundy, Sr.) served as an incredible source of local knowledge, particularly regarding the history of marsh degradation and the subsequent innovative marsh island restorations initiated by the National Park Service and the United States Army Corps of Engineers. Don Riepe, the Jamaica Bay Guardian and director of the American Littoral Society's Northeast Chapter, provided support and access to more than 30 years of his photographic archive of Jamaica Bay imagery. The on-site observation of currents and the mapping of both low and high tide lines in the salt marshes were critical for locating the proposed placement of elevated terraces for sediment traps. Data collection and ground-truthing was best executed in tall rubber boots, Trimble in hand, with the Avenza app activated on a smartphone to track NOAA's Chart 12350. Tide charts proved to be critical

FIGURE 11.6
Fieldwork and a low tide excursion to Little Egg Marsh Island. (Image courtesy of Catherine Seavitt Nordenson/ Structures of Coastal Resilience, Jamaica Bay.)

research tools, enabling walks to certain of the marsh islands during extreme low tides. Visits to the back bay's inlets were timed with the incoming tides for the best access in Don Riepe's research vessel; quadcopter drone overflights were coordinated with mean tides; and frequently flooded streetscapes at Howard Beach were best surveyed during extreme high tides.

Another resource developed as part of the work in the research studio is the Pamphlet Library, consisting of 16 topical pamphlets and three reference manuals, compiled as a structured repository for accumulated data. Inspirations include the *semiotext(e)* model of carefully curated translations of French philosophy published in a small-scale format serving an American pocket, and the low-tech hand-printed-and-folded *cordel* literature and poetry of the northeastern Brazilian troubadours—self-published pamphlets hung from a cord with clothespins and sold in town squares.

Each pamphlet serves as both a reference and a tool. Some compile history through contemporary aerial imagery; some are photographic field notebooks; some contain collections of coastal infrastructural inventions. All address in some fashion the complex cultural, ecological, and infrastructural systems at this dynamic embayment (Black, 1981). As a set, the pamphlets gather the multitude of thoughts and research that influenced and informed the resiliency strategies and designs for Jamaica Bay. The pamphlet library is also intended to be a resource for others, potentially leading to future research directions.

"Experimental Marsh Island Restoration at Jamaica Bay" was the first pamphlet compilation, evolving from a need to understand the techniques, processes, locations, and lessons learned from the USACE New York District's innovative and unprecedented marsh island restorations at the bay. Another pamphlet, "The Air above Jamaica Bay," explores the conflicts between marsh restoration initiatives and Federal Aviation Administration (FAA) controlled airspace, given that high marsh provides ideal breeding habitat for birds that can potentially take down an aircraft. The Port Authority of New York and New Jersey's bird depredation techniques are examined and quantified—the FAA's most-wanted. "The Bottom of Jamaica Bay" showcases bathymetric scans and video stills of benthic habitat, providing a fascinating look at those otherwise unseen species floating within the water column or ambling across the contours of the bottom.

Adaptive Design Strategies for Jamaica Bay

The Jamaica Bay design team of *Structures of Coastal Resilience* has identified a holistic set of strategic design recommendations supporting coastal resiliency at Jamaica Bay, inclusive of the Rockaway Peninsula, the central marsh islands, and the tributary inlets at the back bay's interior perimeter (Figure 11.7). The project seeks to address the region's social, environmental, and infrastructural vulnerability, with the goal of enhancing overall resiliency at Jamaica Bay. These proposals reflect a suite of ideas emerging from USACE that attempt to merge its more recent environmental restoration projects with its traditional flood protection and erosion control practices (USACE, 2013). The Army Corps is a large bureaucratic organization, but the *Comprehensive Study* has initiated the beginning of a notable paradigm shift. For USACE, it is not about flood *control* any longer nor about

FIGURE 11.7
Yellow Bar Hassock USACE marsh restoration, 2012. (Image courtesy of USACE and Great Lakes Dredge & Dock/Gregg Kohl Photography, Inc.)

flood *protection*. At issue now is the *management of risk*. The new terminology, embraced by the Army Corps since the initiation of the Comprehensive Study, is coastal storm risk management (CSRM). Significantly, ecosystem services are included in this thinking— ERDC produced a post-Sandy study arguing that salt marsh island restoration projects at Jamaica Bay effectively reduced wind fetch, and thereby decreased the potential for erosive structural undermining of the Cross Bay Boulevard North Channel Bridge, a critical infrastructural evacuation route for both Broad Channel and the Rockaway Peninsula (Burks-Copes, 2013).

The proposed Jamaica Bay resiliency master plan includes three strategies, each merging novel techniques of ecosystem restoration with layered nature-based features. The first strategy addresses water quality and the reduction of back-bay flooding via a series of overwash plains, tidal inlets, and flushing tunnels at the Rockaway Peninsula and Floyd Bennett Field. The second strategy develops enhanced verges at Robert Moses' Belt Parkway and elevates coastal edges at vulnerable back-bay communities, managing flood risk with a layered system of marsh terraces, berms, and sunken attenuation forests. The third strategy develops novel techniques of bay nourishment and salt marsh island restoration through maximizing the efficacy of the strategic placement of minimal quantities of dredged material. This feature, dubbed the "island motor/atoll terrace," would align with local cycles of maintenance dredging. Dredged material from USACE maintenance dredging may be beneficially placed in order to nourish the Rockaways' coastal beaches, enhance the bay's marsh islands, and create living shorelines at the back-bay perimeter. A resilient marsh ecosystem provides coastal storm risk management services to adjacent communities through wind and wave attenuation, delivering maximum immediate benefits for both vulnerable communities and the degrading salt marsh islands.

Flow and Circulation

An important consideration for any design proposal is the improvement of both water quality and hydrologic flow throughout Jamaica Bay (Figures 11.8 and 11.9). Reducing residence time, particularly in the easternmost reaches of the bay, will support ecological and benthic health, as well as ensure the success of restoration projects within the bay. Improving flushing through the introduction of tidal inlets, overwash plains, and flushing tunnels may also reduce impacts from surge flooding by providing additional outlets for the retreat of floodwaters. In addition, the movement of occasional floodwaters from the ocean to the bay will reintroduce sediment sources to the bay, supporting the nourishment and accretion of the wetland marsh islands.

At Edgemere, a neighborhood with a relatively dense building footprint at the eastern end of the Rockaway Peninsula, both below-grade flushing tunnels and above-grade flow paths are proposed to improve bay-to-ocean circulation, as well as provide a controlled trajectory for periodic floodwaters. At the western end of the Rockaway Peninsula, near Jacob Riis Park and Fort Tilden, both federal lands held by the National Park Service's Gateway National Recreation Area, more open space is available. Here, tidal inlets and overwash plains are proposed as very shallow topographic modifications, allowing for sediment delivery from ocean to bay. Low earthen berms limit the extents of the overwash plains and serve to protect the adjacent residential neighborhoods to the east.

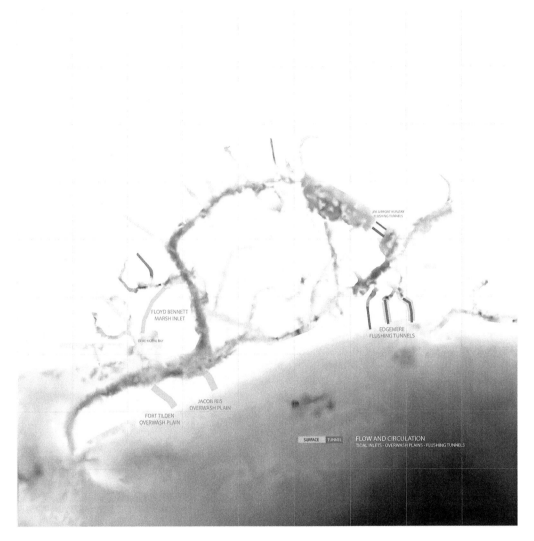

FIGURE 11.8
Key plan: Flow and circulation. (Image courtesy of Catherine Seavitt Nordenson/Structures of Coastal Resilience, Jamaica Bay).

Verge Enhancement

Strategic elevational enhancement through the raising of gradient coastal edges at the vulnerable back-bay communities of Jamaica Bay is addressed through a layered system of marsh terraces, earthen berms, and planted attenuation forests, defined in sum as "verge enhancement" (Figures 11.10 and 11.11). The resiliency design also seeks to connect elevated earthen berms with existing high ground and the linear infrastructure of the Belt Parkway, recasting the verges of this transportation network as a continuous elevated buffer from Mill Creek to Canarsie and Howard Beach. These enhanced verges are foregrounded by low marsh, high marsh, and upland planting zones, providing additional mitigation of

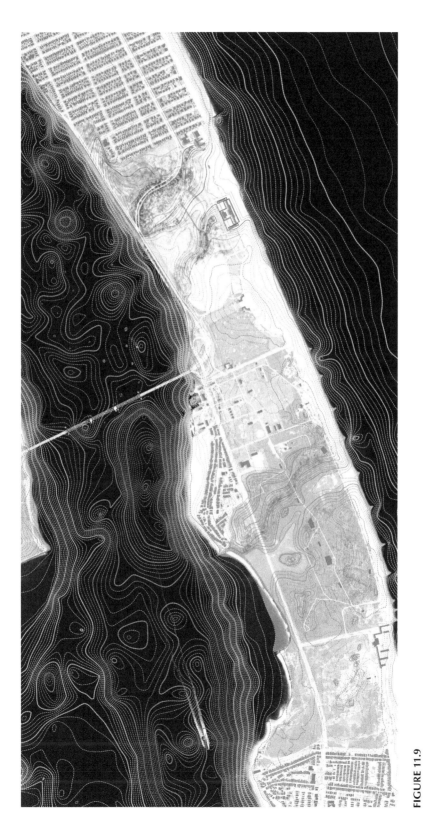

FIGURE 11.9

Design proposal for tidal inlets and overwash plains at Fort Tilden and Jacob Riis Park at the western end of the Rockaway Peninsula. (Image courtesy of Catherine Seavitt Nordenson/Structures of Coastal Resilience, Jamaica Bay.)

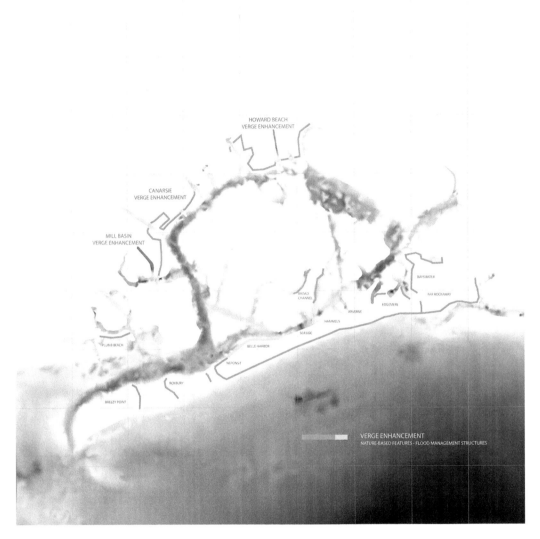

FIGURE 11.10
Key plan: Verge enhancement. (Image courtesy of Catherine Seavitt Nordenson/Structures of Coastal Resilience, Jamaica Bay.)

wave energy, as well as new public spaces at the waterfront. The establishment of densely planted sunken maritime forests are proposed behind the berm.

The neighborhood of Howard Beach is a significantly low-lying back-bay community that was extensively flooded during Sandy and an area that continues to experience street flooding during high tidal cycles (The Nature Conservancy, 2013, 2015). Here, a continuous connection is created from existing high ground at the southwest to the launching of the Belt Parkway's bridge over Spring Creek. A continuous elevated verge is created from the bridge traversing Spring Creek at the west to the Bergen Basin and Lefferts Boulevard at the east. By providing flood closure structures below the Belt Parkway at Spring Creek and near the mouths of Shellbank and Hawtree Basins at 165th Avenue, this continuous

FIGURE 11.11

Design proposal for Howard Beach at Spring Creek, Shellbank Basin, and Hawthorne Creek, establishing a continuous earthen berm tying into high ground at the verges of the Belt Parkway. (Image courtesy of Catherine Seavitt Nordenson/Structures of Coastal Resilience, Jamaica Bay.)

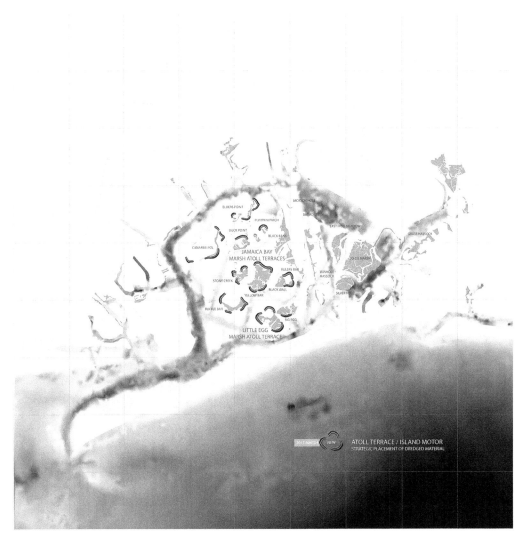

FIGURE 11.12
Key plan: Island motor/atoll terrace. (Image courtesy of Catherine Seavitt Nordenson/Structures of Coastal Resilience, Jamaica Bay.)

enhanced verge has the capacity to provide the entire community with the equivalent protection of FEMA's current 100-year base-flood elevation plus 60 centimeters of freeboard.

Island Motor/Atoll Terrace

Salt marsh degradation and loss at Jamaica Bay is a paradigmatic example of environmental vulnerability, particularly given future sea level rise (Figures 11.12 and 11.13). A resilient marsh ecosystem provides coastal storm risk management services to adjacent communities

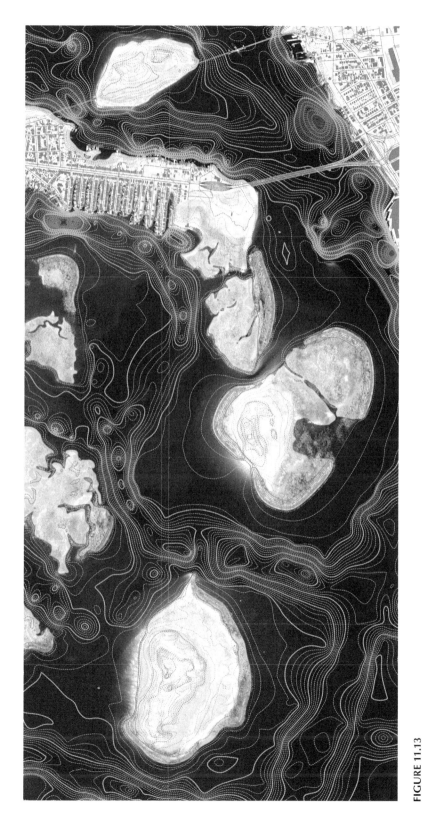

FIGURE 11.13

Design proposal for Ruffle Bar Marsh, Little Egg Marsh, and Big Egg Marsh, incorporating elevated "atoll terraces" at the perimeter of the marsh island footprints. (Image courtesy of Catherine Seavitt Nordenson/Structures of Coastal Resilience, Jamaica Bay.)

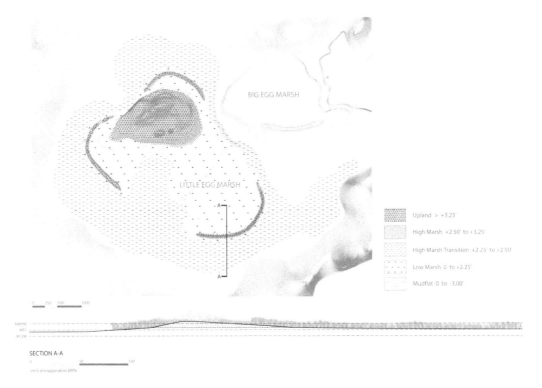

FIGURE 11.14
The "atoll terraces," formed specifically to reduce wave energy, encourage sediment deposition and provide an ideal slope for the establishment and upward migration of low marsh grasses. (Image courtesy of Catherine Seavitt Nordenson/Structures of Coastal Resilience, Jamaica Bay.)

through wind fetch reduction and wave attenuation. After assessing the methods and results of the innovative marsh island restorations at Jamaica Bay undertaken in partnership with the National Park Service, USACE, and local environmental organizations, the project proposes novel strategies of sediment nourishment and salt marsh island restoration at the bay through an "island motor," maximizing the effects of the strategic placement of minimal quantities of dredged material as a perimeter "atoll terrace." This elevated marsh terrace, formed specifically to reduce wave energy and encourage sediment deposition, provides an ideal slope for the upward migration of low marsh grasses *(Spartina alterniflora)* and allows the natural processes of current-driven sediment capture and deposition to nourish the marsh, thus allowing it to accrete upward with sea level rise. Atoll terraces are arrayed at the perimeter of the marsh island footprints, rising just above the high tide datum and positioned to best capture sediment given predominant current speeds, tidal flows, and sedimentary transport. The construction of the atoll terraces would be aligned with the availability of local dredged material and sequenced to provide maximum immediate benefits for the most vulnerable communities and marsh islands.

The atoll terraces, formed specifically to reduce wave energy and turbidity, as well as encourage sediment deposition, will provide both an ideal slope for the establishment and upward migration of low marsh grasses *(Spartina alterniflora)*, as well as enhance the conditions for achieving the water clarity and light penetration required for the successful establishment of submerged aquatic vegetation *(Zostera marina)* (Figures 11.14 and 11.15). Resilient plant selections for the marsh islands given climate change are also specified as

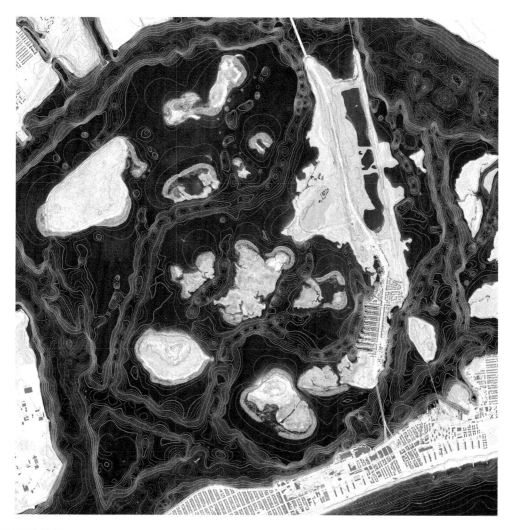

FIGURE 11.15
The "island motor" concept uses natural processes to capture sediment through the use of the "atoll terrace" ridges at the perimeter of the marsh islands. (Image courtesy of Catherine Seavitt Nordenson/Structures of Coastal Resilience, Jamaica Bay.)

part of the design proposal, such as submerged aquatic vegetation that tolerates warmer waters and low and high marsh species that are adaptive to wetter and warmer climates. Ensuring the future success (and indeed, the continued existence) of the salt marsh islands at Jamaica Bay requires the strategic use of dredged material resources, the improvement of water quality at the bay, and a robust palette of plant material.

An Adaptable Framework for Jamaica Bay

The three proposed design strategies for coastal resilience at Jamaica Bay—flow and circulation, verge enhancement, and the island motor/atoll terrace—have been developed

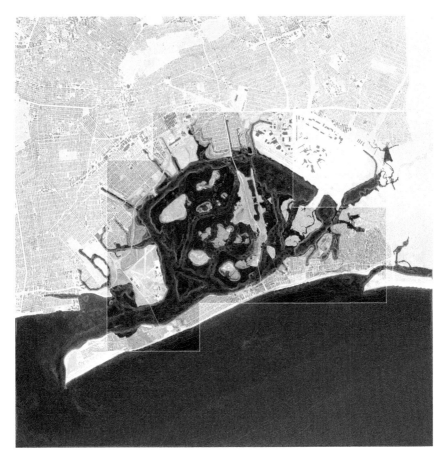

FIGURE 11.16
Resiliency Strategies Design Plan, Jamaica Bay. (Image courtesy of Catherine Seavitt Nordenson/Structures of Coastal Resilience, Jamaica Bay.)

through iterative research using both physical and digital modeling, as well as field investigations (Figure 11.16). These strategies are conceived as complementary, as well as deployable through an adaptable framework over time. Each of the three strategic design components interacts systemically with the others, supporting an enhanced and resilient Jamaica Bay. The bay would thus be enabled to respond to both future sea level rise, as well as ongoing storm and flood risks, while supporting the continued coexistence of an urban ecology of both nature and culture at the bay.

Five years after Superstorm Sandy's landfall, interest in innovative and implementable nature-based proposals for Jamaica Bay's resiliency continues to be high. In February 2017, the New York District Office of the United States Army Corps of Engineers released its *Hudson-Raritan Estuary Ecosystem Restoration Feasibility Study*, incorporating the *Structures of Coastal Resilience* design concept of the atoll terrace as a pilot feature of its proposed restoration of Duck Point Marsh, one of five proposed marsh island restoration projects for Jamaica Bay (USACE, 2017). Another component of the project, the proposal for continuous verge enhancement at Jamaica Bay's Belt Parkway, will be implemented as part of the new coastal resiliency initiatives underway at Spring Creek, adjacent to the back-bay community of Howard Beach (NYS, DEC, 2016). Though often entrenched in bureaucratic processes, the

United States Army Corps of Engineers has been influenced by the collaborative work of the *Structures of Coastal Resilience* research engine. The possibilities of a paradigmatic shift in the policies and practices of coastal resilience, inclusive of the biotic ecologies within the urban ecosystem, appears to be well underway.

Acknowledgments

Structures of Coastal Resilience was fully funded by the Rockefeller Foundation, 2013 through 2015. The City College of New York's Jamaica Bay team was led by principal investigator Associate Professor Catherine Seavitt Nordenson, with research associates Kjirsten Alexander, Danae Alessi, and Eli Sands. A full portfolio of the *Structures of Coastal Resilience* design initiative for Jamaica Bay, New York may be viewed at http://www.structuresofcoastalresilience.org.

References

Black, Frederick R. 1981. *Jamaica Bay: A History.* Cultural Resource Management Study No. 3. National Park Service.

Bridges, Todd and Paul Wagner for U.S. Army Corps of Engineers. 2013. *Use of Natural and Nature-Based Features to Enhance the Resilience of Coastal Systems.* United States Army Corps of Engineers, Engineer Research and Development Center, Vicksburg, Mississippi.

Burks-Copes, Kelly et al. for U.S. Army Corps of Engineers. 2013. *Developing Ecosystem Goods and Service Performance Metrics for Natural and Nature-Based Infrastructure to Support the NACCS.* United States Army Corps of Engineers, Engineer Research and Development Center, Vicksburg, Mississippi.

Hartig, Ellen. 2002. "Anthropogenic and climate-change impacts on salt marshes of Jamaica Bay, New York City." *Wetlands*, Vol. 22, No. 1:71–89.

Moses, Robert. 1938. *The Future of Jamaica Bay.* Department of Parks, City of New York.

New York State Department of Environmental Conservation. 2016. *Spring Creek South: Storm Resilience and Ecosystem Restoration Project.* http://www.dec.ny.gov/about/104426.html (accessed April 1, 2016).

New York City Department of Environmental Protection. 2007. *Jamaica Bay Watershed Protection Plan, Volumes I and II.*

New York City Department of Environmental Protection. 2012. *Jamaica Bay Watershed Protection Plan, Updates 2012.*

Schultz, Martin. 2012. *The Quantification and Evolution of Resilience in Integrated Coastal Systems.* U.S. Army Corps of Engineers, Engineer Research and Development Center, Vicksburg, Mississippi.

The Nature Conservancy. 2013. *Integrating Natural Infrastructure into Urban Coastal Resilience.* https://www.nature.org/media/newyork/howard-beach-report-12-23-2013.pdf (accessed December 1, 2013).

The Nature Conservancy. 2015. *Urban Coastal Resilience: Valuing Nature's Role. Case Study: Howard Beach, Queens, New York.* https://www.nature.org/media/newyork/urban-coastal-resilience.pdf (accessed July 1, 2015).

United States Army Corps of Engineers, Directorate of Civil Works. 2013. *Coastal Risk Reduction and Resilience: Using the Full Array of Measures.* CWTS 2013-3.

United States Army Corps of Engineers, New York District. 2017. *Hudson-Raritan Estuary Ecosystem Restoration Feasibility Study: Draft Integrated Feasibility Report and Environmental Assessment.* http://www.nan.usace.army.mil/Portals/37/docs/Environmental/FINALHRE.pdf (accessed February 1, 2017).

United States Army Corps of Engineers, North Atlantic Division. 2015. *North Atlantic Coast Comprehensive Study: Resilient Adaptation to Increasing Risk, Main Report.* http://www.nad.usace.army.mil/CompStudy (accessed September 1, 2015).

12

Resilience and Coastal Ecosystems: Three Typologies, Three Design Approaches

Steven N. Handel, Gina Ford, and Briana Hensold

CONTENTS

In the aftermath of Hurricane Sandy's devastating impacts on the greater tristate region in 2012, the United States Department of Housing and Urban Development (HUD) launched the Rebuild by Design competition in the summer of 2013—an initiative to produce regional, cross-disciplinary collaboration between state and local governments, international design teams, educational institutions, and the public. As one of the initiatives of President Obama's Hurricane Sandy Rebuilding Task Force, competition partners included the Municipal Art Society, NYU's Institute for Public Knowledge, Regional Plan Association, and Van Alen Institute. Leveraging non-traditional partnerships among designers, scientists, the local communities, and the federal government, Rebuild by Design developed locally contextual, but also regionally scalable resilience solutions. Among 10 teams selected to participate in the initiative, Sasaki, with Rutgers University and ARUP, focused on resiliency for the Jersey Shore.

The Sasaki team's vision for an overall framework is built upon one of the Jersey Shore's most valuable assets: the beach. Months of research, community engagement, and design strategy development focus on the value of the beach, a place of special significance to human memory, state and local economies, and a vital component of coastal ecosystems. In a time of continued climate change and sea level rise, coastal regions are threatened worldwide. We have focused efforts on an area which may have wide relevance. New Jersey's northern shore (Ocean and Monmouth Counties) is an ideal place to study the identity and function of the beach, as it includes the three coastal geomorphic typologies found across the eastern seaboard of the United States: Barrier Islands; Headlands; and Inland Bay.

Practices to support tourism and other development have impacted the underlying ecology and resiliency of the beach and shore communities. The shore's relevancy and geography are ultimately much deeper than the narrow strand of sand that is often defined as the beach. In New Jersey, deep sandy soils extend inland to the expansive, ecologically rich pine forests, the Pine Barrens (Forman, 1998). A series of 22 coastal lakes and myriad rivers and creeks extend into estuarine and other wetlands environments inland kilometers from the coast. Storm surge and coastal flooding pose increasing threats to the coastline. Consequently, the inland environment and patterns of development mean that watershed flooding impacts will also constitute a significant portion of the future threat in the region. Ultimately, the Jersey Shore's future resiliency cannot only be solved through engineering solutions that address the immediate coast; a resilient beach is linked to projects that deepen the physical extent, ecological reach, and cultural understanding of the beach. Considering the economic, ecological, and cultural conditions of the Jersey Shore, we have developed three approaches for its future resiliency: (1) A regional program for environmental improvements along the shore; (2) Local interventions in three specific sites that will serve as replicable "pilot" solutions for the diverse issues that sea level rise and future weather events present; and (3) These design interventions may have wide applicability to other areas confronting similar threats. While many economies are driven by urbanism (linked physically to dense cities) the economies of beaches are an exception, driven instead by the underlying and diverse coastal ecology. Our research focused on understanding the characteristics and vulnerabilities of the coast, considering specifically the links between economy, ecology, and culture along the beach. Since the human experience and impact on the coast is such an integral part of our research, we defined our coastline as much deeper than this literal land-ocean edge, reaching kilometers inland from the ocean to encompass a more diverse ecological landscape, as well as residential and commercial development.

Understanding Ecology and Environment

The shore is defined by the three environmental typologies that comprise all coastlines, including any part of the Atlantic Seaboard, the Sandy-affected region, and—where our research has focused—along the Jersey Shore. These *three coastal typologies are the Barrier Islands, Headlands, and Inland Bay.* The typologies have provided the foundation for the ways that adjacent shore communities have grown and evolved. They also directly inform their vulnerability to sea level rise. This understanding of different coastal types along the Jersey Shore has helped us to shape a regional response to sea level rise challenges. For example, a Headlands community is typically higher and dryer (at an average elevation of 4 meters) than a Barrier Islands community (at 1 meter) (see Figure 12.1). Likewise, a Barrier Islands community, with densely developed, high value land along ocean or bay edges, is highly susceptible to very early levels of sea level rise.

The ecological understanding of the beach is also enhanced by the conception of the shore as deeper than the beach (Burger et al., 2017). Ecosystems here encompass the ocean,

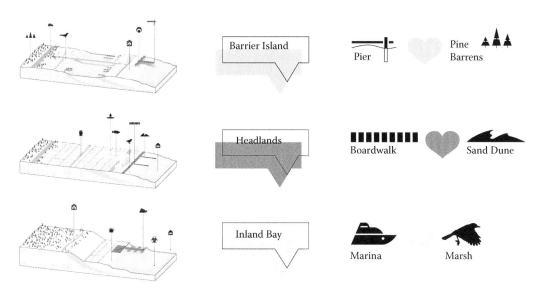

FIGURE 12.1
A typological approach to resiliency and the beach.

the Barrier Islands, the bay, mainland, rivers, creeks, coastal lakes, urban parks and streets, and New Jersey's Pine Barrens forest. Central to a new perspective of the shore, the Pine Barrens include 471 thousand hectares that have been designated as the Pinelands National Reserve. The Pine Barrens extend west from the mainland edge inland nearly to Camden. This deep reach, defined by an ecologically and environmentally sensitive area, is our basis to re-imagine the Jersey Shore as much deeper than the beach.

Protecting Tourism Economies and Social Dynamics

The shore is an economic engine for the state. Monmouth and Ocean counties specifically generate approximately US$28 billion annually in tourism revenue for the state. In effect, the Jersey Shore's economic impact extends much deeper than the immediate beach localities, supporting counties and the state as a whole. As we re-imagine the resilience of shore communities, protecting the shore's ecologically driven economy is one of our primary goals, from lodging to food to beach badge sales, as well as income generation from second home rentals. Certainly not all communities along the Jersey Shore experience the impacts of shore tourism equally. Approximately half of the shore's hotels are concentrated in the Headlands area, while the Barrier Islands have the greatest share (77%) of second homes. The seasonal nature of the shore economy stresses social and community ties, embodied in towns with small populations that swell to ten or more times their year-round population in the summer. The beaches themselves are increasingly preserved and nourished through coastal engineering. Beach nourishment began in the 1930s and continues today as a major part of the Army Corps of Engineers projects along the Jersey Shore. Additionally, much of the shore's edge has been transformed by seawalls, revetments, groins, and jetties. Our survey of our study area's shoreline showed that the dynamic, moving barrier beach is armored with two systems: 99% seawalls on the bay side and 99% groins or jetties on the ocean side. The pattern of coastal infrastructure in such a dynamic environment motivated our search for design opportunities that better integrate ecological processes into beach protection or transformation. Given the shore's need for improved capacity and multi-jurisdictional operations, a regional strategy for a resilient shore requires a connected *Resilience Network*, rather than single project owners or partners. The goal is that these coalitions can easily scale up or down to support projects of different scales and serve as resource sharing platforms, as well as being a mechanism to support disaster response in emergency situations.

Regional Program and Pilot Project Sites

Regional Resilience for the Barrier Islands

> The Barrier Islands are the most vulnerable stretch of the Jersey Shore, the low-lying lands constantly shifting with the energy of tide and storm.

Such ecological dynamism, although beneficial, will be the Barrier Islands' principal threat in the future. Under a 1-meter sea level rise scenario, they are projected to lose half of their land area; in a 2-meter scenario, those popular beachside tourist destinations nearly disappear (NOAA Sea Level Rise and Coastal Flooding Impacts, http://www.csc.noaa.gov/slr/viewer/).

Our project seeks ways to diversify the traditional beach economy, allowing the economic, social, and ecological health of the Barrier Islands communities to persist as the environment of the shore changes.

The classic cultural moment at public piers serves as impetus for design opportunities on the Barrier Islands. There is an opportunity to extend the iconic language of the perpendicular beach pier deeper inland to the New Jersey Pinelands. The new conceptual "pier" will be a transect encompassing beach, dune, marsh, bay, and pineland habitats and communities, enabling a more layered tourism economy to take shape while providing infrastructural functions, including new transit that brings people to the barrier beaches. The inland pier will allow ecological structures to migrate after sea level rise, maintaining habitat and fostering food web security.

Along New Jersey's Garden State Parkway, there are many active and abandoned sand mining pits and other post-industrial sites that present an opportunity to transfer value away from the Barrier Islands to higher and drier development options. The New Jersey Pinelands Transfer of Development Rights program provides a precedent for transferring value from ecologically sensitive areas to more appropriate growth zones.

Barrier Islands Pilot Site

The Toms River–Berkeley area is currently a relatively low-density, suburban complex of mostly middle-income residential communities, with strip commercial abutting highway corridors. We propose to diversify the touristic experience of the Barrier Islands communities and open new opportunities for integrated development inland while providing space for its continued evolution and inland migration over time (Burger et al., 2017).

The pier to pinelands plan can be a mixed-use extension with around 5,000 new housing units for permanent and seasonal residents with eco-tourism opportunities and open areas for natural habitat migration. A new civic node with resiliency center, public education, and entertainment programs can connect into an aerial tram system with new marinas and a water taxi hub on Barnegat Bay, at the mouth of Potter Creek (see Figures 12.2 through 12.4).

Managing risk by transforming access, improving mobility options to the barrier beach, as well as linking in other tourist experiences, creates development value on high-and-dry land. Syncing human and other species, mobility can occur at the disturbed sand mine site, at a land bridge across the Garden State Parkway, and by shared corridors, such as Mill Creek. These restore vital habitat migration corridors long ago severed by development.

Phasing and Implementation

There are several key stakeholders on the state, local, and community levels that are essential to the future resilience of the region. In the Barrier Islands, Ocean County government and local municipalities have coordinated in public workshops and provided information to help understand the current planning environment of the region, as well as offered support to make the Barrier Islands more resilient in the future.

A tourism and travel study is needed to identify existing local and regional values within the study area. In 50 years the master plan suggests a full build-out of the residential component. The 50-year timeframe coincides with an expected sea level rise of about one meter, enabling Barrier Islands communities like Seaside Heights the opportunity to migrate activities inland to the newly developed town center.

FIGURE 12.2
Barrier Islands plan—proposed landscape after sea level rise, Toms River.

FIGURE 12.3
Barrier Islands—population migration over time from beach to nearby mainland.

Regional Resilience for the Headlands

> The Headlands are the most exposed stretch of the Jersey Shore, with open ocean views subject to the direct action of wind and waves, but also the highest and driest of the three shore typologies, making the Headlands attractive for continued future use.

New Jersey's Headlands were the first sites of tourist-oriented occupation of the Northern New Jersey coast. The first boardwalk was built in New Jersey to prevent sand from entering beachside buildings and, over time, boardwalks have become an iconic emblem of Jersey Shore tourism. This is similar to many oceanfront boardwalks, worldwide.

Since the 1930s, beach enthusiasm has been accompanied by extensive investment to control beach erosion, and today the Headlands is marked by groins, jetties, seawalls, and revetment. The beach is raked and shaped for human use only, minimizing its ecological functions as a dune landscape and for the ever-changing wrack line, both necessary for biodiverse and vibrant life near the surf (Nordstrom et al., 1990). These protective engineering efforts have hastened erosive processes, leaving the Headlands beaches dependent on human actions for continuing existence.

FIGURE 12.4
Barrier Islands—a vision for a new ecologically based economy incorporating inland landscapes.

Stakeholders in the Headlands region understand the cost to continually nourish beaches. The federal government covers 65% of the cost for beach nourishment and the state and local government split the remainder.* With the high probability of one meter of sea level rise over the next 50 years, local governments must meet the growing costs associated with protecting the unsustainable landscapes.

The communities of the Headlands today contain traces of the socio-economic stresses common in older urban areas: an aging population, a large low- and moderate-income population, older and less valuable housing stock, and lower rates of home ownership. A substantial lodging sector exists in the Headlands, with 130 hotels occupying around 30 kilometers of coastline, most within six blocks of the beach. Many year-round residents are employed in seasonal occupations oriented to tourism; however, while the average income of a shore tourist is US$99,000, the average per capita income of a shore resident is US$39,000.

Mobility is a considerable challenge in the Headlands. Despite periodic transit stops on the New Jersey Coast Line commuter rail, the culture of the car dominates. The Headlands' built environment reflects this with a prevalence of parking lots occupying prime real estate, space that could be programmed for more tourism-centered and ecologically significant uses.

The "Layered Approach" design opportunity provides protection from the ocean and is coupled with improvements to coastal lakes and the introduction of green infrastructure that provides inland protection and connects the beach to the community. The Headlands relationship of water to land is obvious in the variation in hard infrastructure at the water's edge. The edge conditions along the coastal lakes, where the sidewalks meet the street, and on the beach, present the opportunity for the layered approach to designing resilient infrastructure.

Headlands Pilot Site

The site for this project is the city of Asbury Park, chosen for its iconic boardwalk, coastal lakes, diverse population, and cultural history. Asbury Park's beach naturally captures sand due to a combination of prevailing winds and tidal flow, making it a prime location for investment and a safer location for coastal occupation. While Asbury Park is vulnerable to sea level rise and storm surge, the coastal community has the infrastructure and relatively high elevation to develop projects for long-term protection. Currently, 29% of the land area of Asbury Park is directly impacted by a 100-year storm, but only 2% of the city's population is vulnerable attributable to the high elevation from the beach.

Asbury Park experienced major development booms in parallel to the rise of car culture across the country. The urban form is characterized by unnecessarily wide streets with narrow sidewalks and minimal street landscaping. Existing conditions benefit the summer population swell and encourage vehicular use throughout the town rather than engaging a pedestrian audience through livable, walkable spaces and business districts. Street improvements and landscaping can capture and treat stormwater to further protect the town from long-term flooding. Major east-west corridors with business frontage can have wider sidewalks and incorporate a seasonal street plan for managing parking and summer swell traffic.

Once rich tidal estuaries, New Jersey's 20 coastal lakes are contaminated by urban runoff, vulnerable to storm surge blowouts, and lack edge habitats to support birds and

---

* http://www.npr.org/2013/01/30/170301306/debate-over-rebuilding-beaches-post-sandy-creates-waves

fish needed for higher ecological value. Deal Lake, which borders the edge of Asbury Park, can undergo a thorough restoration process that includes dredging and hard edge removal to make way for soft infrastructure and native vegetation to increase habitat for migratory birds, aquatic invertebrates and other native animal species. Boat launches and small piers can be installed, connecting residents and visitors to the improved water ecology. Through this strategy for resilience in the Headlands, the condition and absorptive power of coastal lakes will be improved to manage stormwater and clean watershed runoff (see Figures 12.5 and 12.6).

The historic boardwalk offers an opportunity to rethink today's beach's monoculture through a more organic boardwalk form and topographic section that have a new relationship to beachside development and promotes a healthier ecology (see Figure 12.7). The shape and design of the boardwalk can maintain its important social role while providing a new infrastructure to capture sand and form dunes over time, creating protection and habitat area for beach wildlife. Microtopography could support a variety of plant species and nesting sites for beach breeding bird species, such as Piping Plover and Roseate Terns. Swales can be dug for ponding which supports invertebrates used by these bird species (Maslo et al., 2011, 2012). Segments of boardwalk can be redesigned to include more experimental dune/habitat creation strategies, protecting the community and acting as a new park. In spatially constrained sections, the boardwalk-dune design will consider a "compressed dune" typology, which includes a newly forming dune, tidal pool, and primary dune (Sasaki et al., 2015).

FIGURE 12.5
Headlands—aerial of Asbury Park, with proposed design solutions on beach, lakeshore, and streets.

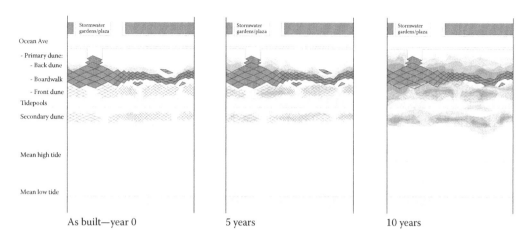

FIGURE 12.6
Headlands—dune creation over time enhanced by boardwalk and fencing infrastructure.

Phasing and Implementation

Within five years the boardwalk-dune project can be complete, and the first signs of natural dune nourishment can begin to occur. Dune maintenance and sand replenishment will naturally occur within the dunes, and the boardwalk would be near its projected useful life of 25 years. Plans for boardwalk retrofits and reconstruction then could begin.

Regional Resilience for Inland Bays

The Inland Bay is the most complex region of the New Jersey shore, with a legacy of industrial uses, densely populated maritime communities, increasing levels of integration into the Greater New York City economy, and a rich estuarine environment.

FIGURE 12.7
Headlands—a vision for a boardwalk as a social and ecological generator of resilience.

Over 50% of the Atlantic coast can be categorized as an Inland Bay condition. Along the Atlantic seaboard and Hurricane Sandy–affected areas, Inland Bay communities support the greatest population, with 62% and 79%, respectively. Only along the Jersey Shore does this pattern switch, with just 23% of the Jersey Shore population living in the Inland Bay. Inland Bay communities in the northern Jersey Shore are located within the New York City metropolitan area commuter shed.

The Inland Bay coastline encompasses diverse habitats conditioned by the fine gradient of salinity and low energy water found along its edge. In comparison to the ocean, the bay contains lower and varying salinity in water at small scales, generally ranging from 15 to 30 parts per thousand. This lower salinity is critical for survival and perpetuation of key estuarine species and has made the bays popular for recreational and commercial fishing. Species, such as blue crabs (*Callinectes sapidus*), require the highly productive marsh habitats that thrive in bay waters. Yet, despite a reliance on water for livelihood, the Raritan Bay's water quality suffers from an influx of contaminants, limiting fish and crab resources for eating and commerce. Contaminants from adjacent industrial uses or other human-dominated land raise nutrient levels in bay waters and reduce species persistence.

Common Inland Bay Vulnerabilities

While seemingly not as vulnerable as the Barrier Islands to sea level rise, Inland Bay communities suffered from both the ocean and the inland rivers during Hurricane Sandy. Routine flooding persists in between storm events, leaving communities perpetually at risk. In fact, sea level rise will present only an incremental increase in vulnerability for Union Beach and Keansburg, for example, because so much of these communities is already vulnerable: 43% of the population is now impacted by a 100-year flood event.

Topographic changes also affect exposure to flooding and future weather events. The Inland Bay is characterized by low-lying communities between a ridgeline and higher, inland communities. While some communities, often lower density, are located on bluffs, the higher-density, low-lying communities along the water's edge are both socially and physically vulnerable.

Resiliency Solutions for Inland Bay Communities

The Inland Bay's complex, mixed use conditions create different risks for storms and sea level rise, making New Jersey's Inland Bays resistant to a singular form of intervention. Water culture in the Inland Bays is centered on the bay, for commercial fishing, commuting, and recreation. A system of creeks, wetlands, and small lakes lines the shore; today these creeks lead to flooding, but the community still has memories of previous recreation use. Building on their recreational and commercial role, there is an opportunity to restore and reuse these water bodies to enhance coastal protection while providing new sources of recreational value for adjacent ecosystems and communities.

Inland Bay Pilot Project: Natco Lake Park

The Natco Lake district is an ideal site for a multi-layered approach to water culture opportunities. The Natco Lake district encompasses dense residential neighborhoods, post-industrial lands, and an increasingly diverse ecosystem with the potential to be an ecological and community asset, as well as offer storm protection and flood mitigation (see Figure 12.8).

FIGURE 12.8
Inland Bay—aerial of Natco Lake area with proposed design solutions.

Both Union Beach and Keansburg are waterfront towns that border Natco Lake and were heavily impacted by Hurricane Sandy, but in vastly different ways. Keansburg's topography represents a bowl-like condition where the berms and dunes act as protective barriers against storm surge. The bowl topography also means that during normal rain events, the community is prone to localized flooding as the storm sewer outlet is below high tide sea levels. In Union Beach, the town is on a higher elevation that has unprotected views to Raritan (Figure 12.8). After Hurricane Sandy, Keansburg suffered no initial building destruction; however, major long-term flooding failed to drain for days and ultimately ruined entire buildings along the waterfront. Union Beach encountered limited long-term flooding but suffered from severe storm surge that destroyed several buildings instantly.

Through this project, Natco Lake, created by adjacent industrial uses, and the surrounding burgeoning marshland can be nurtured and transformed into an ecological system that helps manage storm surge and water inflow, as well as provide a destination for recreational boating and wildlife viewing (see Figure 12.9). In combination with new marina functions, marsh landscapes will be designed to mitigate contamination, contributing to a cleaner bay and supporting the future economic health of estuarine occupations such as fishing. The objective is to simultaneously reduce localized flooding from heavy storm events, protect communities from storm surge risks, and reconnect residents and visitors to a regional recreation amenity that provides better drainage into watersheds and wetlands.

The state and local municipalities (Union Beach, Keansburg, and Hazlet) can encourage the creation of a public-private partnership to transfer property rights from a major

FIGURE 12.9
Inland Bay—a vision for Natco Lake as an ecological and recreational center.

land-holding company to the public, allowing for the design of absorbent recreational green space and educational centers on the remediated land along Raritan Bay.

Natco Lake is surrounded on the east and west sides by creeks. These creeks caused significant flooding during Hurricane Sandy. The creek watersheds can be cleaned through dredging. Non-native vegetation will be removed to make way for native, salt tolerant species (Figure 12.10) and to expand the wetland surface area to absorb more floodwaters from neighboring towns (Handel, 2015a,b). Trails can be constructed to serve as ties between the community and the new regional recreation sites to the north of Natco Lake. While today the community has expressed little knowledge, and even less use, of the Natco Lake area, through this plan in the future they will have a new major community and regional recreation resource.

Phasing and Implementation

The design portion of this immediate phase will entail strategic engineering studies and designs for creek dredging. Most of the project is based along waterways and along the ocean which require Coastal Area Facility Review Act (CAFRA) permits, as well as United States Army Corps of Engineers permits for boardwalk construction and dune creation. Public meetings are needed to engage residents and interested parties in the future of their city.

The short-term phase (two to five years) of the project entails contamination clean up through constructed and re-vegetated wetlands and dredged water bodies. New Jersey can in turn land bank the post-industrial parcels. On a regional scale, the Henry Hudson Trail can entirely connect along the bay to Sandy Hook State Park and path improvements can be made to increase the safety of the trail and role as a resilient barrier to inland flooding from storm surge.

The midterm phase (20 years) for the Natco Lake project will allow the state to take over contaminated land for remediation purposes prior to construction of the recreation park and resilience center. Habitat terracing and habitat migration landscaping will be incorporated into the development of the new ecological park.

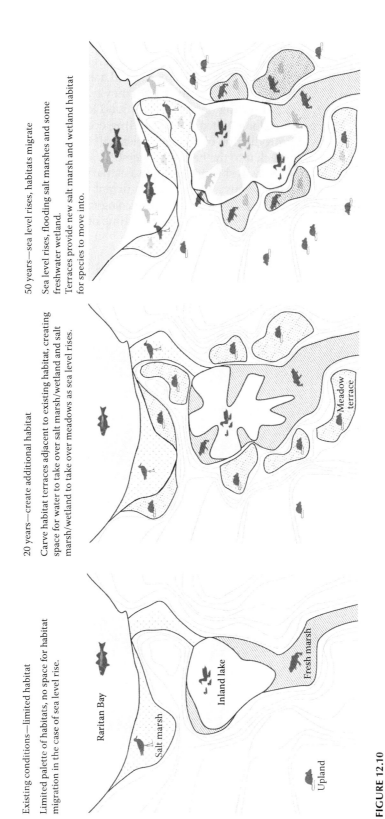

Existing conditions—limited habitat

Limited palette of habitats, no space for habitat migration in the case of sea level rise.

20 years—create additional habitat

Carve habitat terraces adjacent to existing habitat, creating space for water to take over salt marsh/wetland and salt marsh/wetland to take over meadows as sea level rises.

50 years—sea level rises, habitats migrate

Sea level rises, flooding salt marshes and some freshwater wetland.
Terraces provide new salt marsh and wetland habitat for species to move into.

FIGURE 12.10
Inland Bay—habitat engine over time.

Over the long term (50 years) Natco Lake will develop from a strictly freshwater lake to a brackish water body connected to the increasingly saline creeks and wetlands made possible through sea level rise and initial creation of ecological habitat migration corridors and terracing practices along the edge of the lake and creeks moving inland (Handel, 2015a). New development with affordable housing options and ground level retail opportunities can fill out the highway corridor along Natco Lake, building economic health. Wetland maintenance and vegetation reintroduction of native species will be required to challenge invasive species that may move back into the healthy watershed.

The rate of ecological change from freshwater habitat to estuarine habitat is a function of dispersal ecology of plants and animals and can be concurrent with the rate of sea level rise in this area. The land forming "sets the stage" for the ecological transformation that is coming and builds social and economic potential during a time of sea level rise.

Acknowledgments

We thank our colleagues in this project, especially J. Burger and K. O'Neill of Rutgers University, V. Lee of Arup Engineering, and R. Sinclair and N. Chase of Sasaki Associates.

References

Burger, J., K.M. O'Neill, S.N. Handel, B. Hensold, and G. Ford. 2017. "The shore is wider than the beach: ecological planning solutions to sea level rise for the Jersey Shore, USA." *Landscape and Urban Planning* 157:512–522.

Forman, R.T.T. 1998. *Pine Barrens: Ecosystem and Landscape*. New Brunswick, NJ: Rutgers University Press.

Handel, S.N. 2015a. "Jump starting the habitat engine." *Ecological Restoration* 33:1–2.

Handel, S.N. 2015b. "Velocity of climate change and of restoration action: collision course?" *Ecological Restoration* 33:125–126.

Maslo, B., J. Burger, and S.N. Handel. 2012. "Modeling foraging behavior of Piping Plovers to evaluate habitat restoration success." *Journal of Wildlife Management* 76:181–188.

Maslo, B., S.N. Handel, and T. Pover. 2011. "Restoring beaches for Atlantic Coast Piping Plovers (*Charadrius melodus*): A classification and regression tree analysis of nest-site selection." *Restoration Ecology* 19(201):194–203.

NOAA Sea Level Rise and Coastal Flooding Impacts, www.coast.noaa.gov/slr/. Accessed June 13, 2019.

Nordstrom, K.F., K. Nordstrom, N.P. Psuty, and R.W.G. Carter. 1990. *Coastal Dunes: Form and Process*. New York, NY: John Wiley & Sons.

Sasaki Associates, Rutgers University, and ARUP. 2015. "Resilience + the beach." In *Rebuild by Design*. J. Bisker, A. Chester, and T. Eisenberg, editors. New York, NY: Rebuild By Design. Pages 162–177. [J. Burger, S.N. Handel, and K. O'Neill were the Rutgers authors].

13

Resilience of Natural Systems and Human Communities in the Mississippi Delta: Moving beyond Adaptability Due to Shifting Baselines

Craig E. Colten and John W. Day

CONTENTS

Since the stabilization of sea level approximately 5,000 years ago, the Mississippi River formed a vast deltaic wetland complex encompassing about 25,000 square kilometers in the north central Gulf of Mexico (Roberts, 1997; Day et al., 2007, 2014). Before human activities impacted the delta, primarily in the nineteenth and twentieth centuries, it was a healthy functioning ecosystem (Kolb and Van Lopik, 1958; Condrey et al., 2014; Muth, 2014). A hierarchical series of energetic forcing events that occurred over a wide range of temporal and spatial scales formed and sustained the natural delta. These processes included deltaic lobe formation, crevasses, great and annual river floods, hurricanes, frontal passages, and tides (Day et al., 1997, 2000, 2007; Roberts, 1997; Vorosmarty et al., 2009; Roberts et al., 2015). As the delta developed, it also formed a skeletal framework of interconnected natural levee ridges and barrier islands that enhanced sediment trapping and served to protect the delta from storm surge and salinity intrusion (for example, Day et al., 2007; Xu et al., 2016). Although the delta was a dynamic system, it was relatively constant and predictable at a human time scale. This stability presented an apparently secure setting for humans to inhabit the delta by adapting to this an apparent stasis in a dynamic system. From a global perspective, the growth and stability of deltas were due to three primary factors (Day et al., 2016): (1) deltas evolved as open systems with a high degree of interaction among delta plain, river, coastal ocean, and atmosphere; (2) there was a predictable range of input of material (freshwater, sediments, nutrients) and energy from both the drainage basin and coastal ocean; and (3) there was relatively stable sea level. All of these conditions have changed due to human activities.

As a result, during the nineteenth and twentieth centuries (Barras et al., 1994, 2008; Day et al., 2007; Couvillion et al., 2011) land subsidence, exclusion of river sediment by dams and levees, and other hydrologic modifications along the coast, and accelerating sea level rise reduced the wetlands of the Mississippi River Delta by about 28% (Day et al., 2007; Barras et al., 2008; Blum and Roberts, 2009; Couvillion et al., 2011) (Figure 13.1).

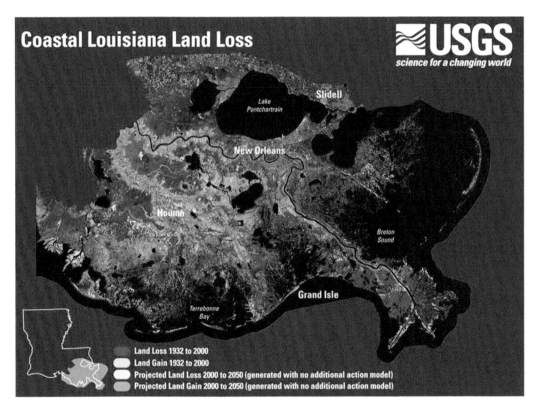

FIGURE 13.1
Past and projected wetland loss in the Mississippi Delta. (U.S. Geological Survey.)

If resilience is considered to be the ability to rebound following a disturbance (NRC, 1999), natural systems and human communities face considerable challenges given the dynamic conditions of this delta setting. Human manipulations of the river's hydrology have fundamentally disrupted the functioning ecosystem, so much so that it will not return to its pre-colonial condition. Human communities, particularly those that are highly dependent on current ecological systems and the resources that exist there, are concerned with chronic disruptions to regional ecosystems and disappearing wetlands that will impact local livelihoods and cultural survival. Acute disruptions, such as hurricanes, can further undermine stressed, resilient capacities.

Human Resources and Community

Louisiana boasts of its deeply rooted coastal communities and their rich cultural traditions (Burley, 2010; Davis, 2010; Colten et al., 2017). Yet, the state has mismanaged its budget since about 2007 and has responded by cutting services and functions rather than restoring revenue. This budgetary approach has severely strained all units of state government, with direct implications for the coastal region. Highway projects, essential to a viable economy and as evacuation routes, have lagged well behind schedule. Funding for education has declined, weakening graduate and faculty research on coastal issues and threatening the retention of the state's best and brightest. By reducing access to healthcare and programs that build well-being, these budget changes have also impacted the state's most vulnerable populations, especially in the coastal zone, and have thereby undermined the resilience and sustainability of communities. The shift of dollars away from prevention to triage means the climb of overall costs in the long run. The state's current budget situation, influenced in part by low global oil prices and combined with the legislature's rejection of revenue-building actions, illustrates little concern for long-term social resilience and sustainability in a precarious location.

Rising sea levels, more intense tropical weather, increased Mississippi River discharge, and more extreme and variable weather events are presenting escalating threats to coastal communities. Repeated exposure to extreme hazard events—stemming from both natural and human causes—contributes to stress and what sociologists call corrosive communities. This condition strains social networks and likely will increase in the already deteriorating communities outside the structural protection system. As more young people move from the coastal area to the cities, an older and often poorer population remains. The departure of this most resilient population leaves the most vulnerable to face storms and floods (Colten et al., 2017). Hemmerling (2017) reports that in general the coastal population is not impoverished, but many residents are just above the poverty line. In the event of a storm or major oil spill, these marginal residents may face economic challenges they cannot afford and may fall below that line. Although they appear to be solvent by standard measures, they are only a disaster away from losing resilience. Over time, the departure of younger people and the marginal economic condition of those left behind to face changing environmental conditions will seriously erode resilience, community viability, and local cultures.

The Environmental and Human Setting

Pre-historic humans migrated southward following the gradual development of the Louisiana coastal plain over the course of several millennia. Relict, pre-historic mounds populate coastal marshes and reveal that people followed closely behind the creation of the delta lobes in order to maintain seasonal access to both marine and freshwater resources (Figure 13.2). Many of the prehistoric societies were obliterated by exposure to diseases unleashed by European colonists, and the indigenous population declined. Europeans who arrived at the Mississippi River Delta in the 1700s were not as inclined as prehistoric societies to populate the marshes, but nonetheless some did. With the arrival of Europeans, numerous Native American communities that sought refuge from conflict with colonists moved to the ends of the bayous (Kniffen, 1987). There they established remote and insular communities reliant on local natural resources and several remnant Native American communities remain in the coastal wetlands today.

While the French clustered on the natural levees, the subsequent colonial power, the Spanish, sent small groups of Isleños (natives of the Canary Islands) into the wetlands of St. Bernard Parish and areas in Ascension Parish in order to keep an eye on potential entryways to their territories by the dreaded English (Din, 1999). They became deeply rooted in St. Bernard and Plaquemines Parishes and were active in trapping and fishing activities. The Spanish also recruited Acadians displaced from Nova Scotia and assisted their settlement along the Mississippi River. After the Louisiana Purchase in 1803, American

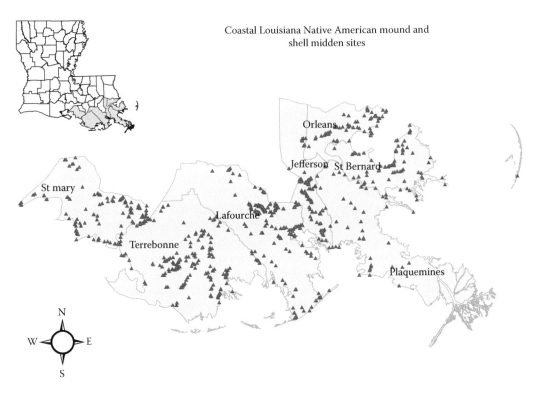

Coastal Louisiana Native American mound and
shell midden sites

FIGURE 13.2
Native American mounds in coastal southeast Louisiana. (Cartography by Rachel Watson, Louisiana Department of Culture, Recreation and Tourism.)

planters assembled larger land holdings for sugar plantations and dislodged the *petite habitant* Acadian farmers. Both levees and raised houses mitigated the risk of river flooding and provided a degree of resilience (Figures 13.3 and 13.4).

After the Louisiana Purchase, some of the French-speaking farming families moved into the Atchafalaya basin and became swamp dwellers who fished and trapped. Others moved to the prairies of southwestern Louisiana, or down the bayous to fish near the coast, and engaged in largely sustainable livelihoods (Estaville, 1986; Brasseaux, 1987; Farragher, 2006). They harvested shrimp and other seafood from the bays and participated in the extension of shrimping to open water in the second half of the twentieth century (Marks, 2012). Many fished for marine species such as oysters, crabs, and a variety of fin fish. They have proven to be a resilient society by virtue of extensive family ties and traditions of economic adaptation both to changing seasons and environmental conditions (Colten et al., 2015). They also proved adaptive, and by the early twentieth century they had added working in the oil fields to their annual round of livelihoods (Comeaux, 1972; Gramling and Hagelman, 2005). Croatians became another prominent ethnic group in the late nineteenth century and harvested oysters from Barataria Bay and Breton Sound. After nearly exhausting the natural populations, they adapted to the conservation policies instituted by the state in the early twentieth century and continue cultivating oysters today (Wicker, 1979; Maass, 2014). Asians also arrived and worked the coast, beginning in the nineteenth century, followed by a late twentieth century infusion of Vietnamese and Cambodians (Airriess and Clawson, 1991). They, along with other shrimpers, have participated in the investment in larger boats

FIGURE 13.3
Raised house on Mississippi River floodplain. (Glendale Plantation, St. John the Baptist Parish, ca. 1933, Library of Congress.)

and fishing offshore waters in recent decades, which is of questionable sustainability. Each of these groups has been part of the re-peopling of the coastal wetlands that produced insular societies, tied to natural resource-based livelihoods.

Within the Parishes that front the gulf coast and contain sizable areas of coastal wetlands, the human population clings to narrow slivers of relatively high ground along the banks of the bayous that snake through the coastal marshes toward the Gulf of Mexico. They have endured natural and technological hazards and remain rooted in place anchored by family, faith, and ethnicity. They are protected by levees and resilient housing and supported by livelihoods based in the rich natural resources of the coastal plain, the adjacent estuaries, and outer continental shelf mineral reserves. Well into the twentieth century, several ethnic groups maintained linguistic traditions and resisted the adoption of English. Foodways and other traditional cultural practices remained defining traits. Census data indicate that they are one of the more geographically stable populations in the United States with a modest standard of living—the vast majority of current residents live in the Parish of their birth (Aisch et al., 2014; Hemmerling, 2017). Their endurance and resilience suggest a stubborn immobility, and indeed this notion of rootedness has become a powerful narrative in the face of coastal land loss (Burley, 2010). However, changing biophysical and environmental conditions are severely testing this rootedness and overwhelmingly traditional building practices.

These changing conditions have resulted in a protracted movement inland from precarious locations, eroding the distinctive local cultures and leaving behind a more vulnerable older and poorer resident population (Bailey et al., 2014; Hemmerling, 2017). Hurricanes and

FIGURE 13.4
Levee repairs Plaquemines Parish, ca. 1939. (Farm Security Administration, Library of Congress.)

lesser tropical storms move across the coastal communities with some frequency. Winds, wave, and surge have prompted the abandonment of numerous small communities, such as Chenier Caminada beginning in 1893 and the current decline of the Isle de Jean Charles Native American community. Mapping of post office closures (Hemmerling, 2017) shows a gradual retreat inland since the nineteenth century due in part to extreme weather events. Recent studies also indicate a decline in coastal populations over the past 60 years (Hobor et al., 2014; Hemmerling, 2017). Oil spills since the 1930s along with periodic opening of floodwater diversions have disrupted shrimping and oyster cultivation. Despite these natural and human-caused shocks to resource-based livelihoods, local societies have adopted new techniques, often energy intensive ones, and remain present on the coast and active in resource-based livelihoods even though diminished in number.

There is also a significant urban population in the wider coastal zone. Three major metropolitan areas are in the Parishes south of I-10. Within their respective metropolitan areas, New Orleans has over a million people, and Lake Charles and Houma over 200,000 each. Baton Rouge and Lafayette, which are not as proximate or as fully tied to the coast, have another 1.3 million residents. The area north of Lakes Pontchartrain and Maurepas, the "Northshore," has another half million people. The state commonly claims that some 2 million people—over half the state's population—reside in the coastal region (CPRA, 2017) and the vast majority of that group resides in urban settings. New Orleans and most of its vicinity is armored behind river and hurricane protection levees. A sizable levee project, termed Morganza to the Gulf, is currently under construction to protect Houma and much of Terrebonne Parish (although some have questioned the reliability of this system). Additional levees protect several smaller communities in lower Lafourche Parish

(Colten, 2009). Given the success of river levees protecting New Orleans since the nineteenth century, local leaders have relied on structural protections against coastal threats. Although hurricane protection levees failed dramatically in 2005, structural protection remains a principal component of the state's current coastal restoration plans, despite its impact on the local ecosystems (CPRA, 2017). The social process known as the "levee effect" tends to foster population growth behind structural barriers, as was the case in New Orleans before Hurricane Katrina (2005). This process increases the number living in high-risk areas.

Unlike other coastal zones in the United States that have experienced recreational, development, and population growth, there has been a gradual decline in rural populations in the already sparsely settled Louisiana coastal zone outside urban areas. This is primarily due to the lack of extensive sandy beaches. Small towns along the highways leading toward the coast have witnessed the decline of local businesses, and boarded up storefronts line the main streets. A corresponding expansion of national chain stores and truck-stop casinos has altered the landscape of the smaller communities in the coastal zone. Despite the trend of rural and small town decline, the state's population has been concentrating closer to the coast in protected urban areas, and most of the near-coastal Parishes have seen moderate population increases (CPRA, 2017). Consequently, greater numbers of people will be at risk as rising sea levels and more frequent category 4 and 5 hurricanes make design heights of fixed levees obsolete and the shift waterfront northward.

Agriculture is present on narrow slivers on the natural levees in southern Plaquemines and Terrebonne Parishes, and along Bayou Lafourche. Sugar cane, too expensive to compete with international supplies, is protected by public policy and a dominant local crop, except in lower Plaquemines Parish where citrus orchards are common. Cattle ranching is prominent on the chenier ridges in Cameron Parish and in the coastal plains of Vermillion Parish, but rising seas and diminishing land area suitable for cattle threaten this activity. While locally significant, coastal farms produce only about 13% of the state's agricultural income (LSU Agcenter, 2015). Petrochemical and mineral extraction and processing are present in the coastal Parishes. The vast majority of oil and gas extraction takes place in the deep outer continental shelf, although onshore ports facilitate the transfer of crews and equipment to the rigs. Pipelines deliver most of the oil and gas to onshore facilities, with much of the processing activity concentrated along the banks of the lower Mississippi River and near Lake Charles. Marine and freshwater cargo transport are intimately linked to the river industries, but they also transport grain from the Midwest, coal from the northern plains, and timber products from the southern forests (Batker et al., 2014). Access to navigation arteries in the coastal zone is a key locational trait desired by most processors and all shippers (LSU Agcenter, 2015).

The economy of coastal Louisiana, despite long-standing foundational activities, has been turbulent in recent decades. Oystering and shrimping have suffered repeated turmoil due to a host of shocks, such as freshwater flushing into oyster beds and oil spills. High fuel prices and global competition threaten the viability of shrimping, and many practitioners are moving out of this livelihood (Marks, 2012). In 2005, citrus farming took a beating from hurricane-induced flooding in lower Plaquemines Parish. Near exhaustion of onshore and near-shore oil and gas deposits has shifted the activity to deeper outer continental shelf areas where drilling costs are higher. A short-lived moratorium on new oil exploration during the BP oil release, and subsequent and more persistent low oil prices since 2014, have greatly impacted offshore mineral extraction and derailed a once expansive onshore shale oil fracturing boom. However, the long-term trend will be to higher oil prices. Cumulatively, the reliance on finite natural resources and even renewable marine life has imperiled the state's economy. In 2001, revenue from oil and gas extraction for the state stood at US$17 billion, and payrolls for petrochemical processing added nearly US$4 billion in 2013, while the total value of marine fisheries accounted for about US$328 million of the state's economy in 2014 (LSU Agcenter, 2015; Upton, 2016). The relative value of the two resource-based activities is clear, and this drives much of the political discourse in the state.

Coastal Louisiana has endured repeated environmental and economic shocks since 2005: Hurricanes Katrina and Rita in 2005; Gustav and Ike in 2008; the BP oil release in 2010; a Mississippi River flood in 2011; Hurricane Isaac in 2012; dramatic decline in oil prices in 2014; and the inland river floods in 2016. In addition, the state has declared the protracted land loss its "coastal crisis" and the annual appearance of the hypoxic zone in the Gulf of Mexico to be ongoing, chronic shocks. With adequate time between major disruptions, a location can respond and move through a long-term recovery—climbing up the resilience curve (Kates et al., 2006). But in coastal Louisiana, the long tail ends of recovery curves overlap. Adequate recovery funding, particularly in tight fiscal times, becomes a greater political challenge, and with limited funds each recovery lasts longer. Louisiana is still rebuilding New Orleans, more than a decade after Katrina. Funds continue to arrive for restoring coastal wetlands from the BP oil disaster but less than a third of the funding for the ambitious 50-year, US$50 billion coastal master plan has been identified. Worse still, the real cost of coastal restoration and protection is likely almost double US$50 billion (Davis et al., 2015). Oil prices currently (summer 2017) are low, contributing to an austere state budget picture. The floods of 2016 merely added another recovery to an already overloaded social and economic system and made competition for limited funds more intense. With an ever-growing budget dedicated to coastal restoration, the state's resilience capacity will be strained to respond to future shocks without external aid. All the while, the national political culture is pushing back against disaster relief and other social programs. Since the hurricanes in 2005 and the oil prices decline in 2014, the state has reduced funding for education, a vital social force to build resilience. These circumstances do not reflect robust resilient capacity.

Implications of Climate Change and Energy Scarcity for Coastal Resilience

The forces expected to impact coastal areas during the twenty-first century include climate change, in the form of accelerating sea level rise (SLR), changes in river discharge, increase in the frequency of extreme weather events (including drought, intense precipitation, and tropical cyclones), and the cost and availability of energy (IPCC, 2013; Tao et al., 2014; Karl et al., 2015; Tessler et al., 2015; Balaguru et al., 2016; Day et al., 2016; Prein et al., 2016; Sobel et al., 2016) (Figure 13.5).

Fossil fuel production, especially oil and gas, are projected to peak between 2020 and 2040 (Maggio and Cacciola, 2012; McGlade, 2014; Mohr et al., 2015; DeConto and Pollard, 2016; Sgouridis et al., 2016). This will affect petroleum prices (McGlade, 2014), upon which maritime activities and delta restoration are heavily reliant (Bray et al., 1997). Highly developed deltaic coasts, such as the Mississippi Delta, that rely on energy-intensive management, are at high risk for non-sustainable outcomes with climate change in a high energy price future (Tessler et al., 2015; Day et al., 2016). Much of the current management of the Mississippi Delta is based on large-scale energy intensive activities (Day et al., 2014) that include dredging, maintenance of navigation channels, building and maintaining levees and dikes, transporting dredged sediments in pipelines, and building large water control structures.

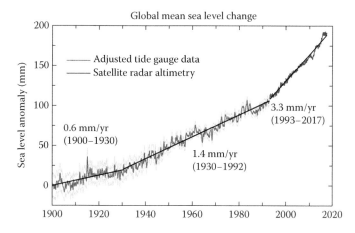

FIGURE 13.5
Twentieth century sea level rise. (From Hansen et al. 2015. *Atmospheric Chemistry and Physics* 15:20059–20179, doi: 10.5194/acpd-15-20059-2015.)

But even without consideration of energy and climate, there are significant financial constraints on coastal management in Louisiana. Only about US$26 billion have been secured for the LACMP, roughly half of the total cost (CPRA, 2016). The actual cost to restore and protect Louisiana's coastline, after including omissions from the LACMP, such as maintenance of existing flood control structures, is estimated to exceed US$91 billion (Barnes et al., 2015). This amount could rise significantly with increasing energy costs. For these reasons, delta restoration plans put in place in the short term will likely have to be largely self-maintaining within a few decades.

Solutions

In addition to actions that could sustain the biophysical setting of the coast, parallel efforts addressing communities and cultures can sustain this component of the coastal ecosystem. Four arenas offer opportunity: education, transplanting communities, integrating formal resilience programs with local inherently resilient practices, and bringing safety to the fore in policy making.

Knowledge is power, and an educated population can be more adaptive to changing conditions. Education in this region should foster understanding of coastal and climatic processes, the rich local tradition of migration that enabled survival, the deeply ingrained community and family practices that fostered resilience, and the comprehension of risks that the coastal region faces. Educators need to examine means to fold natural science, social science, and the humanities into the curriculum in ways that blend with local cultures and social values while not diluting or disputing fundamental concepts that undergird resilience and sustainability.

Louisiana's coastal residents have been mobile and continue to move—generally inland. Yet, there is a vigorous opposition among the older residents in the coastal region who are pursuing natural-resource-based livelihoods (Colten et al., 2017). The current haphazard movement of younger residents to cities in the region and beyond threatens the sustainability of local cultures and communities. It drains communities of the vigorous, educated segment of the population of child-bearing ages. This erodes future population growth and the traditional family networks that have enabled the nurturing of older generations. Programs that foster transplanting entire communities could ensure a degree of community and cultural retention. The state's Office of Community Development has launched a program to engage communities in developing strategies for a safe and sustainable coastal future (Louisiana Office of Community Development, 2017).

Coastal residents have demonstrated an interconnected set of adaptive, resilient practices that have enabled several marginalized and vulnerable populations to persist in place. In response to hurricanes, residents have commonly tapped social and family networks to rebuild modest houses or to provide temporary shelter "up the bayou." Those same networks have been used following oil spills since the 1930s. When oystermen found their leases destroyed, they would turn to an extended family member for employment in shrimping or some other economic activity. Mobility also offered a means to escape damage or to continue their livelihoods. Retreating up the bayou during a storm or moving to safer ground after destruction of homes and communities reflect mobility. Likewise, during oil spills fishermen have worked different areas to sustain their livelihoods. Another fundamentally resilient practice in the coastal region has been economic mobility. Traditional resource-based livelihoods were often seasonal, and families depended on different pursuits at different times of the year. This practice was easily put into action following floods, storms, or oil spills and allowed movement among income-producing activities without relocating geographically (Colten et al., 2012, 2015, 2017). Fostering a connection between formal, top-down government resilience programs that tend to ignore these local networks and practices could yield huge benefits in terms of shifting toward a sustainable coastal society.

Safety and well-being are often relegated to a secondary status in coastal programs. Economic development and the protection of economic interests tend to guide decision making. Yet, there cannot be a healthy economy, for the long term, without a healthy and safe society. While CPRA presents levees and marsh restoration as flood protection measures, they are justified by, and oriented toward, protecting the commercial infrastructure of the mineral and shipping industries. Certainly, safety of their facilities is an essential component of this enterprise, but dedication of huge resources to non-sustainable economic engines demands greater scrutiny. Structures and wetland restoration offer little long-term security for people or social infrastructure in the face of sea level rise. If the question of what provides community safety and well-being is asked first, a different approach would likely be pursued. Investments in helping people make deliberate decisions about shifting inland, along with reasonable expenditures on coastal restoration projects, offers greater safety. It also incorporates safety-oriented planning as part of the resettlement process for coastal residents and for inland communities that will be facing greater risks. Prime consideration needs to be directed toward ensuring that transplanted communities can find a setting where mental health, personal health, economic security, and social justice prevail. These items are not given the same level of attention as the biophysical components of the current coastal master plan. Comments about economic security and cultural heritage appear, but more in terms of how they pertain to large corporations and tourism revenue, rather than for individual or community well-being.

Summary and Conclusions

For centuries, despite a highly dynamic coastal system, both natural ecosystems and human communities coexisted in a sustainable relationship. The riverine system sustained the natural system and human communities developed lifestyles adapted to a dynamic coast and enhanced resilience. But the baselines that existed prior to the twentieth century have changed, and twenty-first century megatrends portend conditions not experienced by either coastal ecosystems or human communities over the past several centuries. These include increasingly severe climate change impacts, most notably sea level rise, stronger hurricanes, and growing resource scarcity. This suggests that sustainability and resilience of both natural and human systems will depend on new visions that take into consideration a future that is increasingly outside of range of conditions that existed when these systems developed.

Acknowledgments

This research was supported by grants from the Louisiana State University Coastal Sustainability Studio Small Projects Fund (Award Number 1512), the Gulf Research Program of the National Academies of Sciences (Award Number 2000005991), and the National Institute of Environmental Health Sciences (Award Number U19ES020676). The content is solely the responsibility of the authors and does not necessarily represent the view of the National Institute of Environmental Health Sciences or the National Institutes of Health.

References

Airriess, C. A. and D. L. Clawson. 1991. "Versailles: A Vietnamese enclave in New Orleans, Louisiana." *Journal of Cultural Geography* 12(1):1–13.

Aisch, G., Befelogg, R., and K. Quealy. 2014. "Where we came from." *New York Times*, 19 August. http://www.nytimes.com/interactive/2014/08/13/upshot/where-people-in-each-state-were-born.html?hp&action=click&pgtype=Homepage&version=LargeMediaHeadlineSum&module=photo-spot-region&region=photo-spot&WT.nav=photo-spot&_r&_r=0 (accessed March 2016).

Bailey, C., Gramling, R., and S. Laska. 2014. "Complexities of resilience: Adaptation and change in human communities of coastal Louisiana." In: J. Day et al. (Eds.), *Perspectives on the Restoration of the Mississippi Delta* 125–140. Amsterdam: Springer

Balaguru, K., Foltz, G. R., Leung, L. R., and K. A. Emanuel. 2016. "Global warming-induced upper-ocean freshening and the intensification of super typhoons." *Nature Communications* 7, doi: 10.1038/ncomms13670.

Barnes, S., Bond, S., Burger, N., Anania, K., Strong, A., Weilant, S., and S. Virgets. 2015. *Economic Evaluation of Coastal Land Loss in Louisiana*. Louisiana State University and the Rand Corporation. http://coastal.la.gov/economic-evaluation-of-land-loss-in-louisiana/.

Barras, J. A., Bourgeois, P. E., and L. R. Handley. 1994. "Land loss in coastal Louisiana, 1956–1990." *U.S. Geological Survey* National Wetlands Research Center Open File Report 94-01. 4 pp. 10 color prints.

Barras, J. A., Bernier, J. C., and R. A. Morton. 2008. *Land Area Change in Coastal Louisiana—A Multidecadal Perspective (from 1956 to 2006): U.S. Geological Survey Scientific Investigations.* Map 3019, Scale 1:250,000. Reston, VA, USA: United States Geological Survey, 2008.14.

Batker, D., Mack, S., Sklar, F., Nuttle, W., Kelly, M., and A. Freeman. 2014. "The importance of Mississippi delta restoration on the local and national economies." In: J. Day et al. (Eds.), *Perspectives on the Restoration of the Mississippi Delta.* New York, NY: Springer. 141–153.

Blum, M. D. and H. H. Roberts. 2009. "Drowning of the Mississippi delta due to insufficient sediment supply and global sea level rise." *Nature Geoscience* 2:488–491.

Brasseaux, C. A. 1987. *The Founding of New Acadia: The Beginnings of Acadian Life in Louisiana, 1765–1803.* Baton Rouge, LA: LSU Press.

Bray, R. N., Bates, A. D., and J. M. Land. 1997. *Dredging: A handbook for engineers.* 2nd edition, p. 448.

Burley, D. 2010. *Losing Ground: Identity and Land Loss in Coastal Louisiana.* Jackson, MS: University Press of Mississippi.

Coastal Protection and Restoration Authority of Louisiana (CPRA). 2017. *Louisiana's Comprehensive Master Plan for a Sustainable Coast: Draft.* http://coastal.la.gov/wp-content/uploads/2017/04/2017-Coastal-Master-Plan_Web-Book_Final-Compressed-05162017.pdf.

Colten, C. E. 2009. *Perilous Place, Powerful Storms: Hurricane Protection in Coastal Louisiana.* Jackson, MS: University Press of Mississippi.

Colten, C., Hay, J., and A. Giancarlo. 2012. "Community resilience and oil spills in coastal Louisiana." *Ecology and Society* 17(3), doi: 10.5751/ES-05047-170305.

Colten, C. E., Grismore, A. A., and J. R. Simms. 2015. "Oil spills and community resilience: Uneven impacts and protection in historical perspective." *Geographical Review* 105(4):391–407.

Colten, C. E., Simms, J. R., Grismore, A. A., and S. A. Hemmerling. 2017. "Social justice and mobility in coastal Louisiana, USA." *Regional Environmental Change* 1–13, doi: 10.1007/s10113-017-1115-7.

Comeaux, M. 1972. *Atchafalaya Swamp Life: Settlement and Folk Occupations.* Baton Rouge, LA: Louisiana State University, School of Geoscience, Geoscience and Man 2.

Condrey, R. E., Hoffman, P. E., and D. E. Evers. 2014. "The last naturally active delta complexes of the Mississippi River (LNDM): Discovery and implications." In: Day, J., Kemp, P., Freeman, A., and D. Muth (Eds.), *Restoration of the Mississippi Delta.* Netherlands: Springer. 33–50.

Couvillion, B. R., Barras, J. A., Steyer, G. D., Sleavin, W., Fischer, M., Beck, H., Trahan, N., Griffin, B., and D. Heckman. 2011. Land area change in coastal Louisiana from 1932 to 2010. *U.S. Geological Survey Scientific Investigations.* Map 3164, scale 1:265,000. 12 p. pamphlet.

CPRA. 2016. "Integrated Ecosystem Restoration & Hurricane Protection in Coastal Louisiana: Fiscal Year 2016 Annual Plan." *Louisiana Coastal Protection and Restoration Authority.* Baton Rouge, Louisiana.

Davis, D. 2010. *Washed Away: The Invisible People of Louisiana's Wetlands.* Lafayette, LA: University of Louisiana Press.

Davis, M., Vorhoff, H., and D. Boyer. 2015. *Financing the Future: Turning Coastal Restoration and Protection Plans into Realities: How Much Is Currently Funded.* New Orleans, LA: Tulane Institute on Water Resources Law & Policy.

Din, G. 1999. *The Canary Islanders of Louisiana.* Baton Rouge, LA: Louisiana State University Press.

Day, J. W., Martin, J., Cardoch, L., and P. Templete. 1997. "System functioning as a basis for sustainable management of deltaic ecosystems." *Coastal Management* 25:115e153.

Day, J. W., Britsch, L. D., Hawes, S., Shaffer, G., Reed, D. J., and D. Cahoon. 2000. "Pattern and process of land loss in the Mississippi delta: A spatial and temporal analysis of wetland habitat change." *Estuaries* 23.425–438.

Day, J. W. et al. 2016. "Approaches to defining deltaic sustainability in the 21st century." *Estuarine, Coastal and Shelf Science* 183:275–291.

Day, J. W. et al. 2007. "Restoration of the Mississippi delta: Lessons from hurricanes Katrina and Rita." *Science* 315:1679–1684.

Day, J. W., Kemp, G. P., Freeman, A. M., and D. Muth. 2014. *Perspectives on the Restoration of the Mississippi Delta: The Once and Future Delta.* New York, NY: Springer. 195.

DeConto, R. M. and D. Pollard. 2016. "Contribution of Antarctica to past and future sea-level rise." *Nature* 531(7596):591–597.

Estaville, L. 1986. "Mapping the Louisiana French." *Southeastern Geographer* 26:90–113.

Faragher, J. 2006. *A Great and Noble Scheme: The Tragic Story of the Expulsion of the French Acadians from Their American Homeland*. New York, NY: WW Norton & Company.

Gramling, R. and R. Hagelman. 2005. "A working coast: People in the Louisiana wetlands." *Journal of Coastal Research* Special Issue No. 44:112–133.

Hansen, J. et al. 2015. "Ice melt, sea level rise and superstorms: Evidence from paleoclimate data, climate modeling, and modern observations that 2°C global warming is highly dangerous." *Atmospheric Chemistry and Physics* 15:20059–20179, doi: 10.5194/acpd-15-20059-2015.

Hemmerling, S. 2017. *A Louisiana Coastal Atlas: Resources, Economies, and Demographics*. Baton Rouge, LA: Louisiana State University Press.

Hobor, G., Plyer, A., and B. Horwitz. 2014. *Coastal Index: The Problem and Possibility of Our Coast*. New Orleans, LA: The Data Center.

IPCC (Intergovernmental Panel on Climate Change). 2013. "Climate change 2013: the physical science basis." In: Stocker, T. F., Qin, D., Plattner, G. K., Tignor, M., Allen, S. K., Boschung, J., Nauels, A., Xia, Y., Bex, V., and P. M. Midgley (Eds.), *Contribution of Working Group 1 to the Fifth Assessment Report of the Intergovernmental Panel on Climate Change*. United Kingdom and New York, NY: Cambridge, 1535.

Karl, T. R., Arguez, A., Huang, B., Lawrimore, J. H., McMahon, J. R., Menne, M. J., Peterson, T., Vose, R., and H. M. Zhang. 2015. "Possible artifacts of data biases in the recent global surface warming hiatus." *Science* 348(6242):1469–1472.

Kates, R. W., Colten, C. E., Laska, S., and S. P. Leatherman. 2006. "Reconstruction of New Orleans after Hurricane Katrina: A research perspective." *PNAS* 103(40):14653–14660.

Kniffen F. 1987. *The Historic Indian Tribes of Louisiana*. Baton Rouge, LA: LSU Press.

Kolb, C. and J. van Lopik. 1958. *Geology of the Mississippi River Deltaic Plain, Southeastern Louisiana. vol Technical Report No. 3-483*. U.S. Army Engineer Waterways Experiment Station, Vicksburg, MS.

Louisiana Office of Community Development. 2017. LASAFE. http://lasafe.la.gov/.

LSU Agcenter. 2015. Louisiana Summary: Agriculture and Natural Resources. http://www.lsuagcenter.com/portals/communications/publications/publications_catalog/money%20and%20business/louisiana_summary_agriculture_and_natural_resources/2015-ag-summary.

Maass, A. 2014. Adapting Resilience to a New Hazard: Oil and Oysters in Coastal Louisiana. *M.S. thesis*, Louisiana State University.

Maggio, G. and G. Cacciola. 2012. "When will oil, natural gas, and coal peak?" *Fuel* 98:111–123.

Marks, B. 2012. "The political economy of household commodity production in the Louisiana shrimp fishery." *Journal of Agrarian Change* 12(2–3):227–251.

McGlade, C. E. 2014. Uncertainties in the outlook for oil and gas (*Doctoral dissertation*, University College London).

Mohr, S. H., Wang, J., Ellem, G., Ward, J., and D. Giurco. 2015. "Projection of world fossil fuels by country." *Fuel* 141:120–135.

Muth, D. P. 2014. "The once and future delta." In: Day, J. W., Kemp, G. P., Freemen, A. M., and D. P. Muth (Eds.), *Perspectives on the Restoration of the Mississippi Delta*. Netherlands: Springer. 9–28.

NRC (National Research Council). 1999. *Our Common Journey: A Transition toward Sustainability*. Washington, DC: National Academies Press.

Prein, A. F., Rasmussen, R. M., Ikeda, K., Liu, C., Clark, M. P., and G. J. Holland. 2016. "The future intensification of hourly precipitation extremes." *Nature Climate Change* 7(1):48–52.

Roberts, H. H. 1997. "Dynamic changes of the Holocene Mississippi River delta plain: The delta cycle." *Journal of Coastal Research* 13:605627.

Roberts, H. H., DeLaune, R. D., White, J. R., Li, C., Sasser, C. E., Braud, D., Weeks, E., and S. Khalil. 2015. "Floods and cold front passages: Impacts on coastal marshes in a river diversion setting (Wax Lake Delta Area, Louisiana)." *Journal of Coastal Research* 31(5):1057–1068.

Sgouridis, S., Csala, D., and U. Bardi. 2016. "The sower's way: Quantifying the narrowing net-energy pathways to a global energy transition." *Environmental Research Letters* 11(9):094009.

Sobel, A. H., Camargo, S. J., Hall, T. M., Lee, C. Y., Tippett, M. K., and A. A. Wing. 2016. "Human influence on tropical cyclone intensity." *Science* 353(6296):242–246.

Tao, B., Tian, H., Ren, W., Yang, J., Yang, Q., He, R., Cai, W., and S. Lohrenz. 2014. "Increasing Mississippi river discharge throughout the 21st century influenced by changes in climate, land use, and atmospheric CO_2." *Geophysical Research Letters* 41:4978–4986.

Tessler, Z. D., Vörösmarty, C. J., Grossberg, M., Gladkova, I., Aizenman, H., Syvitski, J. P. M., and E. Foufoula-Georgiou. 2015. "Profiling risk and sustainability in coastal deltas of the world." *Science* 349(6248):638–643.

Upton, G. B. 2016. *Oil Prices and the Louisiana Budget Crisis: Culprit or Scapegoat?* Baton Rouge, LA: LSU. Center for Energy Studies. https://www.lsu.edu/ces/publications/2016/Upton_10-2016_Oil_ and_Gas_and_the_Louisiana_Economy_FINAL.pdf.

Vorosmarty, C., Syvitski, J., Day, J., Sherbinin, A., Giosan, L., and C. Paola. 2009. "Battling to save the world's river deltas." *Bull. At. Sci.* 65:31–43.

Wicker, K. 1979. The development of the Louisiana oyster industry in the 19th century. *(Doctoral dissertation*, Louisiana State University).

Xu, K., Bentley, S. J., Robichaux, P., Sha, X., and H. Yang. 2016. "Implications of texture and erodibility for sediment retention in receiving basins of coastal Louisiana diversions." *Water* 8(1):26.

14

New Orleans, Coastal City

Elizabeth Mossop

CONTENTS

From Delta to Coast

The very existence of New Orleans has historically been defined by its relationship to water. The city was founded at a key strategic location on the Mississippi, where it controlled military and trade access to the river. Nevertheless, today the city has a deeply troubled relationship with water. Within the city, continuing subsidence increases the city's vulnerability to flooding. Outside the city limits, the coastal land loss threatens the very existence of the city in the next 100 years. As the coast moves inland, it creates confusion about what is coast and what is lake and where the city could exist. While not yet clearly perceived, the city is now truly coastal and connected directly to the Gulf of Mexico through its eastern shore, on the formerly enclosed Lake Borgne, now open to the Gulf. The entire map of the coast has changed and our understanding of the relationship between land and sea needs to catch up to reality (Figure 14.1).

Much of the city was once a watery landscape of swamps and marshes, with limited access routes, characteristic of the Mississippi Delta landscape. Today within the city, despite being surrounded by and threaded through with water bodies, the water is largely invisible, drained from the surface, and hidden by levees and flood walls. Unless one makes a specific effort to find the water, it is possible to traverse the city without ever seeing it. This erasure of the landscape's reality is both symptomatic of, and contributing to, the lack of cultural awareness around the city's contemporary vulnerability to floods and storms.

Until the late twentieth century, the city's relationship with water was defined in two ways: by the commerce of the port and shipping activities, vital to its economic existence, and by the business of flood control and protection, vital to its continuing physical existence. Now in the post-Katrina era, the city is in the process of rethinking its relationship to the water within and around the city, to come to a more holistic, integrated understanding of how water interacts with the city form and function.

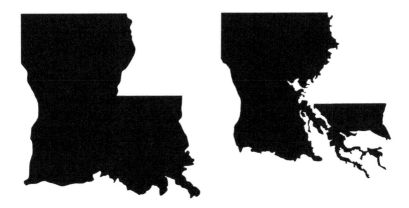

FIGURE 14.1
Louisiana's "boot" has already disappeared and the city of New Orleans is open to the Gulf of Mexico through Lake Borgne. These two maps compare the state's historical outline with the land outline today. The maps, drawn by Matthew Woodson, are from an excellent article that discusses coastal Louisiana's land loss. (Brett Anderson, "Louisiana Loses Its Boot." *Matter*, September 9, 2014 https://medium.com/matter/louisiana-loses-its-boot-b55b3bd52d1e.)

The impacts of Hurricane Katrina and the city's failed levy system, which devastated the city in 2005, have forced new ways of thinking about the city's water issues, generally moving away from the idea of armoring the city with hard infrastructure to keep all water out, toward a greater accommodation of water and flooding within the city. At the regional scale, there has also been a shift in thinking, a move away from the imperative of flood control to address the need to restore the delta's function and to use controlled flooding for land building and ecological restoration. Within the city there is an idea that it would be beneficial to be more open to the river and lake, to engage them as a part of the urban landscape. Important in this new paradigm is the attitude to manage drainage and flooding that calls for accommodating water within the city and design "blue/green" infrastructure into the urban form, based on an acceptance that floodwaters will enter the city.

All of these new ideas remain in a state of flux; with the recent observation of the 12-year anniversary of the storm and the city impacted by major flooding in August 2017, the reality of climate change is becoming a more pressing challenge, and there is an increasing urgency to find strategies that will allow the coastal city and its delta a healthy future grounded in a mutually sustainable relationship.

This essay begins by addressing the development of the delta over time and the ways in which New Orleans's geography and the Mississippi's engineering have influenced the development of the city's urban form from its earliest settlement by the French until the present day. The aftermath of Hurricane Katrina has seen significant upgrading of the city's protection systems. An appraisal of post-Katrina New Orleans and the way in which attitudes to water and risk have shifted in the hurricane's aftermath, looks at the impacts at coast, delta, and city scale. A series of major planning projects undertaken in the last ten years illustrates the new paradigm. These urban and regional projects are reshaping the city's relationship, not only with the river's course through the metropolitan area but also with the delta and coast, whose vitality and restoration is key to New Orleans's longer-term viability.

Landscape Evolution and Human Engineering

Over thousands of years the delta landscape of southern Louisiana has been built by the Mississippi River. Draining 41% of the land area of North America, the river historically deposited sediment over a broad swath of land in annual flooding caused by upstream precipitation and snow melt. This sediment deposit built new land fanning out from the river's course. Periodically the river would change course and begin the build-up of land in another fan, or delta lobe, and the old lobe would gradually begin to degrade as it became more susceptible to the assault of the Gulf Coast wave action. Thus, the coast advanced and receded, gaining and losing land over long spans of time.[1] Until recently, the land gains and losses stayed in balance, generally with a net gain in land building. Since 1930, however, there has been significant deficit, with the addition of only 250 square kilometers and the loss of 3,100 square kilometers. Projections combining the ongoing land loss with sea level rise show a scenario for 2110 where all of southern Louisiana is gone and New Orleans is almost surrounded by ocean.[2]

Beginning in the nineteenth century, expanding human settlement on the river plains, the growth of river-based transport, and the need to protect agriculture, commerce, and housing from the river's annual flooding led to a complex reengineering of the river to support human activity and settlement. Today the river is highly engineered, and although it is generally confined by levees, it has a considerably higher water level and less capacity to hold additional water with, therefore, a significantly higher risk of catastrophic flooding and damage from system failures.[3] Within New Orleans, the river is confined by a combination of earthen levees and concrete flood walls. And at the broader scale of the Mississippi Delta, the river is effectively disconnected from the surrounding delta, so that the natural cycle of land nourishment and building has been broken.

The loss of wetlands off the coast of Louisiana has been attributed to both the engineering of the river and to the impacts of coastal industry. The canals carved through coastal wetlands by industry, most notably oil and gas, permit the intrusion of salt water, destroying the wetland vegetation over time (Figure 14.2). Reduction in land building has been caused by the disconnection of the delta plain from the river distributary network, as well as by the reduction in sediment loads due to upstream reservoir construction. In combination, these have reduced the natural wave attenuation capacity of the coastal shelf, increasing the risk of flood hazard from storm surge.[4] Additionally, in New Orleans similar interventions, for example the construction of the Mississippi River–Gulf Outlet and other smaller industrial canals, have led to the destruction of the cypress swamps and wetlands of the Central Wetlands Unit. In combination, this has resulted in the destruction of the expansive swamps and wetlands to the east and south of the city that formerly separated it from Lake Borgne and the Gulf of Mexico, leaving the city significantly more vulnerable to storm action from the east and effectively open to coastal influence.

The flood events of recent years have caused many people to question New Orleans's location and therefore its viability.[5] In *Bienville's Dilemma*, geographer Richard Campanella clearly articulates the conundrum of the city's location as having an enviable situation but a terrible site. He concludes a certain locational inevitability, balancing being far enough downstream to control the river's mouth and being far enough upstream to be high enough to be resilient to floods and storms. The problems with the city's location for human habitation, which were encountered from its founding, were related to its low-lying terrain and its position between river and lake. The city's terrain flooded in heavy rains

FIGURE 14.2
Industrial canals have contributed to the destruction of protective wetlands over time. (Image courtesy of KD Cheramie.)

from its inception, and also flooded from storm surges driven from the river and the lake by high winds or abnormally high volumes of water.[6]

The city's topography has been a key determinant of its form and performance. The first settlers occupied the highest dry land along the natural levee of the Mississippi, spreading from the French Quarter and thence downstream to the Marigny, Bywater, and Holy Cross neighborhoods and upstream to the Irish Channel, Garden District, and Carrollton. Until the twentieth century, settlement was confined to high stable ground along the river. "Back-of-town," to the north toward the lake, cypress swamps lay just above sea level, but were too wet to inhabit permanently (Figure 14.3).

By the turn of the twentieth century, population pressure had spread development into the back-of-town swamps, where pumps were needed to drain the ground. The construction of a comprehensive drainage system began in 1900 and by 1950 the entire area to the lakefront was completely drained and developed for residential neighborhoods. Stormwater was conveyed to the lake through drainage canals.[7] Flood events were reduced in severity, but they nevertheless continued to plague the city.

This mechanical drainage of the former swamps had a major unintended consequence. The drying out of the ground caused the peaty layers below the surface to shrink and the ground to subside. As the elevation became lower, the ground became even more vulnerable to flooding, and, with the evolution of a bowl between the higher elevations along the lakeshore and riverbank, the water no longer drained naturally to Lake

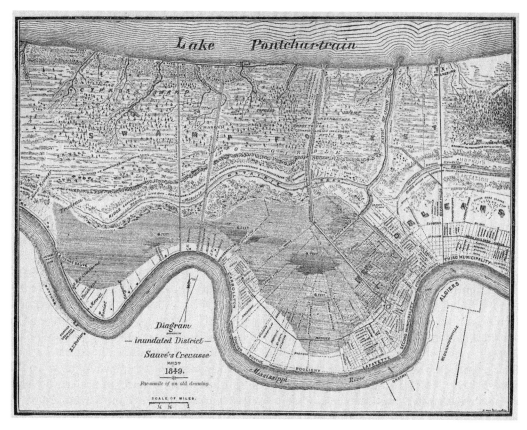

FIGURE 14.3

Map of New Orleans's east bank, showing flooding from Sauve Crevasse in May of 1849. It illustrates the early development on the high natural levee along the river and the original swamps to the north, which were drained in the early twentieth century. (Facsimile of an old drawing, May 3, 1849, Louisiana Digital Map Library, http://usgwarchives.net/maps/louisiana/citymap/1849neworleansflood.jpg.)

Pontchartrain. Mechanical pumping of stormwater became necessary to lift it up to the level of the lake. In the mid-twentieth century, levees were built along the lake, and flood walls were constructed along the drainage canals to protect the subsiding city. By the end of the twentieth century most of the city was below sea level. It is often described as a bathtub, to illustrate its bowl form between the higher boundaries of river and lake. When it rains, the water enters a conventional stormwater system that takes the water from the bottom of the bowl, through pipes and drainage canals to a series of pump stations, where it is mechanically lifted and sent through outfall canals to Lake Pontchartrain. This system can store half-an-inch of rainfall, and it can pump half-an-inch of rain in one hour, after which the city begins to flood.[8] In an average year, two-year storms commonly bring over three inches of rain over three hours.

Historically the flood protection approach has been to armor the city with structures to keep water out. While the city can be protected against floods from the river to a great extent, which can be predicted in advance, it cannot be protected from hurricanes in the same way. Storms are by their nature unpredictable, uncontrollable, and often fast-moving. The impact of Hurricane Katrina in late August 2005 is well documented.[9] Following the many catastrophic failures of New Orleans's system of protection during Katrina, it has been substantially upgraded using significantly more sophisticated modeling and greater transparency in its implementation and management. New flood walls have replaced substandard ones, levees have been made higher, pumping stations have been upgraded and moved to the lake, and improvements are underway to increase the capacity of underground drainage infrastructure. To reduce the risk of storm surge from Lake Borgne, a giant Inner Harbor Navigation Canal Surge Barrier has been built at the confluence of the Intracoastal Waterway and the (recently closed) Mississippi River Gulf Outlet, all of which have contributed to an increased level of protection from both river flooding and flooding associated with storms.[10] But these improvements and interventions only achieve a 90% reduction in flooding from a 100-year storm event (compared to protection in the Netherlands addressing 4- to 10,000-year storm events). There will still be overtopping of levees in some areas, pumping will not keep up with rainfall in many storms, low areas will be flooded (although to shallower depths), and power will be lost. And the modeling on which the new system is based does not effectively account for slow-moving storms like Hurricane Harvey in 2017, or unpredictable storm events. So there is a continuing and significant risk.

The catastrophic impacts of Hurricane Katrina were in part due to the failure of the flood protection system, but the extensive flooding that the city experienced was also due to the fact that more than 80% of the city is now below sea level. As the city continues to occupy former swamp land and continues to pump out groundwater, the ground will continue to subside. This land is already very low, and as it continues to sink further below sea level, it becomes increasingly vulnerable. While it is problematic to raise subsiding areas back to their original elevations, restoring the flows of groundwater in former swamp areas could prevent their sinking further. This would require a major rethinking of the city's water management strategy and the development of new kinds of drainage infrastructure.

New Coast Sustainable Delta

The impact and aftermath of Hurricane Katrina has led to profound changes in our conceptualization of the relationships of city, river, delta, and coast and of how to manage these going forward. This consideration of big questions and looking to the future has become easier as people have moved beyond the pressing questions of survival that took precedence in the years immediately following the hurricane. The enormous coastal land loss that was caused by Katrina and its notable and ongoing impacts on coastal communities and industries has forced a reevaluation of river engineering and much greater consideration of river diversion strategies to reconnect the river to the coastal basins of the delta, thereby allowing the redistribution of sediment to rebuild the coastal land mass. While there was a loss of population caused by the impacts of Katrina, the migration away from the coastal areas of southern Louisiana has been ongoing as land loss has continued. Over the last 10 years, a series of unprecedented broad-scale planning strategies has been proposed for the delta and the Lower Mississippi in a series of initiatives by the state, as well as through influential cross-disciplinary collaborations among academic institutions, foundations, non-profit and community organizations, and industry. Within the city there has also been a profound rethinking of the approach to urban water management.

Another advance born out of hurricanes Katrina and Rita has been a ground-breaking planning effort by the Louisiana Coastal Protection and Restoration Authority (CPRA) in the development of a series of Coastal Master Plans.[11] The CPRA was formed in December 2005, in the immediate aftermath of the hurricanes, to be the central state authority accountable for planning and oversight of all coastal protection and restoration activities, including the development every five years of a Coastal Master Plan. In 2005, the state developed Louisiana's Comprehensive Master Plan for a Sustainable Coast, known as "Louisiana Speaks," and in 2007 developed the first Coastal Master Plan. The problem of land loss and the potential of land-building strategies for the delta were clearly identified. The subsequent 2012 Coastal Master Plan represented an important step in acknowledging the scale of intervention needed to achieve coastal sustainability in Louisiana.[12] The plan selected 109 high-performing projects that could deliver measurable benefits to communities and coastal ecosystems over the coming decades. Moreover, the plan showed that if these projects were fully funded, at a price tag of US$50 billion, flood protection could be substantially increased for many communities, and Louisiana could move toward a more sustainable coast.[13]

In concert with this, in 2012 the United States House and Senate passed the Restore Act that dedicates 80% of all administrative and civil penalties related to the Deepwater Horizon Spill to a Gulf Coast Restoration Trust Fund. The act outlines a structure by which the funds, expected to be tens of billions of dollars, can be utilized to restore and protect the natural resources, ecosystems, fisheries, marine and wildlife habitats, beaches, coastal wetlands, and economy of the Gulf Coast region. And so we are at a historic juncture where there is both the public impetus and the potential to fund a sweeping restoration effort.

While these plans represent an enormous step forward in even acknowledging the need for comprehensive regional strategies for coastal restoration, they still fall far short of what is required for a more sustainable coastal Louisiana. Nothing can reverse the land loss that has taken place, and massive investment is required in the near term to begin rebuilding land in such a way that the delta and coast could become sustainable in the future. Substantive criticisms of the 2017 plan focus on its overly optimistic view of the context and lack of emphasis of the critical impacts of climate change. The plan also takes an overly restrictive view of the coastal zone, ignoring its integration in to the delta and delta systems. Thus it ignores possible upstream strategies and does not address the Gulf's "dead zone" (a low oxygen area deadly to marine life, caused by high nutrient runoff from the Mississippi). This lack of an integrated approach to restoration means that the plan does not emphasize the most promising strategies for rethinking river management using sediment diversions and suggests it will not be sufficiently impactful in changing the tide of coastal loss.

The Changing Course Design Competition, initiated by the non-profit Environmental Defense Fund, with support from the Van Alen Institute, brought together three teams of engineers, scientists, planners, and designers to create innovative visions for a self-sustaining Mississippi Delta ecosystem. The explicit goal of the competition was to influence the 2017 State Coastal Master Plan. It was hoped that the competition proposals would be sufficiently developed, tested, and considered by stakeholders to be eligible for inclusion in the plan, to drive significant new government spending in the future.

The competition facilitated creativity and flexibility that could not be achieved in a formal government process, but that could garner the support and participation of the state and the U.S. Army Corps of Engineers. Teams were asked to address two issues: the restoration of the Mississippi River's land-building capacity as part of a self-sustaining ecosystem, and the support of the high-functioning navigation system in the lower river. The primary geographic focus for the competition was the Lower Mississippi River, from New Orleans south to the Head of Passes in the Bird's Foot Delta, but teams were able to consider a broader geography socio-economically, as relevant to strategies proposed. Schemes to be considered successful were those that maximized the use of the river's freshwater and sediment for land building while achieving the best outcomes for sustainable community and socio-economic development, ecosystem restoration, navigation, and flood management. The competition's end product (it concluded in August of 2015) was a series of compelling scenarios for a more sustainable and effective lower Mississippi River Delta. However, the big question remains: to what extent will these be influential? They have the potential to be profoundly significant in changing how people view the questions of planning for both the river and the delta.

The solutions included major proposals for river reengineering through river diversions and/or distributaries to promote land building and to address the associated changes to both human and ecological systems of the deltaic plain. They also addressed the form of the Port of New Orleans, the future of the energy industry in the delta, strategies for transport infrastructure, and the form of resilient human settlements.[14]

The Baird Team's proposal was perhaps the most radical of the three finalists, proposing to capture 100% of the river's flow and deploy it for land building, using a series of river distributaries that would change location over time, operating in the different delta basins. This would effectively remove the river and its sediment from the current river bed, leaving a navigation channel less prone to flood risk and the need for constant dredging. The scheme proposes a multi-modal logistics hub (including the port) much higher up the river and a smaller sustainable delta with an emphasis on eco-tourism and fishing. Benefits to communities include greater certainty of future safety, a transitioning economy with new investment, and new jobs in a restoration economy. The scheme also articulates issues of change over time and proposes new governance structures for the delta restoration zone (Figure 14.4).

FIGURE 14.4
The Baird Team's proposal uses the total land-building capacity of the Mississippi to create a smaller and more sustainable delta region. (Image courtesy of Sasaki Associates.)

FIGURE 14.5
Hydro-Infrastructure Corridors proposed as part of the "Misi-Ziibi Living Delta" submission for the Changing Course Design Competition by Studio Misi-Ziibi. (Image courtesy of John Hoal, from Misi Ziibi Framework Plan Main Report, Studio Misi-Ziibi, p. 64.)

In the Moffat & Nichol plan (discussed in Chapter 15), a close understanding of the delta processes led to a scheme driven by ecological restoration, with a new system of river diversions to restore the delta's five main basins using pulsed flooding. This will build new wetlands to reinforce the smaller delta and buffer it from storms. Further coastal restoration is proposed with the use of a sand engine at the southern extremity of the new delta. The lower delta region is reconceptualized with a zone of recreation, fishing, and key port and oil industry infrastructure. There are three protected "delta cities," Morgan City, Houma-Thibodeaux, and New Orleans, below the I-10 corridor. The delta's crucial infrastructure is renewed with a shorter navigation canal to the Gulf of Mexico with new deep-draft port facilities at Port Sulphur. The proposal is multi-generational and closely calibrated to the region's cultural and administrative realities.

The Misi-Ziibi competition entry also proposed land building through river diversions, as well as dredge-siphons and conventional dredge and dump. This included the creation of new barrier islands, bay islands, reefs, wetlands, swamps, and sand engines as part of a smaller sustainable delta. The scheme also included strategies to rationalize industrial movement through the delta with the creation of a consolidated hydro-infrastructure corridor network as part of a more sustainable energy strategy for the region (Figure 14.5).

This scale and ambition of integrated planning is exciting and heartening in the face of the region's challenges. The competition's focus on real solutions generated by professionals guaranteed a body of work that should not be ignored in the discourse moving forward, both in public and also within the various levels of government and the Army Corps. The competition also provided material to those within key institutions and organizations advocating for more progressive solutions to the problems of navigation, land building, and flood management.

Adapting to a Water City

Within New Orleans, the new approach of working with the city's connections to its watery environs is expressed at different scales. There are many small-scale projects throughout

FIGURE 14.6
The landscape design for the Rosa Keller Library in Broadmoor stores water in bioswales, soft surfaces and the Louisiana iris rain garden seen here to slow its entry into the city's stormwater system, effectively increasing its capacity. (Image courtesy of Spackman Mossop Michaels.)

the city, such as the renovation of the Rosa Keller Library in the Broadmoor neighborhood or the Deslonde Rain Garden in the Lower Ninth Ward that have been designed so that their landscapes maximize their sites' water-holding capacity through underground storage, bio-swales, detention basins, and rain gardens (Figure 14.6). The New Orleans Redevelopment Authority, which owns the city's vacant land holdings, has undertaken a number of landscape experiments to develop water-holding strategies on their unbuilt lots using grading and planting.

At the metropolitan scale, the Greater New Orleans Urban Water Plan presents a holistic water management strategy for the city. In 2010, Louisiana's Office of Community Development Disaster Recovery Unit funded Greater New Orleans, Inc., a regional economic development alliance, to develop the Greater New Orleans Urban Water Plan for St. Bernard Parish and the east banks of Orleans and Jefferson Parishes using Federal Community Development Block Grant Disaster Recovery Funds from the Department of Housing and Urban Development. A large team of local and international designers, engineers, scientists, and planners (many of them Dutch), led by the architectural firm Waggonner & Ball, was responsible for developing the plan.[15]

The Greater New Orleans Urban Water Plan proposes a rethinking of the city's approach to water management by finding ways to accommodate water within the city using green infrastructure to achieve greater resiliency. Within the context of the flood protection system, the plan proposes two major principles to guide water management: (1) to slow water down and store it wherever possible, taking pressure off the pumping system, and (2) to infiltrate as much water as possible back into the natural cycle to circulate water through the system and recharge the groundwater. Key elements in the new system are (1) making the drainage system more robust by pumping from the northern side of the city to the lake and from the southern side to the river, (2) maintaining the city's network of canals full of water to replenish the groundwater and prevent further subsidence, and (3) at low points and in vacant areas, building substantial areas of water storage to create strategic parklands and integrated wetlands (Figure 14.7). Another key aspect of the

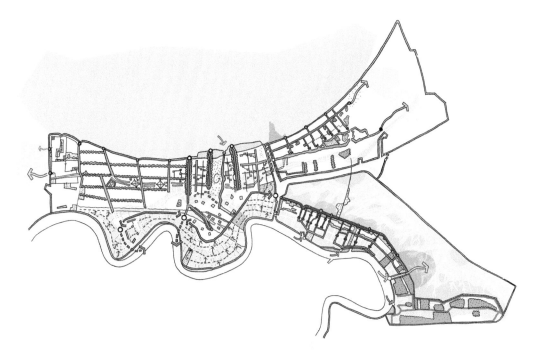

FIGURE 14.7
The Greater New Orleans Water Plan proposes a new urban structure of green infrastructure to manage water that improves and expands the existing canal system, creates a network of water storage in streets and parks, and pumps to both river and lake. (Image courtesy of Waggoner & Ball Architects.)

plan is to make this infrastructure serve multiple purposes, so that water management is incorporated into the design of new public spaces, streets, and a network of parklands, all contributing to new investment and development in the city. The plan provides useful technical background for the reengineering of the city's water management and espouses many of the pragmatic urban design moves available to the city in developing an integrated network of multi-purpose open space to support more resilient water systems. These strategies have the potential not only to provide a viable framework for future urban development and revitalization but also to increase the quality of urban life in the city. While the plan did not have official status, it was incorporated into the city's first Resilience Strategy launched in 2015 and so is now in some way part of the city's policy framework.[16] The approach it represents is already influencing city agencies, such as the New Orleans Redevelopment Authority and the New Orleans Sewerage and Water Board, in small-scale implementation of green infrastructure. The Water Plan has also led directly to the creation of the Gentilly Resilience District in the city's north, which has been funded by the U.S. Department of Housing and Urban Development as part of its National Disaster Resilience Competition. The Gentilly Resilience District projects currently underway include the Mirabeau Water Garden from the Water Plan and a number of green infrastructure projects involving rethinking neighborhood streets. The Lakeview Drainage Upgrades project illustrates the conversion of residential streets and laneways to green infrastructure with soft drainage solutions and water storage incorporated into the street profiles (Figure 14.8a–d).

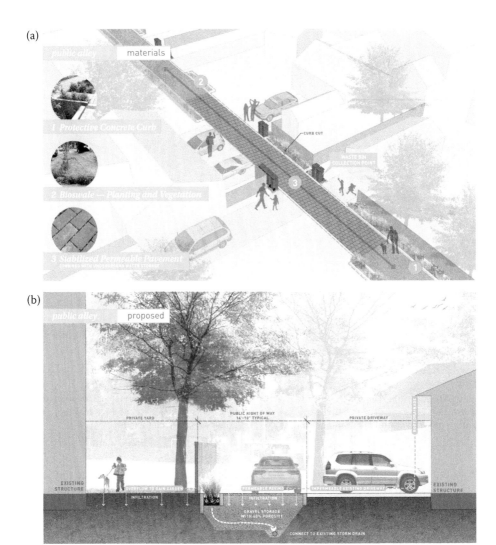

FIGURE 14.8
(a–d) Lakeview Green Streets Drainage Upgrades illustrates soft solutions and water storage designed into residential streets and public alleys in the neighborhood. (Image courtesy of Spackman Mossop Michaels.)
(Continued)

A Way Forward

As in many coastal cities, the questions of resiliency to storms and floods are forcing New Orleans's citizens to change the way they imagine urbanism. In the United States, the impact of Hurricane Sandy, affecting as it did the populous Northeast, has brought these issues much more into the mainstream. Water must be taken into account and designed into cities in a way that it has not been in the past. In the case of New Orleans, if the river can be reengineered then the delta landscape can once again become a dynamic one, with massively increased land building balanced against the inevitable, continuing land loss. In this scenario, the city becomes part of a different delta landscape, perhaps with a highly

FIGURE 14.8 (Continued)
(a–d) Lakeview Green Streets Drainage Upgrades illustrates soft solutions and water storage designed into residential streets and public alleys in the neighborhood. (Image courtesy of Spackman Mossop Michaels.)

developed eco-tourism industry working in concert with a sustainable fishing industry. In the short and medium term, the oil and gas industry would continue to be an important part of the Gulf landscape, but over the long term a different, environmentally sustainable energy and carbon industry landscape would emerge in the region. The form of the city itself would also change substantially if the infrastructure of resilient water management was integrated into its fabric and the city developed all of its waterfronts as important contributors to urban life.

In New Orleans today, there is a rediscovery of the advantages of stronger connections to the river, the lake, and other urban waterways, which offer great potential for associated urban development and improvements in city life. The influence of a newly energized urban population is evident, with new ideas about public space and urban recreation emerging in the public realm. There is also a new understanding of and focus on, the city's urban water challenges, as manifested by communities and many special interest groups. There are some really interesting grassroots educational initiatives, like the Ripple Effect project, that aims to educate elementary school children in the city about New Orleans's water issues through a combination of interactive programs and demonstration gardens on school campuses.[17] This is the beginning of the cultural shift (in many ways more significant than technical solutions) needed to catalyze essential changes. If the city's vested interests—politicians, bureaucrats, corporations—can join its citizens in embracing this approach, it could be possible to remake the urban form in such a way that all streets, buildings, and open spaces will be designed to slow and store water, and the city is framed within a network of green infrastructure.

Although the technical issues are exceedingly complex, at all of these scales there are viable solutions, and there is a substantial body of innovative work already existing and currently in progress that could drive successful new solutions. The cultural questions are more problematic, and the bigger question is whether these ideas can influence state and federal politics and corporate interests quickly enough to make a significant change to investment in coastal restoration and city development. In New Orleans, there are hopeful beginnings, but even in this small city, with a clear direction forward, and an ever-present threat, it is hard to see the concerted planning necessary to allow the city to be viable into the next century.

Endnotes

1. See Charles R. Kolb and Jack R. Van Lopik. 1958. *Geology of the Mississippi River Deltaic Plain, Southeastern Louisiana*, Technical Report 3-483, 2 vols. (Vicksburg, MS: U.S. Army Corps of Engineers Waterways Experiment Station, 1958); see also Michael D. Blum and Harry H. Roberts. 2012. "The Mississippi delta region: Past, present, and future." *Annual Review of Earth and Planetary Sciences* 40:655–683.

2. Michael D. Blum and Harry H. Roberts. 2009. "Drowning of the Mississippi delta due to insufficient sediment supply and global sea level rise." *Nature Geoscience* 2:488–491, doi:10.1038/ngeo553.

3. Craig E. Colten. 2009. *Perilous Place, Powerful Storms: Hurricane Protection in Coastal Louisiana*. Jackson, MS: University Press of Mississippi.

4. See Edward Barbier et al. 2013. *The Value of Wetlands in Protecting Southeast Louisiana from Hurricane Storm Surges*. PLoS ONE 8(3): e58715. doi:10.1371/journal.pone.0058715; and Ty Wamsley et al. 2010. *The Potential of Wetlands in Reducing Storm Surge. Ocean Engineering, A Forensic Analysis of Hurricane Katrina's Impact: Methods and Findings* 37 January : 59–68.

5. While it is clear that the city has real geographic challenges, it is important to remember that San Francisco straddles a major fault line, and many coastal cities, notably New York City and Miami, have demonstrated significant vulnerability to coastal storms and sea level rise, without calling their existence into question.

6. Richard Campanella. 2008. *Bienville's Dilemma, a Historical Geography of New Orleans*. 88–95. Lafayette, LA: Center for Louisiana Studies, University of Louisiana at Lafayette.

7. Ibid. 36.

8. An excellent description of the city's drainage infrastructure can be found at http://www.guttertogulf.com in "Water History: Three Case Studies" and "Water Today: A Taxonomy."

9. The archive of the Times Picayune has a particularly good set of diagrams that describe the hurricane's impact http://www.nola.com/katrina/index.ssf/2015/08/katrina_flooding_map.html.

10. For a full explanation of the measures see Mark Schleifstein. 2013. "Upgraded metro New Orleans levees will greatly reduce flooding even in 500-year storms." *The Times Picayune* August 16, 2013.

11. The Coastal Protection and Restoration Authority is described at http://coastal.la.gov/about/.

12. The current Master Plan and information about past and future planning endeavors can be found at http://coastal.la.gov/a-common-vision/2012-coastal-master-plan/.

13. See the Data Center's Coastal Index http://www.datacenterresearch.org/reports_analysis/the-coastal-index/.

14. Elizabeth Mossop. 2014. "Changing course: Redesigning the Lower Mississippi delta." *Topos* 87:70–75.

15. All of the documents that make up the Greater New Orleans urban Water Plan can be found at http://livingwithwater.com/blog/urban_water_plan/reports/.

16. The strategy Resilient New Orleans can be found at http://www.resilientnola.org.

17. The Ripple Effect program and its projects is described at http://rippleeffectnola.com.

15

*The Giving Delta**

Jeffrey A. Carney, Robert R. Twilley, Claire Agre, Jonathan Hird, Ioannis Georgiou, and Jeff Shelden

CONTENTS

Introduction: Background and Driving Forces

Coasts are tenuous lines where water and land meet. Human settlements occupying these edges reap the many benefits of this location but must also cope with risks associated with coastal storms and increasingly the effects of climate change. For settlements in coastal deltas, the stakes are even higher. Around the world, deltas hold a disproportionate concentration of population working and living in global cities and mega-regions. As many as 500 million people currently inhabit these regions, drawing benefit from their ecological, agricultural, and industrial productivity (Vörösmarty et al., 2009; Giosan et al., 2014). These deltaic environments face significant threat from land subsidence and sea level rise, but from the Nile River delta, to the Mekong Delta, the wealth from navigation, commerce, and energy has outweighed the risk (Syvitski et al., 2009). In the Lower Mississippi River Delta, a century of constricting the river through levees to limit riverine flood risk and to safeguard navigation has resulted in massive land subsidence that threatens communities

---

* Special mention to all members of the team: Moffatt & Nichol | West 8 | LSU Coastal Sustainability Studio | Deltares | RAND | Ioannis Georgiou | Headland & Associates.

across the coast, including the city of New Orleans (Twilley et al., 2016). Coastal Louisiana has reached a breaking point. As the past decade of hurricane damage has made evident, the region is becoming exposed to an ever-increasing risk from the encroaching Gulf waters, increased storm intensity, and rising economic instability. Without direct focused action to transform the conflict between settlement and environment, Louisiana will see a decline in its resilience to recover from events, rapidly losing national standing in economic and social viability (Figure 15.1a–c).

This framework plan, developed as one of three winning proposals for the Changing Course Design Competition, proposes a vision for how to reinforce the physical environment against the threat of accelerated sea level rise, while maintaining—and even expanding—long-term economic and social opportunities that will allow the people and businesses of southern Louisiana to continue to thrive and prosper. A relentless program of river control has guided Louisiana and the U.S. Army Corps of Engineers (USACE) for nearly a century (Barry, 1997; Reuss, 2004). The seasonal flooding from the Mississippi River, that built and nourished the delta landscape over thousands of years, has effectively been contained

FIGURE 15.1
The dynamic deltaic coast is Louisiana's greatest opportunity. Using the power of the Mississippi River, Louisiana's ecosystem, economy, communities, and culture will adapt to increasing uncertainty. (a) Cities connected along the I-10 Corridor. (b) The Mississippi River crosses on its path to the Gulf. (c) The nexus of ecology and infrastructure. (Moffatt & Nichol + West 8 + LSU Coastal Sustainability Studio.)

behind a continuous chain of levees. By challenging this paradigm, Louisiana can realize an opportunity that most other coastal regions do not have: that is, the power of a dynamic river to build and maintain land against the rising seas. While other cities are exposed on fixed coastal edges, Louisiana can leverage the power of the Mississippi River to sustain a wetland floodplain that protects and promotes its industrial economy, sustains ecosystem productivity, and nourishes human occupation.

What If We Maintain the Status Quo?

The map of the Louisiana "Boot" is an icon and represents the pride and living culture of a people who have, for centuries, worked and lived in this unforgiving and unpredictable landscape. Over just a few generations, the people of coastal Louisiana transformed this deltaic environment from harsh and unpredictable, into a valuable working and industrial landscape. The icon seen on road signs across the state shows a solidified coastal boundary set defiantly against the Gulf of Mexico (Anderson, 2016).

Today more than ever, the static icon of the State of Louisiana belies the true appearance of a rapidly transforming landscape. Attempts by past generations to juxtapose a static boundary in this shifting landscape have suppressed the essential dynamism that nourishes and supports the delta ecosystem's health. The coast has lost 466,198 hectares of wetlands over the last century, and we are now poised to see this trend exacerbated due to new threats over the coming decades (Barras et al., 2008; CPRA, 2012). As wetlands erode into open water, communities are witnessing the profound dissolution of this protective buffer that protects them from storm surge. The landscape is changing so quickly that differences are not measured in generations but in five- or ten-year intervals (Ibid.). Even with the State of Louisiana's 2017 Comprehensive Master Plan for a Sustainable Coast, which includes plans for hundreds of large-scale engineering and restoration projects to rebuild marshes and protect communities, this landscape will continue to decline, unable to keep up with accelerated sea level rise. The severe constriction of the river with hundreds of miles of continuous levees starves the deltaic wetlands of needed pulses of sediment and freshwater. Masterplan projects cannot adequately compensate for the significant reduction of sediment in the delta system without a radical redistribution of the river's sediment budget. Without this fundamental building material, efforts to restore coastal land will eventually fail and the processes of coastal erosion, land subsidence, and sea level rise will continue to overtake the coast.

Alternatively, relinquishing control of the Mississippi River by removing the levees, allowing the river to flood freely, is not a viable approach to management of the river either. It is almost certain that the Mississippi River would transition its primary flow to the Atchafalaya River, causing sudden and calamitous effects for stakeholders locally, nationally, and globally. The city of New Orleans would be without drinking water within days and over time the port of Louisiana, the largest port in the world by tonnage, would likely be restricted to a silted up slack water channel. Situated between these two poles of river management paradigms, there must be a measured approach to river engineering that maintains the valuable economic and public infrastructure that supports the working coast while reenabling the ecological function of the system. The river's dynamic processes are key to the coast's long-term survival and adaptation to climate change.

The Opportunity of the River and Delta

The delta cycle is the process where a river builds land through sediment accumulation along its course in a basin, forcing the river to switch course to find a shorter route to the sea (Roberts, 1997). A deltaic basin experiences its highest biological productivity long after the primary freshwater supply from the river that built it has switched to a different primary outlet. In fact, fisheries are most productive in estuarine basins that are adjacent to the main river course, receiving pulsed river flows from secondary distributaries even after the main distributary has switched to another basin. If the secondary distributary completely abandons a coastal basin, the biological productivity of the estuary's peak precedes an eventual habitat collapse that will be experienced when the estuarine wetlands transitions to open water. Across the arc of the delta's lifecycle, the delta cycle balances times of land building, times of peak net biological productivity, and times of estuarine wetland collapse. Over time this process will build land and extend the delta further into open water, although this process is never constant (Roberts, 1997; Paola et al., 2011). The inherent challenge to managing the river and delta to include stable human habitation is to distribute sediment to adjacent estuaries without undermining the coupled social, economic, and infrastructural system's capacities to adapt to change. The Giving Delta project provides a vision for reengaging the Mississippi River's seasonal flood pulse to build and maintain coastal wetlands within a management framework that allows for human adaptation to estuarine transgression over time.

Anticipated climate change induced increases in precipitation in the upper watershed of the Mississippi River Basin will increase river flooding in the Lower River basin (Barros, et al., 2014). The potential negative impacts of increased flood risk actually *enhance* the strategic opportunity for the Lower Mississippi River Delta plain, offering more sediment to the Louisiana coast during flood events. Strategic adaptation to this uncertain "risk" will enable the delta to derive benefit from extreme events, distributing sediment to the landscape to help balance the encroachment of the Gulf from sea level rise. This proposal changes the mind-set from a twentieth century model of risk management through "flood control" to a new paradigm enabling the opportunity for "controlled floods." By mimicking natural river pulses that enhance seasonal salinity gradients in estuaries, this strategy allows for the self-organization of productive fisheries over the long term, as the coastal basins reorganize into distinct estuarine zones.

A Systems Approach to Delta Design

Trade-offs in complex systems at the scale of the Mississippi River Delta require robust systems analysis of interacting natural and social infrastructure that involves the flows of water, energy, materials, people, and money. This analytical approach establishes values that can steer management and planning options related to present and future risks and opportunities (Meadows and Wright, 2008). The problem with systems analysis in practice is that non-market values of natural systems, or ecosystem services, particularly those in large dynamic landscapes such as the Mississippi River Delta, are not appropriately valued by society until a major disturbance draws public awareness to their fundamental importance. Such is the case of the Mississippi River Basin where a history of competing projects designed to control floods and promote the economic development associated with navigation, have incrementally undermined the overall ecosystem stability and with it, the long-term value of the system in its totality (Figure 15.2).

To overcome the dominance of a compartmentalized economic outlook, systems analysis links multiple-purpose needs of diverse stakeholders. The Giving Delta framework is predicated on a robust systems-based analysis that nests scaled analysis of the overall river network, delta plain ecosystem, and specific coastal estuarine basins. This project conceptually modeled three layered frameworks: Ecosystem, Community, and Economy. Through the overlay and cross-referencing of the three, the team arrived at the final framework proposal.

Ecosystem Framework: Connect the River to the Delta

The proposal establishes an adaptable system to reconnect the Mississippi River to its delta through a series of controlled flood outlets. These diversions will move sediment and freshwater to relieve the Mississippi River when it is in flood stage. This approach will reconnect the natural flood pulse to the coastal landscape returning it to a dynamic balance (Day et al., 2007). Controlled flood pulses from a series of river outlets build and maintain wetlands with additional sediment delivery and shifts seasonal salinity regimes within parameters that promote fisheries and retain nutrients to reduce offshore hypoxia. Controlled flood structures supply coastal wetlands with freshwater, sediment, and nutrients that enhance productivity, vertical accretion, and marsh stability. This design provides sediment and reduces salinity to promote wetland adaptation to accelerated sea level rise in these coastal basins. Additional engineered features that move sediment with sand engines, pipeline conveyance, sediment traps, sediment accretion zones, dedicated beneficial dredging, and other techniques create the platforms that emergent wetlands can subsequently colonize (Figure 15.3).

Controlled flood structures are positioned upriver and just downstream of New Orleans. These controlled flood structures will provide increased flood control capacity for New Orleans and provide freshwater and sediment delivery to upper coastal basins. The location of these structures in an upstream location provides more effective management of freshwater delivery from one generation to the next as a forward-thinking adaptation to climate change that will most certainly increase flooding in the Lower Mississippi River Delta plain. This adaptive management approach to river engineering will also accommodate the need for freshwater and sediment in coastal basins in response to accelerated sea level rise. Choosing to operate a series of controlled flood structures is one of the most important decisions that achieves the multiple goals of wetland maintenance, nutrient reduction, and enhanced estuarine productivity in a delta cycle. The strategy is to utilize a flood-pulse operation of all the flood control structures proposed. The restriction on the duration of the flood-pulse is an annual average of 60 days during the spring flood, operated over decadal periods, but also allowing for an annual 'estuarine recovery' period to promote fisheries productivity.

Community Framework: Link Communities and Infrastructure

Human settlement in the Louisiana Gulf Coast region is framed by two overlapping systems of development. First, the historic linear bayous and Mississippi River support development that reaches from the relative high ground of the Pleistocene ridge to the north around Baton Rouge and extends toward the Gulf of Mexico to the south. These traditional communities hold tight to the relative high ground directly adjacent to the rivers and bayous.

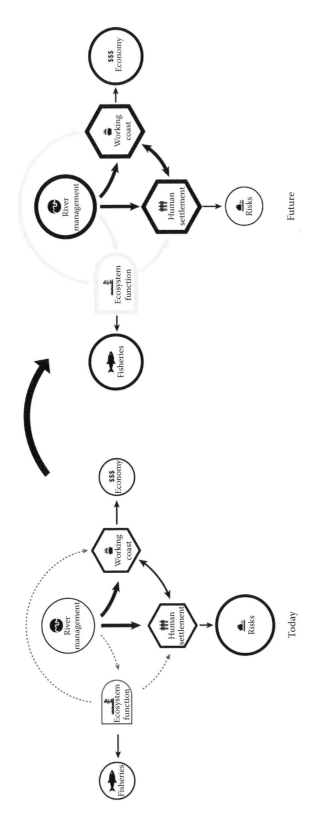

FIGURE 15.2
The Giving Delta proposes a reconnection of ecosystem function as a primary role of the Mississippi River. (Moffatt & Nichol + West 8 + LSU Coastal Sustainability Studio.)

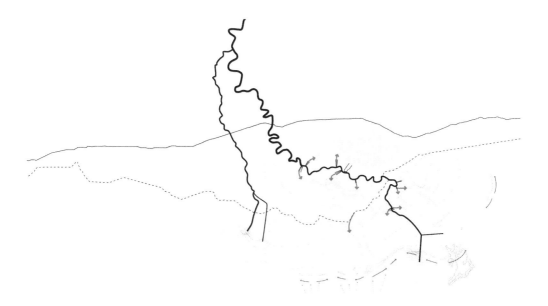

FIGURE 15.3
Ecosystem function: Connecting the river to the basin. (Moffatt & Nichol + West 8 + LSU Coastal Sustainability Studio.)

Second, the contemporary I-10 corridor or Gulf Coast mega-region links communities from Mobile, Alabama to Houston, Texas, and beyond through an east/west linear pattern of contemporary automobile-dominated development. The elevated system of highways has spawned significant development where the I-10 corridor links the traditional high-ground along the coast; areas such as Lafayette and Baton Rouge (Carney, 2013).

The Giving Delta refocuses these two linear land use typologies, concentrating contemporary resilient and adaptable development along a zone bounded by the I-10 and US-90 highway corridors bisected by the industrial river corridor and traditional ridgeline settlements stretching to the south. Emerging at the intersections of these systems are dense urban nodes, "Delta Cities," that thrive behind and within a renewed marsh zone. A rigorous concentration of development patterns closely aligned with transportation infrastructure will allow a "hyper-use" of investment on river and road transportation to serve both commerce and community needs (Figure 15.4).

The I-10 corridor serves multiple significant roles structuring settlement for the next 100 years. The corridor is located along the Pleistocene ridge that runs north of the modern Mississippi River Delta. This ridge makes the corridor a refuge above the rising sea levels and high rates of land loss in the shifting soils of the delta. According to a report by the Federal Highway Administration and the U.S. Geological Survey on the impacts of climate change, "A 20′ (6.096 meter) storm surge applied across the coast would flood 100% of the roads south of the I-10/I-12 (north shore) corridor" (Savonis et al., 2008).

As sea levels rise and much of the delta continues to transgress, communities south of the I-10 will consolidate behind protection and elevate in areas with increased flood risk (Twilley et al., 2016). Maintaining a strong connection to the Gulf of Mexico for international trade and the ongoing development of offshore oil and gas resources will provide critical links for the nation's economy for many decades to come. These communities include the parallel development along the US-90 and the future I-49, connecting Lafayette, Morgan

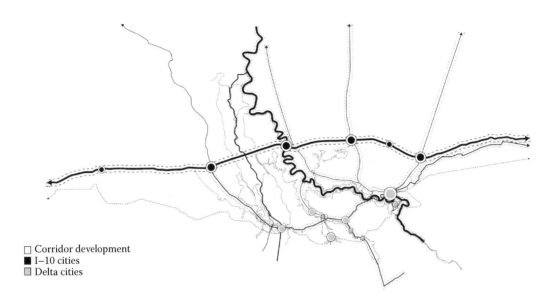

☐ Corridor development
■ I–10 cities
☐ Delta cities

FIGURE 15.4
Community system framework: Link communities and infrastructure, particularly in three delta cities. (Moffatt & Nichol + West 8 + LSU Coastal Sustainability Studio.)

City, Houma, and New Orleans. This highway connects the communities that form the headland of the reinforced wetland zone fronting the Gulf.

Settlement in the wetland zone between the I-10/I-12 corridor and Gulf Intracoastal Waterway (GIWW) will support emergent concentrations of settlement at the interface of east-west highway transport and north-south river transport. Morgan City, Houma-Thibodaux, and New Orleans to the south, and Lafayette, Baton Rouge, and the North Shore of Lake Pontchartrain communities to the north, provide the integral crossing points between the I-10 mega-region and the bayous and rivers linking Louisiana to the ecosystem services of the Gulf of Mexico. These features form infrastructural spines, that provide a foundation for the reemergence of the culturally historic long lot, or "arpent" system, that once allowed an individual family to access transportation along the bayou, a place to build a house on the natural levee, space to farm, places to raise cattle, and, as one continues further into the depth of the rural fabric into the back swamp, places to trap and fish.

This layered framework will help the region to transform and advance over time through a deliberate engagement with the local ecosystem, balanced and integrated with the mega-regional network. Ecologically linked and engaged communities connected by a network of rapid and dependable transportation will provide increased access to jobs, recreation, and flexibility in an increasingly dynamic ecological system.

Economy Framework: Consolidate Investment According to Infrastructure and Geologic Realities

New Orleans is situated at one of the most powerful strategic positions in the United States. Controlling the mouth of the Mississippi River, the terminus of six Class I railroads, and a central section of the I-10 corridor, this region has potential to be innovative in the way it moves and processes goods by water, rail, pipeline, and road. Bulk processing, oil and gas transport, and investing in cargo are all likely areas of continued investment and expansion for ports in this region. Maintained access to the Gulf for future oil and gas exploration will also maintain a competitive advantage for this region. However, as the navigation and petroleum industries continue to change, diversify, even diminish over the next century, the port system must be designed to adapt to a number of different scenarios, even sudden changes due to a storm, or as seen in 2010, a catastrophic oil spill.

In order to capitalize on a strategically advantageous location, the ports of southern Louisiana must look ahead to what physical measures are required to stay commercially competitive. 16.764 m deep draft channels to accommodate post-Panamax shipping, improved rail access, and diversified port complexes including bulk processing and cargo located closer to where people live and to the inland transportation network, will bring the coastal zone into greater global competitiveness (Figure 15.5).

Central to the Giving Delta is the development of an ambitious new port complex north of a redirected Mississippi River alignment, in the vicinity of Port Sulphur. This new port facility will replace structures to the south of the new deep draft river channel that are either abandoned, not replaced at the end of their current design lives, or not rebuilt after catastrophic storm damage. Currently, the infrastructure and associated economic investments along the lower section of the Mississippi River, are located along a lengthy, but relatively narrow, linear band; primarily on the West Bank. As a result, they require more costly linear levees to protect them, and they are susceptible to being inundated if any section of the levee is breached or if flooding occurs in this region. Additionally, the further down the river they are located, the more exposed they are to direct impacts from tropical storms. Hence, they are, and will become, even more vulnerable to disruption from tropical storms and river flooding events. An additional important consideration is that this location is upriver from an area with significantly higher subsidence rates which, combined with the potential increases in sea level, will result in those areas requiring substantial increases in levee heights and building elevations over time. Port Sulphur's location coincides with the area where subsidence decreases significantly, especially as

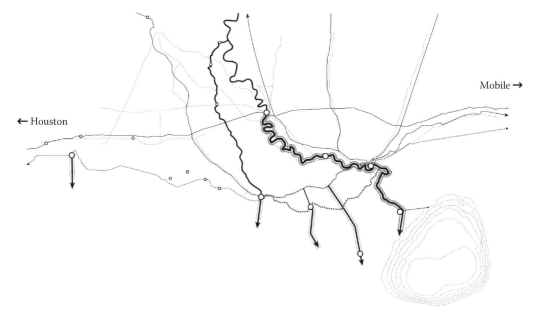

FIGURE 15.5
Economic framework: Consolidate investments according to infrastructure and geologic realities. (Moffatt & Nichol + West 8 + LSU Coastal Sustainability Studio.)

compared to areas further south, which currently incur subsidence of up to 22–24 mm each year (Ayres, 2012; CPRA 2013; Georgiou et al., 2016).

As previous studies have documented, dredging the existing channel at Southwest Pass to a depth of only 15.240 m would have a total economic net benefit to the State of Louisiana equivalent of over US$5 billion over the first generation alone (first 25 years) including the dredging costs to maintain the channel. Additional secondary economic benefit to the nation would be an equivalent of a further US$8 billion per year. It is anticipated that the additional national economic impact of a 16.764 m channel to accommodate post-Panamax vessels as part of this plan would be substantially higher (Ryan, 2013).

To support this more intensive industrial zone, upgrades will need to be made to the rail and highway system along the West Bank to ensure safe, efficient, and protected means of transport of people and materials. These transportation upgrades, though, should be seen as part of the larger need for a resilient transportation network including the GIWW and other shallow-draft channels, linking the various economic zones of coastal Louisiana. These systems must be made less vulnerable to extended disruptions due to storm events in order for these areas to prosper. The economic importance of the Mississippi River Delta region is dominated by the role the river plays in navigation for the state and interior of the nation. While navigation routes would adapt to the changing course of the river, the project design is centered on maintaining and improving, navigability of the lower Mississippi River. Access to the Port of New Orleans will be improved with shorter navigation times upriver and deeper drafts.

Project Phasing

The challenges that we face today are often presented as an unpredicted consequence of progress. However, for over a century, scientists and policy makers have understood the risks of the river's confinement by levees. The decision to prioritize riverine flood protection above all other uses of the river was a decision made with eyes wide open. An 1897 *National Geographic* article puts it plainly, "No doubt the great benefit to the present and two or three following generations accruing from a complete system of absolutely protective levees, excluding floodwaters entirely from the great areas of the lower delta country, far outweighs the disadvantages to future generations from the subsidence of the lower Gulf delta lands below the level of the sea" (Corthell, 1897). Thus, a decision was made, one that generated great prosperity for four generations, but with a validity whose time has come to be reconsidered. The consequences of that choice for this and future generations are now apparent, and the negative impacts will soon outweigh the benefits, if they have not done so already.

The communities along the Mississippi River grew tremendously over the last 100 years. Over the course of four generations, residents and businesses adapted to the negligible risk of riverine flooding introduced by the levee system up and down the Mississippi River. The continuing risk of storm damage tempered these changes along the Lower River, leaving this area more exposed to the impacts of storms, but resulte in the consequence of becoming more resilient and capable of adapting as the coastal environment changed. In a future without action, these communities would face relative sea level rise and suffer increased storm damage, forcing residents and businesses to choose between moving further inland or elevating and reinforcing their homes and businesses. The Giving Delta framework plan lays out a vision for restoring the environment and reintroducing the natural river dynamics into parts of Louisiana's coast. However, by stretching this plan across the next four generations, we leverage the natural ability of human settlement to adapt and change to this long-term vision. While this and future generations will inevitably face considerable challenges in carrying out a plan for changing the River on this scale, the great benefit far outweighs the disadvantages.

First Generation: Connect—From Flood Control to Controlled Flood Pulses

Connecting the Mississippi River and reestablishing historic and surrogate freshwater and sediment pathways back to the delta/estuarine complex is of critical importance and the first order of business. With the Mississippi River draining 41% of the contiguous United States, the strategic opportunity to embrace the river to maximize freshwater and sediment flow into the interior of the basins in order to combat sea level rise must be seized upon immediately. During the first generation, the controlled flood and multi-height spillway structures will be constructed, as well as the settling basins and sediment traps. In addition, the GIWW will be utilized for inter-basin management of flows, and South Pass and Pass A Loutre will be closed to ensure sufficient flow continues through Southwest Pass when the upriver controlled flood and spillway structures are operational. The power of controlled flood pulses will allow the deltaic coast to adapt to sea level rise through input of sediment into the system and as salinity gradients are expanded through increased upriver controlled flooding into the upper basins. The delta cycle of river reoccupation and river abandonment is then harnessed to enhance diverse levels of estuarine productivity across the coastal basins by allowing river floods to organize ecosystem development.

Second Generation: Align—Transform the Lower River

The mouth of the Mississippi River is not sustainable. Uninterrupted deep draft navigation cannot be guaranteed through the Southwest Pass and thus an alternative deep draft navigation channel entrance into the Mississippi River and the inland waterways of the nation is constructed at Port Sulphur during the second generation. The Southwest Pass navigation channel is abandoned, a process the Mississippi River has already begun and which will continue to increase in rate with increasing relative sea level rise, thereby transforming the Lower Mississippi River and its mouth from what we know today. The Southwest Pass delta lobe is mined for sediment and the material placed in the littoral zone south of West Bay. As the mouth of the Mississippi River is reestablished south of Port Sulphur, the material excavated from the new channel is beneficially used to establish interior marsh platforms and to reestablish and nourish the barrier island chain. Establishment of sediment traps in the new navigation channel will fuel sand engines located at the intersection of the new navigation channel and littoral zone allowing littoral processes to transport deltaic deposited sediments to sustain the Barataria Bay shoreline.

Third Generation: Adapt—Consolidation and Economic Dividends

By the commencement of the third generation, the strategies and tactics implemented in order to change the way the Lower Mississippi River is operated and maintained are now fully operational. The natural and human systems are responding and adapting to the implementation of the new paradigm. The basins are adapting seasonal salinity gradients, as a result of the implementation of a closer relationship of the basins to the dynamic state of the river. A broad, dynamic wetland zone has been established across a significantly concentrated delta zone. Port infrastructure and the economic activity and associated support infrastructure is consolidated and solidified, providing for a globally competitive port region, centered on Port Sulphur and aligned along a zone of economic concentration along the Mississippi River corridor north of Port Sulphur. The footprints of the more vulnerable communities located on the extremes of the buffer zone are reduced. There is strategic development of dense communities, concentrated along current and future areas of high ground and infrastructure.

Fourth Generation: Position for Change

In 75 to 100 years from now, the Mississippi River Delta plain and Louisiana coastal zone will look very different from how they look today. If we act boldly, we can turn the risks of the delta into our greatest asset and return the delta landscape into a protective, sheltering environment for generations to come. The fourth generation of this plan depicts a place that has adapted to change and is poised to continue to adapt in the face of changing climate and economies. We have adapted to increased fluctuation in the Mississippi River. We have shortened the length of the river and increased the flow of sediment into our wetlands while improving the dependability and capacity of navigation. We have provided a clear future roadmap for communities, industry, and the workforce (Figure 15.6).

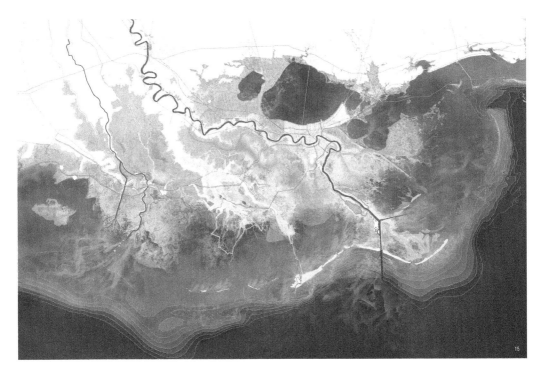

FIGURE 15.6
The 100 year vision for the Lower Mississippi River Delta. (Moffatt & Nichol + West 8 + LSU Coastal Sustainability Studio.)

Stakeholder Conflict and Policy Implications

The dramatic transformations to the management of the Mississippi River proposed in this framework plan will undoubtedly require significant changes to the way that people live in this landscape, how businesses operate, and the way policy is written. To implement such bold actions, it is essential that they are matched with clear directions and options for communities and industries to adapt. Producing a long-term vision for change will reduce the public anxiety over large scale projects, build mutual trust, and enable projects to move forward in an equitable manner.

Increased sea level rise and storm intensity exposes people to higher risk. New approaches to planning and building resilient and adaptive communities are central to this framework. A widespread effort toward enhancing the resilience of whole communities and their regional context and networks will be essential, not only because it will reduce costs and risk in the long term, but because it will enable the state to accomplish the ecological projects with significantly less public anger and legal dispute. Resilience planning augments traditional land use planning with information and expertise from various fields, particularly disaster science, floodplain management, climate, and environmental sciences. The processes and outcomes of resilience planning build stronger communities that are better equipped to respond and adapt to disturbances. The integrated efforts of experienced planning professionals in close collaboration with the Coastal Protection and Restoration Authority will be essential for success (Figure 15.7).

This project suggests a mix of regulatory and incentive-based approaches to both establish a central policy and administrative structure that will reduce community conflict and replication of efforts but will encourage communities to innovate and maximize the benefit of traditional ecological knowledge and local economic development. The goal of planning is to enable the dramatic and large-scale projects proposed in this framework, to be implemented as effectively as possible. Efficient planning will enable communities and citizens to adapt to change given adequate time or to work through state-funded buyouts to relocate to a different place where risk is lower or protections are in place.

The framework plan is laid out over multiple generations in order to allow time for these coastal communities to adapt. While some of that adaptation would have happened naturally during the first generation without the framework plan, the state can take steps to facilitate that adaptation and alleviate the costs in cases where the framework plan requires more immediate action. A trust fund will be developed for residential property purchases, and a first right of refusal for the state will be established for properties south of the new channel and in portions of the Barataria Basin that will eventually be impacted by the influx of freshwater into the basin. Because these properties can be bought out over many years, one way of thinking about those costs in present dollars is to consider establishing a trust fund to acquire residential property so that the investment return of that trust fund in addition to the principal, would provide sufficient funds to buy up a small portion of the properties in each year as they come on the market due to natural turnover and housing churn.

Social upheaval and resistance to major changes in the landscape and human habitation will undoubtedly occur if the plan described herein is conducted. The key to navigating this resistance is awareness of cultural sensitivity, forthrightly assessing and communicating the impacts to humans, and using all means available to mitigate those impacts. All tools available, including financial, social, political and legal will be necessary to achieve public acceptance of this bold plan.

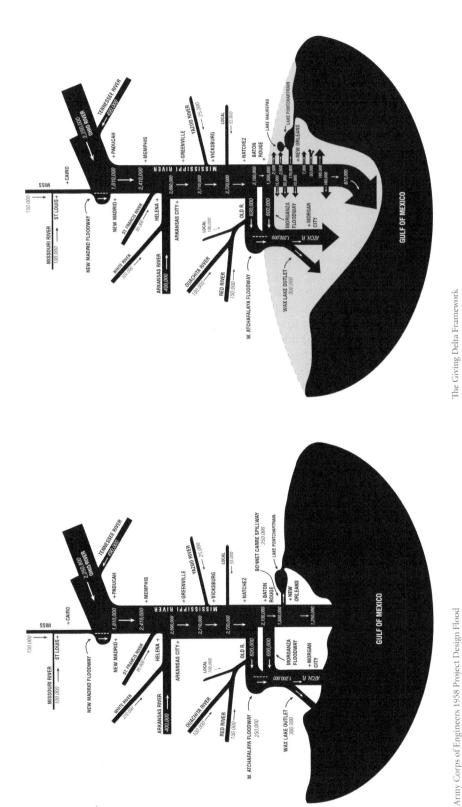

Army Corps of Engineers 1958 Project Design Flood

The Giving Delta Framework

FIGURE 15.7
Building upon the USACE 1958 Project Design Flood, the Giving Delta framework plan reconnects the river to the delta. (Moffatt & Nichol + West 8 + LSU Coastal Sustainability Studio.)

FIGURE 15.8
View from a resilient marsh along the renewed infrastructure of the Lower Mississippi River. (Moffatt & Nichol + West 8 + LSU Coastal Sustainability Studio.)

The Reality of Implementation Must Be Bold

The Changing Course Design Competition asked teams to think outside the box. This charge has led to the bold ideas developed as part of The Giving Delta (Figure 15.8). However, over the arc of this process, this competition entry and the entry of other teams, make the point that there is no single silver bullet available to solve this challenge. Coastal Louisiana has seen numerous singular ideas without focusing on the greater need for a comprehensive framework where projects must work in concert to achieve the ultimate goals.

This project presents a vision of a 100-year future for the Mississippi River Delta, the Louisiana coast, the communities that exist here, and the economy that sustains them. A bold vision is matched with a strategy for implementation that will benefit each generation over the next century. However, the ability to implement bold ideas, to break outside of a deeply ingrained paradigm, will be far more difficult than conceiving of the ideas themselves. The major challenge will be the ability for policy makers to convert this vision and its overarching framework into the policies, tools, and practices to reconnect the many stakeholders in a system in line with the strengths of the Mississippi River.

References

Anderson, B. 2016. Louisiana Loses Its Boot. https://medium.com/matter/louisiana-loses-its-boot-b55b3bd52d1e (accessed December 07, 2016).

Ayres, S. 2012. *Atlas of U.S. Corps of Engineers Historical Daily Tide Data in Coastal Louisiana.* LCA Science and Technology Office Report. 1–22.

Barras, J. A., Bernier, J. C., Morton, R. A. 2008. "Land area change in coastal Louisiana—A multidecadal perspective (from 1956 to 2006)." U.S. Geological Survey Scientific Investigations Map 3019, scale 1:250,000. *Pamphlet* 61(12):127–142. http://pubs.usgs.gov/sim/3019 (accessed August 23, 2010).

Barros, V. A., Field, C. B., Dokke, D. J. et al. 2014. *Climate Change 2014: Impacts, Adaptation, and Vulnerability - Part B: Regional Aspects.* Contribution of Working Group II to the Fifth Assessment Report of the Intergovernmental Panel on Climate Change.

Barry, J. M. 1997. *Rising Tide: The Great Mississippi Flood of 1927 and How It Changed America.* New York, NY: Simon and Schuster.

Carney, J. 2013. "Gaining ground: Structuring settlement in the uncertain economic and climactic landscape of the gulf coast mega-region." Ed. Berman, I., Mitchell, E. *New Constellations, New Ecologies. Proceedings of the 2013 National Meeting of American Collegiate Schools of Architecture,* San Francisco, CA, Yale University Press.

Corthell, E. J. 1897. "The delta of the Mississippi River." *The National Geographic Magazine* VIII(12):351–354.

CPRA. 2012. *Louisiana's Comprehensive Master Plan for a Sustainable Coast.* Baton Rouge, LA: Louisiana Coastal Protection and Recreation Authority, Editor. http://coastal.la.gov/a-common-vision/2012-coastal-master-plan/.

CPRA. 2013. Process for addressing relative sea-level rise in the feasibility studies for the projects in the state comprehensive master plan.

Day, J. Jr. et al. 2007. "Restoration of the Mississippi delta: Lessons from hurricanes Katrina and Rita." *Science* 315:1679–1684.

Georgiou, I.Y., Gaweesh, A., Hanegan, K., T. Yocum. 2016. "*Flow distribution and salt-wedge propagation in the lower Mississippi River and Delta.*" Final Report Submitted to the Coastal Protection and Restoration Authority, as a cost-share contribution to the LCA Mississippi River Hydrodynamic and Delta Management Feasibility Study. Contract No. 2503-12-17.

Giosan, L., Syvitski, J., Constantinescu, S., Day, J. 2014. "Climate change: Protect the world's deltas." *Nature, International Weekly Journal of Science.* http://www.nature.com/news/climate-change-protect-the-world-s-deltas-1.16428 (accessed December 07, 2016).

Meadows, D. H., Wright, D. 2008. *Thinking in Systems: A Primer.* White River Junction, VT: Chelsea Green Pub.

Paola, C. et al. 2011. "Natural processes in delta restoration: Application to the Mississippi delta." *Annual Review of Marine Science* 3:67–91.

Reuss, M. 2004. *Designing the Bayous: The Control of Water in the Atchafalaya Basin, 1800–1995.* College Station, TX: Texas A & M University Press.

Roberts, H. H. 1997. "Dynamic changes of the Holocene Mississippi River delta plain: The delta cycle." *J. Coast. Res.* 13:605–627.

Ryan, T. 2013. Study: Deeper Mississippi River to Add USD 11.5 B to U.S. Economy. http://media.nola.com/business_impact/other/Mississippi%20River%20deepening%20economic%20study.pdf (accessed December 07, 2016).

Savonis, M. J., Burkett, V. R., Potter, J. R. 2008. Impacts of climate change and variability on transportation systems and infrastructure: Gulf coast study, Phase I.

Syvitski, J. et al. 2009. "Sinking deltas due to human activities." *Nature Geoscience* 2:681–686. doi: 10.1038/ngeo629.

Twilley, R. R., Bentley, S. J. Sr., Chen, Q. J., Edmonds, D.A., Hagen, S. C., Lam, N., Willson, C. S. Xu., Braud, D., Peele, R. H. 2016. "Co-evolution of wetland landscapes, flooding, and human settlement in the Mississippi River delta plain." *Sustainability Science* 11(4):711–731. doi: 10.1007/s11625-016-0374-4.

Vörösmarty, C. J. et al. 2009. "Battling to save the world's river deltas." *Bulletin of the Atomic Scientists* 65(2):31–43.

16

A Fine Dutch Tradition in the Noordwaard

Robbert de Koning

CONTENTS

Fine Dutch Tradition

The Netherlands, with two thirds of its land vulnerable to flooding by rivers and the sea, has a very long tradition of struggling against water (Figure 16.1). Levees were already being built here in the Middle Ages, first to channel the water, later to protect the lands that lay behind them. In time, these levees were connected to form levee rings. The areas behind the levees were reclaimed through drainage and later also with the use of wind power. And so, the polder was born.

The Dutch have always lived with water; and they have always had a love-hate relationship with it. Floods have taken many lives, but the water has also allowed the country to make a fortune through trade and fishing. The water also provided the country with fertile lands for its profitable agriculture. The delta was a place for enterprising people to live and work. Besides tulips and cheese, the Netherlands is known for clever water management (Figure 16.2). Engineering has turned into landscape art, even world heritage, such as the windmills of the Kinderdijk and the Beemster polder (Figure 16.3).

The presence of water in the landscape and the need to control it has resulted in a range of characteristic images, such as the long lines of ditches, the clean-cut edges of the levees, the pumping stations. The result are beautiful, rational landscapes.

This way of dealing with water is also called the Fine Dutch Tradition. It refers to the dynamic tradition of creating spatial designs by efficiently integrating hydrological

FIGURE 16.1
The flow area of the de-poldered Noorwaard at average high water. (Image by Robbert de Koning.)

FIGURE 16.2
World Heritage Kinderdijk, with rows of windmills along the water. (Courtesy of depositphotos.com.)

characteristics, soil properties, and existing technology. In the Dutch history of water management and urban planning one can find good examples of how this was done up until the Industrial Revolution. In the twentieth century, manipulating technology was used to sidestep the natural system. Nowadays, the Netherlands is looking for a new relationship between technology, natural systems, and spatial design to be able to respond to the changing climate.

FIGURE 16.3
World Heritage Beemster Polder, with long lines of ditches and tree-lined roads. (Courtesy of depositphotos.com.)

Water Management in the Netherlands

Urgency

It is sad, really, that it almost always takes a disaster to convince people that making changes to the water system is of vital urgency. After major floods in 1916 along the Zuiderzee, a large, shallow inlet of the North Sea in the heart of Netherlands, it was decided to build the Afsluitdijk (enclosure dam, 1932), a 32-kilometer dam between the provinces of Friesland and North Holland. This dam transformed the saltwater Zuiderzee into a large freshwater lake, the IJsselmeer, and reduced the risk of flooding for the cities and polders along the Zuiderzee coast. In the 1953 North Sea flood, nearly 2,000 people died in the southwest of the Netherlands. This led to the drafting and implementation of the Delta works (1954–2010). As part of these Delta works, many dams and levees were constructed in the provinces of Zeeland and South Holland. In many places, the tide disappeared as a result and saltwater areas turned into freshwater ones, leading to ecological decline. At a few locations, a new balance between water safety and ecology was found, such as the Oosterscheldekering (Eastern Scheldt storm surge barrier) with its characteristic moveable sluice gates (Figure 16.4).

In 1993 and then again in 1995, disaster almost struck. The water of the river Waal nearly rose up to the levee crown and 100,000 people were evacuated from the adjacent polders. In the end, not one levee was breached, but it was a wake-up call.

Room for the River

This high water in the 1990s prompted the Dutch government to install various advisory water management committees. In 1991, based on the climate scenarios of the Dutch

FIGURE 16.4
Eastern Scheldt storm surge barrier. (Image from beeldbank.rws.nl, *Rijkswaterstaat.*)

meteorological institute KNMI, the committee for Water Management in the Twenty-first Century assessed the impact on the indicative river water discharge. The committee used the average climate scenario for 2100 as a starting point.[1] In 2015, maximum discharge of the river Rhine was assessed at 16,000 m³/s (which is 1,000 m³/s more than the 15,000 m³/s in the period of 2000–2015) and 18,000 m³/s in 2100. For the river Meuse the discharge requirement was raised from 3,650 m³/s to 4,600 m³/s. The committee also took into account a sea level rise of 500–710 millimeters (mm) between 2015 and 2050.

This assessment demonstrated the need for both a short- and a long-term strategy to prevent flooding. The possibilities for the long term were examined in the Spankrachtstudie (resilience study). The insights of this study are especially important since the river and sea levels are not expected to stop rising after 2015. Climate scenarios show that the Netherlands will face substantial water level rises in the coming century. The study resulted in a catalogue of dozens of possibilities for broadening rivers within and outside the winter bed. These were included as long-term views in the Planologische Kernbeslissing (Key Planning Decision, KPD) Room for the River (2006). The main part of the KPD Room for the River concerned 34 river safety measures for the short term along the major rivers

Waal/Merwede, Nederrijn/Lek, and IJssel. These measures were to be implemented to meet the legal safety requirements along the Rhine and the Meuse in 2015.

Room for the River had two objectives. The main objective was to increase water safety, by limiting the risks of flooding and dike breaches. The second objective was to improve spatial quality. This objective is exceptional. It meant the safety measures also had to improve the user value, future value, and experiential value of the landscape. So, two birds would be killed with one stone.

It was the first time that improving spatial quality was explicitly named as a goal in connection to flood prevention. Room for the River is therefore seen as a break with tradition, and an intentional one at that, for the regular levee reinforcements in the last decades of the twentieth century sometimes faced strong opposition from local residents and interest groups. Levees were sometimes brutally transformed into inelegant "sausages" with no anchoring in the landscape. So, things had to be different with Room for the River.

The Room for the River Program Directorate (PDR) had central control of the implementation of the measures. For each measure, the PDR looked for a government body to organize its planning: municipalities, water boards, provincial governments and, on occasion, Rijkswaterstaat (part of the Dutch Ministry of Infrastructure and the Environment). In addition to the two objectives (increasing water safety and improving spatial quality) three directive principles were formulated for each measure: the water level in the river had to be lowered, a budget had to be met, and the measure had to be completed in 2015.

The implemented measures were diverse. Levees and summer beds were relocated, reclaimed land was "de-poldered," side channels were constructed, obstacles were removed, and groins and smaller levees were lowered. The measures were linked in order to tackle the bottlenecks in the river system, so the water level would not rise as high in events of extreme discharge.

Integrating Spatial Quality in Planning

The PDR offered practical assistance in various ways to support spatial quality initiatives and to ensure the coherence of the measures.

An atlas[2] was made with all kinds of information about the rivers in the area. This included four types of maps. A blue map of the existing water systems, a green map of the river area as an ecosystem, a yellow map of spatial and cultural-historical values, and a red map of land use developments.

A regional spatial framework[3] was set up as well. This set out the spatial quality of the river area in images and assignments.

An inspirational book[4] was made with examples of river designs and river broadening projects. This was a book of completed projects of comparable assignments abroad, both in the field of ecological design and new urban waterfronts. The PDR also included a number of specialists in spatial quality in its organization. They were the point of contact for the initiators should they have any questions.

Perhaps most importantly, the PDR set up a "quality team," or Q-team (described by Dirk Sijmons in Chapter 6). This team consisted of authoritative specialists in landscape architecture, urban planning, river ecology, and hydraulic design. The Chief Government Advisor on Landscape, a direct advisor to the minister, led this team. The Q-team visited all measures at different points in the planning process and assessed their spatial quality. Precisely because the concept of spatial quality is sometimes elusive and cannot be substantiated with facts and figures, this was a unique and effective way to make initiators

FIGURE 16.5
Characteristic image of the Noordwaard before de-poldering, with farmlands and relics of wooded creeks. (From Joop van Houdt, Rijkswaterstaat.)

fully aware of the efforts needed in their planning in order to realize spatial quality. At first, skeptics questioned the need of such a Q-team and thought the team might be expensive and would only slow things down. Eventually, certainly after the measures had been implemented, everyone agreed that the Q-team had been very effective. It played a leading role in the planning process.

De-Poldering Noordwaard

The most extensive and expensive measure in the Room for the River program was the project De-Poldering Noordwaard. This is an area of more than 4,000 hectares (9885 acres) along the River Nieuwe Merwede in the Biesbosch area in the southern province of Brabant. Both the measure and the landscape are unique. First of all, de-poldering on such a large scale does not usually happen in the Netherlands. The Dutch are known for reclaiming land and building levees, not removing them. And the landscape of the Biesbosch is unique because it is a freshwater tidal area at the border of the Dutch river area and its delta. Its nature and landscape values are unparalleled in Europe (Figure 16.5).

De-Poldering Noordwaard *in a Nutshell*

The water safety objective of the De-Poldering Noordwaard project was to lower the water level in the river by 300 mm at the location of the city of Gorinchem (8 kilometers [km] upstream), because there is a bottleneck in the river there, causing the water to accumulate at high discharge events. By creating more room in the riverbed downstream, at the Noordwaard, the water would be able to pass the bottleneck more quickly, thus lowering the water level in Gorinchem.

The essence of the De-Poldering Noordwaard project was to lower the primary levees at the northeastern and southwestern sides of the Noordwaard area. This allowed the river water to flow through the area diagonally when water levels were high as the water level of the Nieuw e Merwede was higher than that of the River Amer, south of the Biesbosch. The starting point was that farmers had to be given a viable perspective in- or outside of the polder, and that residents should be able to continue living in the Noordwaard if they wanted to.

Historical Context

The Noordwaard is an area with a long tradition of living with water. At the end of the fifteenth century, the St. Elizabeth's flood completely submerged the landscape of the current Noordwaard polder and its surrounding area, effectively transforming it into a gigantic inland sea. In the centuries that followed, the interplay between the water dynamics of the Maas and Waal rivers and the ebb and flow of the sea created a unique area, the Biesbosch. This area consisted of sandbars, creeks, reed fields, and willow forests, and was used for the cultivation of reed and withy (willows that are regularly cut off low to the ground) and for fishing. A problem arose when the area increasingly filled up with deposits and the vegetation grew ever thicker. When water levels were high, the water discharge of the rivers was greatly hampered, and the shipping route was hardly usable. The water levels along the levees upstream thus became increasingly dangerous, which led to levee breaches. At the end of the nineteenth century, it was therefore decided to dig new rivers through the Biesbosch. This is how the Nieuwe Merwede and Amer rivers came into being, which are actually extensions of the River Waal and the River Maas, respectively. This divided the Biesbosch into three parts: the Brabantse, Dordtse, and Sliedrechtse Biesbosch.

The area that is now called the Noordwaard borders the Brabantse Biesbosch, and in the nineteenth and early twentieth centuries, it consisted of many small polders surrounded by low levees between wooded creeks. It was sparsely populated, and people lived on mounds. Levees were built around agricultural parcels to limit the consequences of flooding. These parcels were pumped dry with the use of simple windmills, or the water was drained when water levels in the Biesbosch were low. The farmers could only reach these lands by boat, sailing from one polder to the next. They kept enlarging the polders, at the expense of the creeks and woodlands. The willow cultivation also flourished, mainly thanks to the many levees that were built during this period. The withies were used to make mattresses, the bottom parts of a levee.

1970 was a milestone in history of the entire Biesbosch area. As part of the Delta works, the Haringvliet dam was built that year, closing the Biesbosch's direct connection with the sea. The tidal difference decreased from 2 meters to 300 mm. The water dynamics largely disappeared from the Biesbosch, offering opportunities for agricultural development. It was decided to build one main levee around the small polders, thus creating one large one: the Noordwaard. This was followed by land consolidation, road construction, and a decreased water level for sufficient drainage. Many creeks were filled up and people no longer needed a boat to get to the Noordwaard. It was too much effort to fill up the largest creeks, so they remained as relics in the landscape. The pattern of small meadows with low levees around them and meandering wooded creeks transformed into a relatively large-scale landscape with lands for all kinds of agriculture. The characteristically small-scale landscape disappeared. Traces of the old creeks are still visible in the micro relief and composition of the soil. Relics of creeks and farms on mounds remind us of the time when the Noordwaard lay outside the levees.

Nonetheless, the Noordwaard was still a peaceful and natural-looking area, which was partly due to its isolated location. From the eastern side, the Noordwaard could be reached by land, but from the western side one had to take a ferry to get there. People from other parts of the country came to live in the polder for the peace and quiet.

In the context of the national nature policy NURG,[5] three large nature reserves were realized along the major rivers in the 1990s. The aim was to strengthen the existing nature reserves to counter the increasing fragmentation of nature. One of these new reserves was planned in the eastern part of the Noordwaard. This area was called NOP; nature development project Noordwaard. The specific aim of the NOP was to realize a water connection between the Nieuwe Merwede River and the Biesbosch in order to strengthen the aquatic ecosystem.

For that reason, the existing levee was relocated, placing the NOP on the outside of the levee ring. Farmers were bought out and all buildings were demolished. Broad waterways that looked like secondary channels were dug straight through the polder. The cultural-historical layer of inhabitation was completely removed, all in favor of nature development. The implementation of the NOP was completed in 2008 (Figure 16.6).

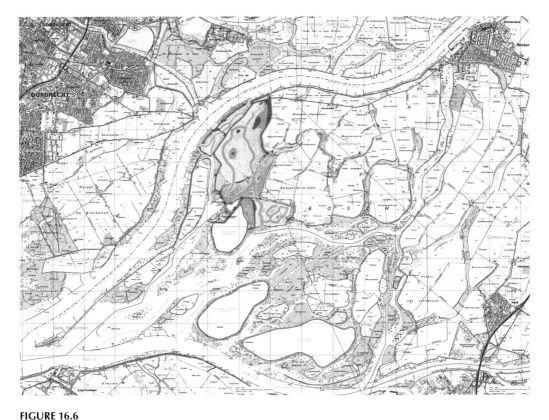

FIGURE 16.6
Topographical map of the Noordwaard and part of the Biesbosch, before de-poldering. The NOP area is shown on the map. (Courtesy of Topografische Dienst.)

Design Philosophy De-Poldering Noordwaard

The Room for the River project Noordwaard is very different from the NOP, even though it started in 2003, before the NOP was realized. In the Noordwaard, the landscape's multi-functional and cultural-historical layers were used to guide and inspire the spatial redesign. On this basis, a coherent plan was made narrating the rich history. The plan aimed to increase the experiential value along with the water safety, while retaining the peacefulness and spaciousness for the residents. It also aimed to increase the ecological diversity and to realize beautiful forms with detail and materialization in the style of the Biesbosch, showing the function and logic of the design.

Water safety and spatial quality had to strengthen each other. For reasons of water safety, large volumes of water would have to be able to flow diagonally through the Noordwaard in a short amount of time. Because the ground level of the Noorwaard was fairly low, the water could flow over the surface. It soon became clear that it was very important to understand the verticality in the landscape and to thoroughly think through the designs for the levees, mounds, and vegetation. This was crucial for the water flow and for the spatial layout. For the spatial plan, the topography of the Noordwaard in 1905 proved to be an important source of inspiration. At the time, the Noordwaard lay outside the levees and was an area made up of both an agricultural landscape of small polders with grazing cows and withy cultivation, and a natural landscape with creeks, woodlands, reeds, and mudflats. More than 100 years ago, there was a balance that also had to be found in the Room for the River project. There were so many relics in the landscape that referred to the landscape before the reclamation, that it made sense to take a good look at the past and use it as a basis for the future (Figure 16.7).

The General Spatial Design

The narrative of the history of the area is visible in the design. De-poldering the Noordwaard reintroduces river dynamics to the area. The most important measure is that the main barriers on the northeastern and southwestern side of the Noordwaard were dug back to the ground level. This allows the water of the Nieuwe Merwede to flow diagonally through the Noordwaard. So, the Noordwaard now lies outside the levee. The creeks are once again connected to the Biesbosch, and as a result the water level on the outside of the levees is now almost 1.5 m higher than it was before the de-poldering. Tidal dynamics of 300 mm now also come into the Noordwaard (Figure 16.8).

Throughout the Noordwaard, the 1905 pattern of the polders and creeks has been redesigned in a contemporary way. In total, 17 polders were made, enclosed by small green levees. These levees have more or less the same locations they used to have, but they are in a straighter line.

The area where the river water flows through when water levels are high is called the *doorstroomgebied*, the flow area (Figure 16.9) This area features polders with very small levees only about 600 mm high. The levees are so small because drainage is central here, with nature development and agriculture taking a back seat. Various types of polders were designed. Analogous to the polders that are characteristic of the Biesbosch, there are polders with shared agricultural use (cows graze here in the summer), there are nature polders with high water levels, and in the lowest parts there are polders with so-called "breached levees," where the water flows freely in and out. In order to facilitate the water flow through the flow area when water levels are high, the polders here are only allowed to have grass on them. Every year during the winter period the water flows through this area and floods the polders for a few weeks.

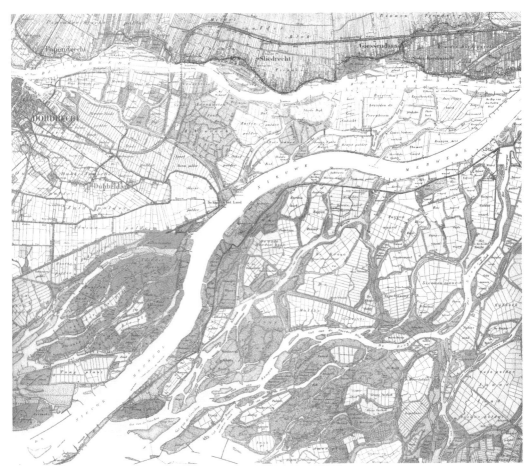

FIGURE 16.7
The 1905 topographical map that served as an inspiration for the spatial structure of the de-poldered Noordwaard.
(Courtesy of Topografische Dienst.)

On both sides of the flow area, there are polders with small levees around them. These are the farmers' polders. Each of these polders has its own, closed levee flanked on the outside with wooded groves. The new levees reflect their function in their shape and height. These polders will rarely be flooded, some only once every 100 years, others once every 1,000 years. On either side of a small-levee polder, part of the levee has slightly lower and extra wide dimensions. This is where the polders will flood first. Because of its wide profile, the levee cannot be breached, which will allow for the polder to be drained quickly after inundation. The inundation segment is marked with beacon trees on the levee to indicate the water level.

New creeks were dug between the polders, about 40 kilometers in total (Figure 16.10). Almost all of them are at the same locations they were in the past. There are wooded creeks in the part of the Noordwaard with little flow. Here, as in the rest of the Biesbosch, the forest comes up to the shore. From the water, one gets that typical Biesbosch experience of sailing through a green tunnel. In the flow area, there are mudflat creeks. The banks here were designed to be long and so gentle that the tide covers the entire shore. This accentuates the movement of the tides (Figures 16.11 and 16.12).

FIGURE 16.8
Spatial design De-Poldering Noordwaard. (From Robbert de Koning, Rijkswaterstaat.)

FIGURE 16.9
The part of the flow area with small levees is drained using small wind-water mills. (From Werry Crone, Rijkswaterstaat.)

FIGURE 16.10
Creek bridge over a partly wooded creek in the flow area. (Image courtesy of Jeroen Bosch.)

FIGURE 16.11
Cross-section of a wooded creek. Because the forest comes up to the water, one can see a characteristic "tunnel" effect" on narrow creeks. (Image by Robbert de Koning.)

FIGURE 16.12
Cross-section of a mudflat creek. Making the tidal (0.3 m) bank long and gentle gives full play to the ecological processes of a freshwater tidal area. (Image by Robbert de Koning.)

To regulate the drainage of the polders, pumping stations were built (Figure 16.13). Each small polder used for agriculture has its own pumping station, some even have two. In total, one pumping station was installed for a levee ring and 10 for agricultural polders, plus 26 small wind-water pumping stations. The pumping stations for the agricultural polders are shaped like a masonry "box" and have a sturdy look. The stations are "waterproof,"

FIGURE 16.13
Pumping station with a viewing platform right behind a small levee. (Image by Robbert de Koning.)

the pump is on ground level height, and the electrical part is at the top, so high that the water will never reach it. They can function even when they are inundated. The brickwork shows how high the water can reach. For the portion above Maatgevend Hoogwater[6] (norm for high water) bricks were used from old farms that had to be demolished because of the de-poldering. Cultural history is being "reused" here for new functions.

Houses built after the levee around the Noordwaard was closed in 1970 are all at low ground levels. These are too low for people not to get their feet wet if the river water flows through the area. However, Maatgevend Hoogwater is used, as agreed with PDR, to determine that an inundation of up to one meter, measured from the threshold, does not create a life-threatening situation and would thus be acceptable. If such houses are also strong enough to withstand waves, they do not have to be demolished. Residents can make use of a claim settlement if the water indeed enters their house. Houses where the water level rose above one meter were demolished. The owners of these houses were given a reasonable settlement based on the estimated value of their house and property, and they were given the opportunity to build a new house on top of a mound elsewhere in the Noordwaard. Around half of the residents opted for this. The other half moved elsewhere. About 15 low-lying houses were rebuilt on mounds at the outer edge of the new polders. These houses have a private area on the creek side, where the residents can moor their own boats. The rebuilt farms were situated inside the levee. This was to underline that the agricultural land inside belonged to the farmers (Figure 16.14). Interestingly, the old houses and farms, built before 1970, were usually already on a mound. These buildings were often at sufficiently high locations and could be maintained. All of these mounds did have to be enlarged and reinforced to meet the contemporary water safety requirements.

The new pattern of polders and creeks was reason to adjust the road pattern and to place 34 bridges. The architects of the contractor designed a family of bridges. Some of these are long bridges spanning the in- and outflow openings of the levees at the edges of the Noordwaard. These bridges are in line with the levees, and the pillars are parallel to the direction in which the river water flows. Because the water can flow quite fast here, they were built to be robust. Creek bridges were placed between the polders. They form a slender

FIGURE 16.14
New house on a mound. (Image courtesy of Jeroen Bosch.)

connection and offer beautiful views over the landscape and the water. They are arched to make crossing the creeks a real experience. The Steurgat bridges are a third type. These have a narrow passage because they have to be able to be closed with shot beams in the winter. This is so the water will keep flowing in the desired southwestern direction. (Figures 16.15 through 16.17). At Fort Steurgat, a fortification of the New Dutch Water Defence Line from the end of the nineteenth century, a new levee had to be constructed right next to the fort. This was necessary to prevent river water to flow into a small residential area, a business park, and the fort.

The levee was not easy to design, because this side of the fort had never had a levee before and had to have a clear line of fire. What made the plan particularly difficult is the fact that the fort is currently privately owned. To limit the effect on the residents' view as much as possible, two measures were taken. The gentle inner slope of the levee has a ratio of 1:15. This way, the ground level is "lifted" and the crown of the levee seems visually lower. The river side of the levee is planted with willows that greatly reduce the wave forces on the levee. The crown could therefore be lowered by almost a meter.

The addition of withy beds to the main levee is an innovative measure, which was applied in the Noordwaard for the first time in the Netherlands. The withy beds consist of two different types of willow that are planted in strips. One year one type of willow is cut, the next year the other, so there are always withy beds of sufficient length and roughness. The use of two types of willow limits the vulnerability to damage and diseases. The twigs of the two willow types differ in color, one is greener, the other more yellow, and the cutting regime means the color of the withy bed is slightly different every year. A study was done to see what direction of planting would be most effective. When the plan was first made, the idea was that the best way to plant the strips would be at right angles to the tidal waves from the southwest. Based on later calculations, the strips are now planted parallel to the

FIGURE 16.15
One of the levee bridges near the inflow opening. These bridges are built in a robust style. (Image courtesy of Jeroen Bosch.)

FIGURE 16.16
Creek bridges connect the small polders and elegantly cross the creeks. There is a pumping station on the right side. One can clearly see that the upper part (lighter color) that always remains dry is made from bricks from demolished farms. (Image courtesy of Jeroen Bosch.)

FIGURE 16.17
The Steurgat bridges have a narrow passage so that they can easily be closed off. (Image courtesy of Jeroen Bosch.)

levee to protect it from waves from all directions. Three lines of sight were created through the withy beds, so people can see two old farmhouses and a new bridge from the fort. By including the bridge in the lines of sight, a link was created between the levee and the flow of water through the Noordwaard (Figure 16.18).

It is expected that water sports and outdoor activities in the Noordwaard will increase as a result of the de-poldering. This will affect the residents of the Noordwaard for whom the tranquility of the area is one of its best qualities. Therefore, the recreation was proactively zoned by creating a recreational gateway and hub at the edges of the area. Visitors can have a drink and a bite to eat here, and the trails and facilities are just a little better than they are elsewhere, in order to prevent visitors from roaming through the entire area. The headroom of the new bridges is such that not all boats can go everywhere. Farmers are allowed to build small-scale tourist accommodations. The new trails and bicycle paths will make the Noordwaard an adventurous destination for the residents of the cities of Gorinchem, Dordrecht, and Rotterdam.

The topographical map from the beginning of the twentieth century was a great inspiration for the design. The spatial mix of polders, small levees, and creeks was beautifully balanced at the time. At several locations, this structure was redesigned in the de-poldering plan in a contemporary way, with twenty-first century resources and style. With beacon trees on levees to indicate the locations the river might flood and with gentle slopes of mudflat creeks in the tidal zone to allow optimal tidal movement, the dynamics and processes of the river and tidal system flow together (Figure 16.19).

The Noordwaard is (once again) a landscape in which living with water is central. Different levee heights lead to different "levels of protection against flooding" and thus to different possibilities for use. The flow area is inundated every year; the polders with small levees on the other hand rarely are (Figure 16.20a–f). The system of channels and

FIGURE 16.18
The innovative levee with withy beds to reduce the tidal pressure at Fort Steurgat. (Image by Robbert de Koning.)

FIGURE 16.19
Aerial photograph of the de-poldered Noordwaard. (From Yourcaptain, Rijkswaterstaat.)

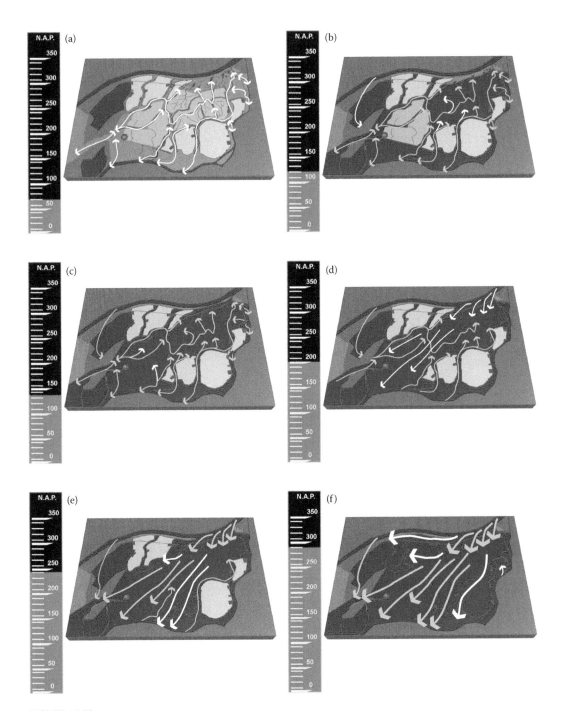

FIGURE 16.20
(a) through (f) What happens when the river water rises? The daily tidal movement is between 40 and 70 centimeters above NAP.[7] When it reaches 70–120 cm above NAP it floods the nature polders. When it rises above 135 cm the water flows over the embankments into the agricultural polders with low levees. At 200 cm above NAP the inlet dam is submerged and the direction of the main flow changes. If the water reaches 240 cm the polders with higher levees are also flooded. At 290 cm the whole Noordwaard is under water. (From Rijkswaterstaat.)

creeks between the old polders is revived. Reeds, withy beds, and willow thickets flank the creeks. The creeks are spatially enclosed whenever possible (at locations with limited flow) and open where they have to be (in the flow area). The landscape is strongly determined by the "empty" polders, lined by green levees and wooded creeks. Farms and barns are iconic elements in the landscape that accentuate the emptiness. Small and large levees form both borders and connections between polders and creeks. Many new bridges were built, new pumping stations were placed, and mounds were raised and fortified. All in all, a palette of culture and nature, of old and new, of functions for both people and animals was created.

To be able to guide the contractor during the realization phase in terms of spatial cohesion in the Noordwaard and to create a fitting identity, the so-called "Biesbosch style" was developed. This style is of a robust landscape, where nature and culture alternate, where one can find space and tranquility, and where new architecture is sober and subservient to the landscape. It is a landscape with many shades of gray due to the willows and shades of green in the meadows and fields, where the open fields and wooded creeks make for great spatial contrasts.

Planning Process and Design Plan: The Essence of the Design

A good design needs a good planning process. The planning process had three phases: the planning study, the contract phase, and the implementation phase. The design process was always dominant but different in each phase.

The planning study (2003–2010) started with an exploration. The first act was the signing of an administrative agreement with the aim to examine how the Noordwaard could best be incorporated in the Room for the River program. Research by design was used to explore the brief, and several design workshops were organized with the residents and local authorities (the stakeholders). This was also called the regional process. It led to broad outlines for several variations.

Parallel to the regional process, an environmental impact assessment was made, for which variations were also drawn up, and which explored the effects on nature, soil, and archeology.

As it turned out, the regional process greatly accelerated the planning process. At first, all the stakeholders were very critical, but they were included in an open planning process that changed over time into a dialogue with much input into the design. This resulted in the selection of a preferred variation in 2007: The Design Vision. The most important choices were made at that time, including the polders with the small levees, the functions, and roughly where the new creeks would be located. This was essential, because it was established that the Noordwaard had to be redesigned as a relatively small-scale area. This plan was subsequently elaborated in the 2010 Design Plan. This Design Plan is an integral spatial plan. Additionally, a Rijksinpassingsplan (National Government Zoning Plan) was drawn up to include all the legal requirements.

Then it was time for the contract phase (2010–2012). In this phase, all contracts were drawn up, including the brief specification. A contractor was selected using a Design & Construct contract. To ensure the spatial quality during the realization phase, Rijkswaterstaat also drew up an Ambition document. This outlines the requirements the work would have to meet.

In 2012, the implementation phase started, and the Noordwaard was effectively de-poldered at the end of 2015. During the implementation, the landscape architect gave advice and evaluated the work.

Lessons Learned

The Noordwaard has been de-poldered since the end of 2015. The area has gone through an enormous metamorphosis. A layered landscape has been laid out in which water, polders, residential, agricultural, and nature areas form a coherent whole. It is a distinctive area for people to live and recreate in, in which the history of the area is referenced in many different ways.

The area has been reconnected to the Biesbosch, just as it used to be. From now on, the Biesbosch can be experienced both from land and from the water. The layout of the Noordwaard with its diversity and contrasts forms the basis for the spatial quality of the area. In short, the Noordwaard has been given a more varied character. There is still room for agriculture, but nature has been given a boost, and water dynamics once again help to shape the landscape. The daily freshwater tidal flows have returned, and the river will flood large parts of the Noordwaard a few times per year (Figure 16.21a and b).

What lessons have been learned? The evaluation of the process concerning the bridges showed that architects on the contractor's side had not been involved in the way it was established in the official documents. So, contractual guarantees do not necessarily lead to the actual realization of spatial quality. It is also important here that at crucial moments during the approval of the final designs, the technical managers have to completely understand the spatial quality aspects of the designs. In the course of the implementation phase, the informal, non-contractual process got better and better. As a result, so did the working process in which the spatial quality was realized.

Looking back, it can be said that many time-consuming and sometimes costly discussions about the realization of the project and the spatial quality could have been prevented if the client and the contractor had gone over the entire process right after it had been commissioned. All in all, it is clear that it is people who determine whether spatial quality is realized or not.

Fine Dutch Tradition in the Noordwaard?

Can the design of the de-poldered Noordwaard be seen as part of the Fine Dutch Tradition? Hydrological characteristics of the river and the sea were used as a firm basis for the plan, soil characteristics and geomorphology were utilized, and existing and innovative technologies such as the levee at Fort Steurgat were integrated. The different components of the plan were geared to one another to achieve spatial coherence. The layered design very subtly plays into the verticality of the landscape and the water levels that come with it. It is a contemporary, robust landscape with ingenious references to the past, such as the new structure that was derived from the historical topography.

The jury of the Landschapsarchitectuur en Stedenbouw in Nederland Jaarboek 2016 (annual book with highlights of Dutch landscape architecture and urban design) wrote: "It is an unpretentious, pragmatic plan with a sensitivity and poetry that is unusual in the Netherlands." So, one might even conclude the plan goes beyond the Fine Dutch Tradition.

(a)

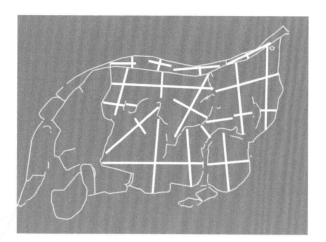

(b)

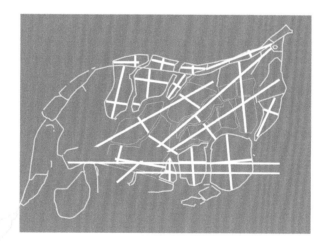

FIGURE 16.21
(a) and (b) The visual experience before and after the de-poldering. The contrast between the sheltered edges and the open central area has been heightened. (Images by Robbert de Koning.)

Project Data De-Poldering Noordwaard

Surface area	• 4,450 hectares (10,996 acres)
Flood risk	• 1/2,000 years in levee ring 23 with a high levee (±5 meter)
	• 1/1,000 years (in 2015) in four of the seven polders with small levees (±3 meter)
	• 1/100 years (in 2015) in three of the seven polders with small levees (±2.5 meter
	• 30 days per year in the summer polders in the flow area with very small levees (±0.5 meter)
	• 60 days per year in the nature polders in the flow area
Brief	• 0.30 meter at river kilometer 955 (Gorinchem, ±8 kilometer upstream)
Initiators	• Rijkswaterstaat, Projectbureau Ontpoldering Noordwaard
Plan study	• Projectbureau Ontpoldering Noordwaard

Realizer	• Rijkswaterstaat, Projectbureau Ontpoldering Noordwaard
Spatial design	• Robbert de Koning landscape architect BNT with Rijkswaterstaat and DLG
Implementation	• Combinatie Noordwaard (Boskalis bv, Martens en Van Oort, Van Hattum en Blankevoort and Gebr. Van Kessel bv.) Architects involved: Ipv Delft creative engineers and West 8 urban design & landscape architecture bv.
Planning	• Plan study: 2003–2010 • Contracting: 2011 • Implementation: 2012–2015 • Management: from the end of 2015

Endnotes

1. Published in 2001, the third climate report of the IPCC was confirmation for the KNMI that the earlier climate scenarios used by the committee for Water Management in the twenty-first century—the scenarios that Rhine and Maas discharge requirements were based on—were correct. Source: final report Spankracht (2002).
2. Atlas Ruimte rond de rivier, PDR (Program management Room for the River). (2003).
3. Regional spatial framework Room for the River. 2004. PDR.
4. Rivers and inspiration. 2008. PDR.
5. NURG, further detailing river landscape, implementation period 1992–2015, 55 projects with a total surface area of 60,000 hectares (148,263 acres).
6. Maatgevend Hoogwater is the theoretical water level at a certain location based on the maximum discharge. For the river Rhine, it is 16,000 m^3/s.
7. N.A.P. = Normal Amsterdam level. This is the reference level to which height measurements in the Netherlands are related. In general, it is equal to the average seawater level.

References

Hendriks, M. 2016. *Yearbook Landscape and Urban Design in the Netherlands*. Stichting Lijn in Landschap, Wageningen.

Hooimeijer, F. L. 2012. The Fine Dutch Tradition, *Architecture and The Built Environment Delft*. Faculty of Architecture, Silpakorn University, Thailand.

Koning, R. de. 2010. *Ambition report De-poldering Noordwaard*. Rotterdam: Rijkswaterstaat Projectbureau Noordwaard.

Koning, R. de and L. Eshuis. 2008. *Rivieren en Inspiratie*. Utrecht: Programmadirectie Ruimte voor de rivier.

Kors, A. and F. Alberts. 2002. *De spankracht van ons rivierenland: eindrapport Spankracht. Ministerie van Verkeer en Waterstaat, Rijkswaterstaat, Rijksinstituut voor Integraal Zoetwaterbeheer en Afvalwaterbehandeling (RWS, RIZA), Projectgroep Spankracht* (The resilience of our river land: final report Resilience study. Ministry of Transport, Public Works and Water Management, Rijkswaterstaat, National Institute for Integrated Freshwater Management and Wastewater Treatment (RWS, RIZA), Project group Spankracht (*Resilience*)). Lelystad: RWS, RIZA.

Projectbureau Noordwaard. 2007. *Ontwerpvisie Ontpoldering Noordwaard*. Rotterdam: Projectbureau Noordwaard.

Rijkswaterstaat Project Office De-poldering Noordwaard. 2007. *Design strategy De-poldering Noordwaard*. Rotterdam: Project Office De-poldering Noordwaard.

17

Thresholds and Contingencies: A Design Process for Regional Coastal Resilience

Penny Allan and Martin Bryant

CONTENTS

Introduction

Adaptation for climate change and resilience in coastal regions outside urban centers will be a protracted program. Cities, with their high population densities and capital investment, are deemed most vulnerable to the impacts of climate change, and hence this is where most research and money is invested. However, the impacts of climate change in underfunded coastal regions are likely to be no less significant. From a resilience perspective, these regions provide a back-up service to the cities. They provide biodiversity and recreation opportunities and, with worldwide food and water scarcity likely to be one of the most significant threats of climate change, and as producers of a high percentage of these resources, they have the potential to contribute to global solutions.

Although the climate change impacts associated with rural, urban, and suburban sites in coastal regions are similar, the challenges are somewhat different. For coastal farms to remain viable, access to land is critical, but inundation from flooding and sea level rise is likely to effectively diminish arable land in low lying areas by up to 50% in some cases. And, because land is relatively cheap, extensive suburban development continues to occur along the seafront with road networks that prioritize waterfront views for individual houses over neighborhood amenity and community networks. Perhaps most significant is that there are very few financial incentives to alter the status quo. This chapter addresses a way of thinking about adaptation to climate change in regional coastal areas that relies not on financial incentives and government policy, but rather on a community-based approach.

Through design explorations in two regional case study sites, we describe how ecosystem thresholds can be linked with specific actions in a contingent relationship to transfer agency from government bodies to local communities, thereby encouraging effective and lasting adaptation to climate change impacts. The two sites are located on the coast northwest of Wellington, New Zealand, one suburban, the other rural. The suburban case study provides place-based solutions to the global problems associated with the anti-communitarian and ecosystem-threatening subdivision practices. The rural case study offers a unique window into the values of distinctive land-based indigenous practices that might link culture and community in non-traditional ways. Both uncover opportunities for local solutions to address global issues in a meaningful way. They address the vulnerability of coastal landscapes to the impacts of climate change, the relative lack of resources to cope with this vulnerability, and what design can do to address these problems.

The Kapiti-Horowhenua Coastal Plain

New Zealand's land mass is long and narrow, with an extensive coastline, often visually connected to the spine of snow-covered mountains that stretches from the south of the South Island to the Bay of Plenty in the north. North of Wellington, the Kapiti-Horowhenua lowlands extend from the coast to the foothills of the Tararua ranges (Figure 17.1). This is wet and fertile country fed by a network of rivers. The ground is composed of layers of sand and peat. The sand is washed down from the coast to the north and then drifting inland in waves of parabolic dunes. The peat is the result of the cycles of growth, flooding, and decay in wetlands navigable by *waka* (canoe), which once stretched in a continuous line for at least 100 kilometers behind the coastal dunes. The extensive forests in the area were felled in the early twentieth century to make way for dairy farms. The coastal plain is now a landscape of artificial drains and ditches, straightened rivers, and polluted waterways that lacks the capacity to deal with large storm events. The water table sits just below the surface and when it rains and the rivers flood, its capacity is quickly exceeded, resulting in widespread flooding.

There are three typical conditions on the coastline: Low density residential in the coastal towns; clusters of lifestyle blocks scattered across the coastal plain; and farmland, where dairying is marginal in the wet ground and farmers plant exotic pine in an attempt to stabilize the shifting dunes. In each of these areas, the endemic riparian vegetation has been stripped as a way of increasing the amount of arable or developable land, and the ground has been heavily engineered to improve drainage. These conditions expose the area to the challenges of climate change effects.

The Typical Response

In a recent brief for the government's National Science Challenge, the National Institute for Weather and Atmosphere (NIWA) suggested that New Zealand is good at climate change science but not good at communicating it to the public. And so, people may not see the need to adapt. There are several possible reasons for this.

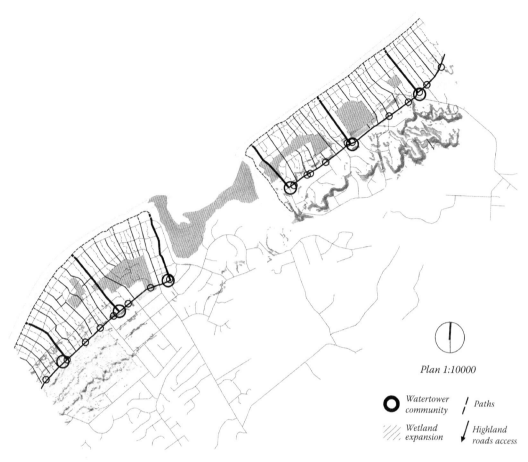

FIGURE 17.1
The coastal suburb of Waikanae: Plan of realignment of coastal infrastructure to encourage neighborhood reorganization in smaller units around new community facilities on the high ground. (Image courtesy of Caitlin Wallis.)

The first might be a "head-in-the-sand" scenario: the information is good, just falling on deaf ears. The New Zealand public seems to be heavily resistant to accepting the realities of climate change and doing something about it. Change is costly. In the Kapiti region, for example, residents with property on the coast ejected the standing local government for mapping a red line through private property to indicate the vulnerable coastal edge (Shand, 2013) and the limits of development. And in the Horowhenua district to the north, dairy farmers continue to engineer artificial drainage solutions because addressing climate change impacts through ecosystem restoration might significantly impact the viability of their farms.

A second reason may be that climate change is typically translated into top-down policies and regional land use planning instruments that address mitigation rather than adaptation. They focus on stopping climate change (often through economic incentives Helm, 2015) rather than on what anyone might do at a local level to deal with its impacts; on mitigating emissions and dealing in carbon credits, rather than the design of space or how the spatial configuration of settlements might accommodate climate change impacts.

The communication of climate change and its impacts focuses at a meta level and seems to bear little relationship to the land or the people that live on it.

A third reason is that most spatial solutions are just too expensive, all-encompassing, and disruptive, their fixed and monumental nature no match for the slippery uncertainties of climate change. Even projects that rely on catalyzing existing environmental systems rather than working against them require significant investments of capital.[*] Perhaps climate change scenarios will not play out as predicted. Why bother spending a lot of money for something so unpredictable?

An Alternative Methodology: Speculative Design

Design-led speculative projects play a crucial role in presenting spatially integrated design, bridging the gap between local scales and regional planning and expanding the public conversation about possible futures in a way that scientific papers never can (Bryant et al., 2006). Design processes enable land-based decision making on the growing availability of fine-grained geographic, climatic, social, and cultural data. This data availability provides an opportunity to deinstitutionalize centralized broad-scale planning processes and conceptualize settlement-making with finite, local, place-specific, and formal solutions. In this context, design becomes a useful tool to test cultural, social, and ecological processes and "address the complexities of everyday life" (Speaks, 2006). Design also facilitates the empowerment of communities through its inherently legible visual language, providing a concrete and visually accessible projection of settlement.

Because they are speculative, design research projects present no direct threat to government, landowners, or community groups, but they are nevertheless real enough to raise genuine questions, contribute to important debates, and envision possible futures. They help to develop a collective understanding of what is possible and act as something to work toward.

But while distinctly spatial, design research visions are nevertheless abstract propositions, too far in the future to act as catalysts for change. Sometimes a vision, even one clearly articulated and visually rich, is not enough to overcome the head-in-the-sand inertia brought on by talk of climate change and its impacts, particularly among those who have a stake in maintaining the status quo. When we undertook workshops and meetings with regional communities we realized that communicating a vision—the "what"—was not enough. We also needed to communicate the "when" and the "why." We subsequently expanded our methodology to include the documentation of key environmental thresholds—tipping points, when a relatively small change or disturbance in external conditions causes rapid ecosystem change (Walker et al., 2004). We communicated these thresholds as experiences that, when recognized, would require an urgent adaptive response. And we linked these in a contingent relationship to design solutions that could be easily implemented, thereby connecting "what," "when," and "why."

---

[*] For example, see "The Sand Motor in the Netherlands," http://www.dezandmotor.nl/en/home (accessed November 8, 2017).

A Possible Solution

The thinking around this issue was stimulated by three things: a paper written about ecosystem thresholds on coastal plains in Pacific Islands of Hawai'i (Kane et al., 2015); speculative design research collaborations with researchers at Victoria University of Wellington (VUW) about climate change and the urban/suburban condition; (Allan and Stutterheim, 2007) and firsthand experience of climate change on the coast of Wellington. The thinking was further developed during recent design research with members of the local indigenous community, artists, and climate change scientists, which looked at how design might assist Māori coastal farmers adapt to climate change.

The paper on ecosystem thresholds in Hawai'i describes how low lying coastal plains in the Pacific are at risk of inundation, not just from the sea but from a combination of sea level rise and changes in groundwater levels. Because of the complex interaction of both systems, inundation can be separated into slow and fast phases: water from the sea slowly inundates the land, but when hydrostatic pressures start to impact groundwater levels, inundation accelerates and is much more difficult to manage. The slow and fast phases of inundation are separated by a threshold, and the length of the slow phase can be approximately calculated. During the slow phase, there is time for management plans to be prepared to anticipate the loss of critical wetland habitat and either protect or relocate it to higher ground.

Anticipation is an important feature of the collaboration with research associate Nick Jones in the 2007 MOVED to Design studio, a joint initiative between VUW and the Royal Melbourne Institute of Technology (RMIT). Nick's project, "xyt:", anticipates climate change impacts in a series of design moves that, implemented over time, might allow the seawall at Lyall Bay in Wellington to be ultimately removed, thereby stimulating the regeneration of sand dunes and enhancing coastal resilience. The strategy acknowledges the contingent relationship between inevitable but unpredictable sea level rise and the thresholds precipitated by very specific urban impacts. For example, in anticipation of sea level rise above the invert level of stormwater outflow at the seawall, the open space network was redesigned to accommodate detention; and in anticipation of storm surge topping the seawall and eroding the road linking the airport to the CBD, the road was rerouted further back through the urban grid. The combination of moves suggests a contingent relationship between climate change impacts, thresholds, and adaptive responses. The project is interesting for a number of reasons: it accommodates certainty *and* uncertainty; it emphasizes the importance of the *experience* of thresholds, shifting the focus from planning at a distance to localized agency based on firsthand experience; and for this reason, it has the potential to overcome the head-in-the-sand inertia that prevents lasting adaptation to climate change.

Our own firsthand experience of climate change was in Wellington, New Zealand, where the narrow road separating our house from the sea was regularly affected by winter storms. Twice a year we would wake up after a particularly bad storm to find debris from the sea on the road. We accepted this as part of living there: the road was soon cleared and things returned to normal. But during the 10 years we lived there, the storms became more frequent and more intense. We often wondered where the threshold was: what frequency and severity might indicate an irreversible shift that showed us once and for all that climate change was real and that it was happening now. We often thought about what we might do in anticipation of that shift. We designed a narrow concrete tower, to be constructed floor by floor over time. We thought we might start by using the ground floor as a garage and the flat roof as a viewing platform, then slowly build a second story and, at a designated time, move in. Each time we felt a threshold had been reached we could build another floor, always in anticipation of the next one.

In these examples, design imagines possible futures, allowing a designer or ecologist to map a pathway or a sequence of possible pathways that might deliver that future over time. The pathway, one of many that might be taken, is contingent on the recognition of thresholds, and specifically connected to the firsthand experience of climate change impacts as they occur. And the awareness of possible futures ensures that the path taken keeps options open rather than closing them down.

This chapter now examines our two regional case studies in light of these ideas. In the first: *High Ground Low Ground*, we worked together with VUW designer and research associate Caitlin Wallis, to address the suburbs in Kapiti's lowlands beside the sea. In the second, *Wai o Papa/Waterlands*, we propose evolving solutions for farmland in the Horowhenua.

Case Studies

High Ground Low Ground

In the Kapiti town of Waikanae, coastal suburban development is aligned in rows parallel to the beach. This is a seemingly efficient way to use land. Streets and services and freestanding houses run along the contours of the foredunes and the dune slacks, which has minimized extensive earthworks, except for some filling-in of wetlands. The fenced housing lots are exclusively private: apart from the natural landscape of the beaches, community facilities are conspicuously absent in this subdivided dormitory suburb. The back dunes, 200 meters away, have been avoided as a locus of development because they are too far from the sea and, being on high ground, are more exposed to the coastal winds.

There are several problems with this condition. First, there are immediate threats from storm surges, earthquakes, and tsunamis, and the opportunity for residents to escape is circuitous because the roads have not been arranged to allow easy access to high ground. Second, the houses are at risk: the Kapiti Coast District Council's Coastal Hazard report, released to the public in 2012, determined that 1,800 of Kapiti's beachfront properties are threatened by coastal flooding and storm surges. And finally, there is a lack of well-defined community: the low density, the fences, and absence of community facilities, such as local schools, denies neighborliness, which in turn threatens a community-based approach to resilience.

High Ground Low Ground (Wallis, 2015) looks at the shifting nature of the dunes and wetlands and suggests that, rather than retreating inland, residents should stay and adapt, reconfiguring the urban pattern to accommodate wetlands and create more legible connections to high ground.

The project addresses flooding and coastal hazards as constraints, but also sees these impacts as opportunities to positively alter the way communities have settled coastlines. It proposes retrofitting a linear infrastructure, perpendicular to the coast (Figure 17.1). This post-urban intervention would not require the removal of houses, but it would enable quick access up to high terrain in case of disaster events, and generate a community neighborhood structure that relates to the landscape of foredunes, wetland, and back dunes. Extant backyards would be partially amalgamated to provide a combination of public and private space, allowing wetlands to reestablish, providing biodiversity, options for water storage in peak events, and places of communal recreation and community gardens when water levels are low. New infrastructure, including community facilities (schools, churches, community centers, water towers) would be built on the high ground of the back dunes. The design moves would encourage the currently uncontextualized detached housing units to evolve into small, complex, but manageable neighborhood units (Allan and Bryant, 2011).

An important feature of the project is that it shows the benefits for a coastal settlement, vulnerable to sea level rise, storms, and tsunami, to retrofit and self-organize slowly over time so that it is not only resilient to inundation but also refocuses the existing community toward environment and place. It suggests how the ambiguous public/private landscape between houses in suburban communities can be used to accommodate overlapping environmental, hydrological, and social systems. The realignment of the infrastructure of streets, front yards, and backyards improves the everyday community facilities of residents, while providing an enhanced opportunity for access to high ground.

The project also shows how design can encourage local agency via a process of adaptation that is tied to evidence of indicators. Communities can adapt slowly and accelerate when thresholds are reached and adaptations become urgent, for example when groundwater levels rise in backyards and they become hard to manage individually, when storm surges or earthquakes repeatedly close off roads, or when tsunami warnings become more frequent. The adaptive retrofit of the suburb evolves as a series of irregular-paced, localized, community-based responses, triggered by thresholds, if and when they occur. The design provides a conceptual framework, something to adapt toward, and the act of adaptation is catalyzed by a threshold experience, reinforcing what Nick Jones identified in "xyt:" that thresholds and adaptive responses exist in a contingent relationship.

Wai o Papa/Waterlands

Wai o papa/Waterlands is part of the Deep South challenge, a much larger initiative emerging from the national government's current concern about the impacts of climate change on coastal New Zealand. The Ministry of Business Environment and Innovation (MBIE) has established 11 cross-disciplinary, mission-led challenges requiring collaboration between researchers from universities and other academic institutions, Crown Research Institutes, businesses, and non-government organizations to achieve their objectives. In their brief, MBIE asked that solutions communicate the urgency for adaptation from an indigenous perspective. With research partner Huhana Smith, we established a team of landscape architects, climate scientists, a fluvial geomorphologist, representatives from the local *iwi* (tribes), and artists to show how design could help Māori coastal farmers adapt to the impacts of climate change.

We used a case study site on a 500-hectare coastal farm in the Horowhenua (Figure 17.2). Tahamata Incorporation is a shareholder, collectively owned coastal farm with a Board of Directors governed by the 1993 Te Ture Whenua Māori Act. The land is low-lying coastal plain: part land, part water, with a high-water table, extensive wetlands, and a meandering river, which makes it particularly vulnerable to a wide range of impacts, including pollution from dairy farming and flooding caused by storm surge and groundwater inundation. As part of our speculative design approach, we proposed a series of land-based spatial strategies that could be implemented by farmers over time, together with exhibitions in a variety of venues to capture the design work in progress, raising awareness of the issues among farmers, politicians, and the local community while receiving feedback that encouraged local ownership for the project.

FIGURE 17.2
Bird's-eye view of the farm showing the Ohāu River, parabolic dune landscape, and Tararua Ranges in background. (Image courtesy of Aerial Photography, June 6, 2009, by Laurie Cairns, Palmerston North.)

Unlike urban areas, which have an economy that can build a seawall or a levee, a farm has a limited economy but a lot of fertile land, good for growing food and harvesting water. Climate change impacts including groundwater salinity, loss of productive land, flooding, coastal and riverbank erosion, and a loss of habitat and biodiversity will challenge farming practices and potentially disrupt livelihoods. But these impacts have a positive side too. They potentially act as a catalyst for adaptation toward more resilient, productive, and viable practices, which can contribute in a positive way to the potential food and water crisis as the climate starts to change.

In terms of a methodology that would suit this kind of research, we saw the potential in working through the multiple, knowledge-creation lenses of *Matauranga Māori* (traditional cultural knowledge) and Western science, looking for parallels or slight disjunctions between the two worldviews, so that the resulting tensions expanded the possibilities inherent in each perspective. Three Māori methodologies: *whakapapa*, *hīkoi*, and *kōrero tuku iho* proved to be particularly useful. *Whakapapa* (to layer) is the definitive genealogical reference system for *hapū* (sub-tribe) and *whanau* (family) and their connections to each other, to lands, waterways, ecosystems, and areas of spiritual importance, layered and embodied over time. *Whakapapa* has important resonances with ecosystem thinking, particularly in its codification of the relationship between ecological, meteorological, and human events. *Hīkoi* (walking and talking on the land) encourages firsthand knowledge, and *kōrero tuku iho* (oral narratives) tell stories of the past, the present, and the future, building on *whakapapa* understandings to establish spiritual and physical associations between lands, waterways, and people. The stories often highlight transitional moments and contingencies, and the oral narrative structure was key in communicating the significance of climate change related thresholds and contingencies to local communities.

We began the research with a combination of *wānanga* (workshops) and *hui* (meetings) to consult the farm's shareholders about the project, and give local *iwi* a chance to develop a long-term vision, something to adapt *toward*. These were distilled into a series of principles:

Whanaungatanga ki te whenua: Bringing *whanau* (family) back to the *whenua* (the land)

Puāwaitanga o te whenua: Ensuring the farm is economically viable

Whakahokia ngā kai o te awa: Practicing traditional resource management

Kaitiakitanga mō āpōpō: Protecting the farm for future generations

Tiakitanga o ngā wāhi tapu: Acknowledging and protecting cultural/ancestral sites and burial areas

The research team's climate science focus was on understanding and communicating the climatic and hydrological rhythms that have affected the site for the last 400,000 years—from the melting and freezing of the ice caps to the diurnal fluctuations of the tide and the moon. Land and water on the Horowhenua coastal plain is constantly shifting to these rhythms. In some places, the groundwater still rises and falls with the tides. It is difficult to appreciate this flux because much of it is invisible or it happens too slowly. However, measuring and mapping it, as we did for the first exhibition, highlights the magnitude and pulse of systems, allows us to expect and anticipate change, communicates the very large timescales involved in climate change science, and establishes a relative scale between human and geological time (Figure 17.3).

Perhaps the most pressing and complex issue was how to elicit an adaptive response. We were reasonably certain about the impacts, but the weather and the tides are always volatile. Our challenge was to communicate to farmers, struggling in the short term to make ends meet, that the evident environmental changes were not simply part of seasonal flux. It is very clear if you live on a cliff by the sea and the cliff is eroding, that something needs to be done. It is less clear—but no less pressing—on a coastal plain where change is often less dramatic. Nevertheless, taking steps early to adapt to irreversible shifts might mean the difference between economic viability and financial ruin.

We began by designing a series of possible responses: a toolbox of actions that the shareholders and farmers could implement over time (Figure 17.4). They were framed around a number of ideas. The first involved understanding and valuing existing natural systems, how they were expressed at a local scale, and the role they played in buffering the extremes of climate change impact. On the farm, because of financial pressures and the need to maximize arable land, these systems had become degraded. The coastal dunes had eroded, and most of the wetlands, which absorb flooding and prevent salt-water intrusion, had been drained. The first design responses dealt with protecting and restoring these first lines of defense.

The second set of responses related to imagining an adaptive future, based on the principles of the long-term vision, establishing core interventions, and thinking about how, once constructed, that core might catalyze a series of adaptive activities. So, if the principle

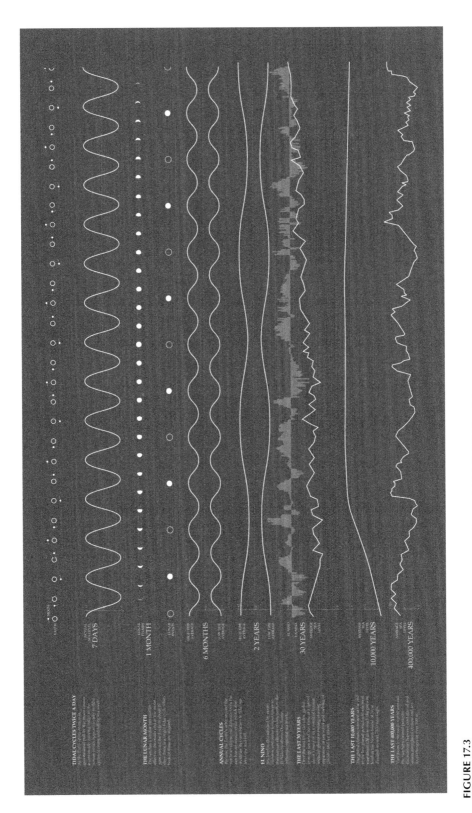

FIGURE 17.3
Cycle of weather and water. (Image courtesy of Penny Allan and Abdallah Richards.)

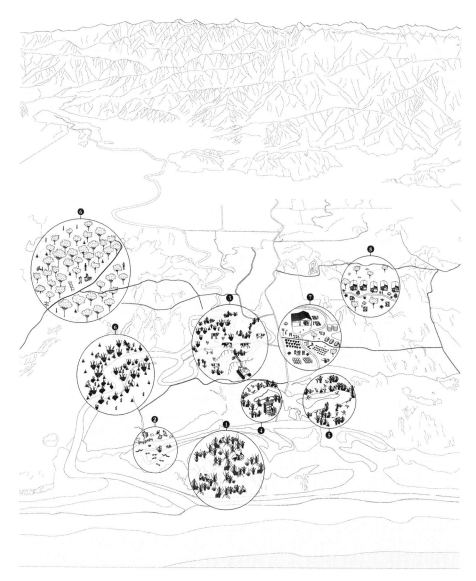

FIGURE 17.4
The toolbox strategies: 1. Protect coastal dunes. 2. Protect wetlands. 3. Protect the most arable land. 4. Protect habitat and biodiversity. 5. Make room for water. 6. Diversify farming practices. 7. Develop adaptive infrastructure. 8. Prepare and settle high ground. (Image courtesy of Penny Allan, Martin Bryant, Miguel Guilarte, and Charlie Curtin.)

was to bring *whanau* (family) back to the *whenua* (land) and settle on high ground, the core might be infrastructure, for example all-weather tracks, tank water, temporary power, and simple platforms. These could support cultural festivals *now* as possible sources of alternative income, while this intermittent adaptation might signal and support the development of longer-term settlement in the future. This approach keeps options open, which is an important corollary of the contingent approach to managing climate change impacts.

The third set of moves related to the idea of diversification. Monocultures like dairy farming close options down. They tend to focus on short-term fixes rather than long-term adaptive solutions. The long-term goal was to ensure the farm was economically viable while also protecting the farm for future generations, so the final set of strategies were conceived around diversification. We identified a range of low-risk farm practices that not only accommodated climate change but also reflected cultural values, for example the harvesting of freshwater and marine species (eels were once plentiful on this land and are considered a great delicacy) and the cultivation of flax. Dairying would continue, but the community might begin to experiment with alternative practices, thereby developing a range of options to choose from should dairying ultimately fail. However, these three types of design response were ultimately ineffective on their own, particularly since the short-term economic viability of the farm was at stake. The designs helped visualize solutions. If implemented now they would benefit the farm, and ensure long-term economic viability but although site-specific, they remained abstract: nice ideas but why bother? Until the immediacy of climate change was effectively communicated, the farmers would remain unmoved.

Insight into this conundrum may lie in the narrative about our house on the coast in Wellington. Firsthand experience of the frequency and severity of climate change impacts showed us that climate change was real, that it was happening now, and that it demanded a response. All we needed was an understanding of what constituted a threshold in that particular environment, the sense, based on firsthand experience of the event, that a threshold had been reached and some idea of possible adaptive actions that would give us a sense of agency should a critical, irreversible event occur. It is difficult to communicate the urgency of climate change through scientific reports, policies, or even a toolbox of site-specific strategies. It takes the personal experience of a storm, a severe flood, or a hurricane for the reality to sink in.

Thresholds are typically assessed through quantitative analysis such as scientific monitoring. But on a farm, scientific monitoring is too expensive and there is no-one to interpret it. So, we determined thresholds from a qualitative perspective, based on the severity and frequency of impacts and their associated implications for the farm. Thresholds included events such as the collapse of coastal dunes, significant change in the river morphology, the failure of pasture species to thrive, and standing water in the paddocks. All would dramatically affect the amount of available arable land. Each signaled an irreversible shift: slow to start and then fast once the threshold had been reached.

Identifying thresholds from a qualitative perspective allowed us to draw parallels between Western science and Māori worldviews of environmental flux. For example, the *maramataka* (moon calendar) (Figure 17.5), developed as a drawing for our second exhibition based on material found in the national archives, represents traditional ecological knowledge. It is an expression of *whakapapa*, the genealogical relationship between people and the complex web of relationships between culture and the environment. It codifies traditional knowledge gained through *direct experience* of the cycles of the moon and the rise and fall of the tides and connects this in a contingent relationship to specific actions, such as fishing, hunting, and harvesting. This kind of relational thinking is a feature of both ecological science and the Māori worldview.

The challenge then, was to harness this thinking and connect it with what we knew from Western science about climate change to create new *kōrero tuku iho*. The third and fourth exhibitions described climate change thresholds as physical events, communicating the experience using images and text to ask the question "what would you do if…?" (Figure 17.6), tapping into traditional knowledge systems to connect future climate change impacts to adaptive responses, and embedding the experience in the rhythms of everyday life. By doing so we linked the cycles of the first two exhibitions with the threshold-contingency "call to action," so when thresholds occurred, locals might be in a position to recognize them and act accordingly.

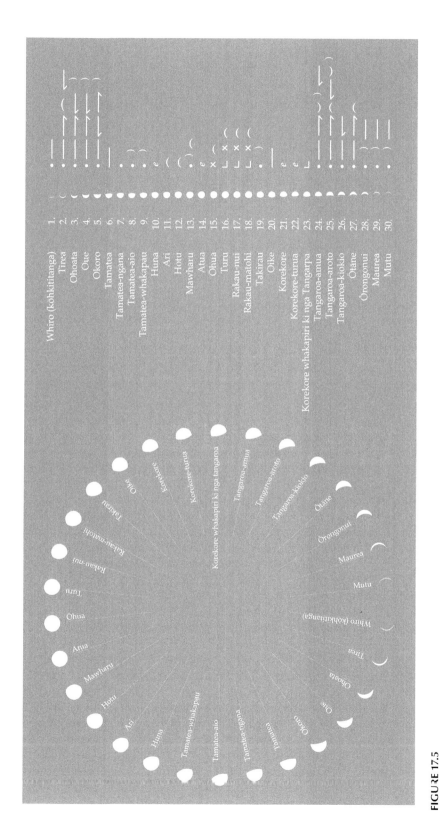

FIGURE 17.5

Mararataka prepared for second exhibition. Each entry on the right represents a lunar day, for example *Huna*: bad weather, food products suffer; *Ari-roa*: favorable for spearing eels; *Maure*: a fine, desirable day; *Māwharu*: crayfish are taken on this day; *Ohua*: a good day for working. For a full key, refer https://teara.govt.nz/en/maramataka-the-lunar-calendar/page. (Image courtesy of Huhana Smith and Abdallah Richards.)

FIGURE 17.6
The Kōrero Tuku Iho shed showing banners asking the following threshold questions: "What would you do if: … Coastal dunes erode during storms and high tides? … You have trouble getting pasture species to thrive? … You lose arable land to riverbank collapse? … Floods restrict access to the property for several days? … You notice standing water in paddocks normally dry in winter?" (Image courtesy of Penny Allan.)

Conclusion

Several key ideas emerge from this design research. These are concerned with the role of design in the adaptive process and how design might address certain obstacles to effective local and community-based adaptations, such as inertia, the tendency toward planning and policy based solutions, and in coastal rural areas particularly, the prohibitive cost of undertaking adaptations.

Design is useful in this context, but we need to be aware of the type of design process that is most effective here. Sometimes design methods can be reductive, overly programmatic, stylized, or driven by a Western scientific rationale. None of these design processes suit this problem. In this project, design's role is to anticipate and encourage adaptive change and ensure that critical decisions are always open-ended. Keeping options open means that there is always a multiplicity of potential ways forward, and the risk associated with making decisions when it is too late is minimized. This kind of design is speculative and agile rather than prescriptive, and it embeds thresholds and contingencies as part of its process.

Thresholds are powerful stimuli to action, perhaps because by their very nature, they are indeterminate. Knowing what constitutes a threshold, how thresholds "look and feel," and what one might do about them can empower communities to adapt, transferring climate change agency from governments to individuals and from the global to the local. It means there is something one can *do*. Designers can assist by identifying and communicating the nature of thresholds, their implications, and the contingent relationship between thresholds and a suite of possible adaptive responses. The methodologies employed in this research show that the relationships can be enabled by a clear understanding and interpretation of culture. The indigenous culture and its worldview provided the link between thresholds and design actions in the farm case study. In the coastal suburban case, a more traditional Western approach to neighborliness provides the foundation for understanding of thresholds and the responsive design actions.

Identifying, through design, what is core to a vision of the future allows local interventions to be small, slow, targeted, and responsive to suit the scales of economy of local farmers and collectives of residents. Slow and small progress still has the potential to generate transformative outcomes. Given the uncertainty of climate change, the value of a suite of possible spatial solutions like the design toolbox is that it allows individuals and communities to adapt if and when necessary and in a way that is specifically tailored to local impacts. A local, collective response to climate change adaptation that is site-specific and synthesizes a number of issues together with climate change, encourages the goodwill of the community to take advantage of these improvements in a collaborative way. The economic benefits of this kind of slow and adaptive approach may in fact be the most realistic way to begin to deal with enhancing resilience.

References

Allan, P. and C. Stutterheim (eds). 2007. *MOVED to Design*. Melbourne: RMIT University Press.

Allan, P. and M. Bryant. 2011. Resilience as a framework for urbanism and recovery. *Journal of Landscape Architecture* 6(2):34–45.

Bryant, M., Allan, P., and S. Kebbell. 2017. A Settlers' Guide: Designing for Resilience in the Hinterlands. *Buildings Buildings*, 7(1):23; doi: 10.3390/buildings7010023.

Helm, D. 2015. *Natural Capital: Valuing the Planet*. New Haven, CT, USA: Yale University Press.

Kane, H. H., Fletcher, C. H., Frazer, L. N. et al. 2015. Critical elevation levels due to sea-level rise in Hawai'i. *Reg Environ Change* 15:1679. doi: 10.1007/s10113-014-0725-6.

Shand, R. 2013. *Assessment of the Kapiti Coast Erosion Hazard Assessment Report 2008 and 2012*. http://www.kapiticoast.govt.nz/contentassets/a933446e8c094de8a946d20b9f36a1de/assessment-of-the-kapiti-coast-erosion-hazard-assessment-report-2008-and2012.pdf. See also Blundell, K. 2014. Residents win hazard line challenges. Dominion Post Wellington. Available from http://www.stuff.co.nz/dominion-post/news/kapiti/9854214/Residents-win-hazard-lines-challenge (accessed November 8, 2017).

Speaks, M. 2006. Theory was interesting but now we have work. *Archit. Res. Q.*, 6:209–212.

Walker, B., Holling, C. S., Carpenter, S. R., and A. Kinzig. 2004. Resilience, adaptability and transformability in social–ecological systems. *Ecology and Society* 9(2):5–13. doi: org/10.5751/ES-00650-090205.

Wallis, C. 2015. High Ground, Low Ground: Explorations in Topography and Neighbourliness in Coastal Dune Settlement. *Victoria University of Wellington. Masters thesis*.

18

The Mekong Delta: A Coastal Quagmire

Bruno De Meulder and Kelly Shannon

CONTENTS

The future of the Mekong Delta, specifically in relation to the contemporary challenges of climate change, must be grounded in an understanding of both its geology and its context-embedded cultural practices of agriculture, settlement, and floodwater management. This chapter will first briefly review the physical and socio-cultural formation of the Mekong Delta and then introduce the Asiatic concepts of "hydraulic civilization" and *desakota* in order to conceptually expand both past and possible futures of humankind's organization and settlement of the territory. Finally, the chapter will focus on a plan that was co-produced by a team of international consultants (led by the authors) with the Southern Institute for Strategic Planning for the Ministry of Construction. The Mekong Delta Regional Plan 2030, Vision 2050, was envisioned to respond to warming waters, rising seas, and saline intrusion, and yet not fall back on tried-and-true, global models of adaptation and resilience. It sought to create a contemporary continuum to the region's material and cultural practices; to update its "hydraulic civilization" and *desakota* settlement tendencies. The main impetus was to accentuate, intensify, and differentiate its regional assets and maintain its vital importance as a fertile, productive urban delta.

Domestication of the Mekong Delta is relatively young, yet has been dramatic and profound. Revolution and rupture have always been more a part of the territory's history than simple evolution or change. During the last three centuries—through both Viet and French colonization—the Vietnamese component of the delta (40,576 square kilometers (km^2) of the 49,500 km^2 catchment) (Biggs, 2010) transformed from a quagmire, a "heavy water-saturated mud" or "marshy swampy ground,"[1] into a monumental cultural landscape. The original, "as found" landscape was systematically and almost completely altered (Figure 18.1). The frontier river-water civilization (*van minh song nuoc*) consisted of minimal dispersed occupation by Khmers and Chams, marginal Chinese trading posts among the (as majestic as endless) mangroves, and Viet villages resulting from those who migrated from the North (Brocheux and Hémery, 2009).

Today, all appears as radically inverted. The extensive canal mesh made every inch of the hardly penetrable quagmire integrally accessible. The Vietnamese component of the delta accommodated 17.59 million people in 2015 (GSO, 2016). From an almost inaccessible region characterized by an immensely rich and dense tropical biodiversity, the Mekong

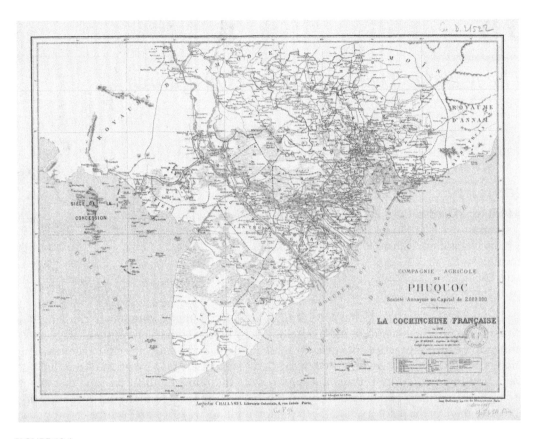

FIGURE 18.1
Cochinchine Française 1891. Throughout the Mekong River Delta, there was intense settlement along rivers and along the coast. The rapid domestication of the water-sick territory radiated from Saigon and anchored itself on the riverbanks (the alluvial higher land banks resulting from sedimentation) and the coast in relation to economies of agriculture and fisheries. (Bibliothèque nationale de France, GED-3663; FRBNF40727469.)

delta has become an icon of endless monoculture, methodically unlocked by an enormous and fine-meshed network of canals and rivers. Successive enterprises of hydraulic works—extensive canal and irrigation networks—have been inscribed in the terrain, allowing a green revolution to flourish (beginning with fertilizers and pesticides for agriculture and more recently with antibiotics for aquaculture). The region has historically been the rice basket of Vietnam. During the process of modernization, urbanization, and densification, the marginal local economy initially grounded in subsistence has been overwhelmed by an escalating race of ever-increasing economies of scale and productivity gains. Vietnam is presently the third-largest exporter of rice in the world (USDA, 2014), with just over half hailing from the Mekong Delta.

The territory, 0–4 m above mean sea level, is incredibly varied and rich, due to its complicated system of nine estuaries and its long history of a densely interlaced river and canal network and submerged forests. Tidal pulses from the Gulf of Thailand in the west and the East Sea penetrate up to 100 km upstream in the tributaries of the Mekong, complicating irrigation and transportation. The highly developed water management system includes over 7,000 km of main canals, 4,000 km of secondary canals, and more than 20,000 km of protection dikes to prevent early floods (Le et al., 2007), in addition to floodgates, saline protection dams and dikes, sluices, and pumping stations. Channel density is estimated at 20–30 m/ha, and the channel area occupies 9% of the delta area (Nguyen, 2002). Canals and channels are part-and-parcel of the Mekong Delta's mutation from a quagmire into rich agri- and aquacultural mosaic.

Most interesting is the very notion of "the coast." The geomorphology of the Mekong Delta is in a constant state of flux due to the intense interaction between river and sea waters. The delta coast has prograded more than 200 km seaward from Cambodia over the past 8,000 years as a result of the voluminous supply of sediment from the Mekong River. The Mekong is among the top ten rivers in the world in terms of sediment discharge—known for an astonishing 145 million tonnes (U.S. metric tons) per year into the East Sea. However, it should be noted that this volume was recorded in 1993 and has been declining due to various reasons, the primary one being the extensive dam building upstream but also forest clearing, arable land expansion, and mere water diversion. According to some, in 2009, the number was less than half, down to 68 million tonnes per year (Kummu et al., 2010). The complex bathymetry and the delta's different tidal circulation systems result in the fact that the delta consists of a 150 km wide subaerial plain and a subaqueous portion. There is a complex movement of sediments from the Mekong that is both wave- and tide-driven, in addition to being controlled by the seasonal monsoon system, which, in the East Sea, shifts direction twice annually in May (from southwestward to northeastward) and October (from northeastward to southwestward) (Xue et al., 2014). There are multifaceted processes of both sedimentation and erosion along the delta's coastline. Presently, sedimentation still prevails near the estuarine inlets, but due to natural and human-induced causes, erosion is occurring away from the inlets and it is anticipated that erosion will increase in the future (Phan et al., 2015).

At the same time, one might read the Mekong Delta as one giant mud plain, a conceptually thickened coastline, one enormous quagmire: "a largely waterlogged world of black mud and mangrove trees, bordered by thick tropical forests where the land rose away from the flooded plain" (Osborne, 2000:21). In such an immeasurable quagmire, gradients of wetness and slight, even minimal differences in topography and soil differentiations and water conditions, lit by the tropical sun, generate a wide variety of ecological conditions, resulting in an astonishing biodiversity where everything is, to a certain degree, about simultaneity: land and water, forest and salt marshes, flora and fauna, birds and fish. The land is saturated with water and the water is full of land. One could say that the mangroves, which covered substantial parts of the "as found" landscape, function as a relatively dark interior that houses an amazing combination of living things. During the domestication process, many of the variations and differentiations were replaced by a simple and categorical trichotomy: water (in rivers, canals, and irrigation ditches), lowlands (paddy fields), and highlands (natural and man-made, including river or canal embankments that are generally stabilized by tree planting, landfill areas for urbanization, and naturally higher elevations of alluvium with orchards and vegetable production). Over time, gradients and transitions in the landscape were systematically exchanged for borders: organic for geometry, curves for orthogonal. Porous environments, where different habitats interpenetrated and overlapped, mutated in a juxtaposition of clear categories: water, paddy fields, and highlands that are stabilized with trees. Nowadays, an enormous wave of fish and shrimp ponds have invaded the territory and extend the water landscape, with what one could consider a mosaic of reconstructed shoreline conditions. Studies have also shown that shrimp farming, which replaces rice culture at an accelerating rate in many parts of the delta, is not only an opportunistic reaction to salination, but also a far more lucrative one than that of rice. Shrimp farming has gained an irresistible momentum as the way forward, out of poverty for many, while causing rapid soil exhaustion and generating many other thorny ecological issues.

"Hydraulic Civilization" and *Desakota*

According to renowned professor of political science and anthropology James C. Scott, the Mekong Delta has become, like so many other deltas, a "high-modernist space" (Scott, 1998). It is a highly ordered, legible, and standardized landscape. As mentioned previously, through imperial and colonial engineers, American advisers, and successive waves of socialist hydro-technicians, the landscape was transformed from a virgin, almost uninhabited jungle into a densely occupied and productive territory (Figure 18.1). Traditionally, populations in the Mekong Delta settled on the natural levees of waterways, creating the river-water civilization. Drainage canals enabled the "river-water" or "canal-creek" civilization to develop linear settlement structures, following the alluvial, unsalinated, high-land banks of waterways while the swampy area of the delta was progressively transformed into fertile alluvial plains for wet paddy cultivation (Figures 18.2 and 18.3). Traditionally, life in the delta was highly dependent on the rhythms of river water regimes. Nuances in the Vietnamese language underscore the sensitivity of *living with water*: it distinguishes between several kinds of floods: *lu* (flood), *lut* (inundation), and *ngap* (submergence). Floods occur when excess water upstream flows over riverbanks or dikes to floodplains. Inundation is effected by floods. Submergence is a phenomenon that occurs when water levels in the plain (fields or land of the inner dike or depression areas or wetlands) are higher than the ground surface area. Inhabitants of the Mekong Delta call the yearly September to October flood period *mau nuoc noi* (water rising season), and *nuoc son* (reddish water) refers to the water color of heavily silt-laden water from upstream. *Nuoc bac* (silver water) refers to greenish and transparent water that flows from fields with acid surface soils (Le et al., 2007).

FIGURE 18.2
Water-based urbanism. A series of French colonial images (1919–1935, by L. Busy and R. Tetard) reveal vibrant urban and rural water-based life throughout the Mekong Delta, including in Rach Gia. The overwhelming productive activity of the time was wet rice cultivation. (Archives nationales d'outre mer, FR CAOM 30Fi107/102.)

FIGURE 18.3
Paddy Landscapes. Busy and Tetard also captured the overwhelming productive activity of the time was wet rice cultivation. (Archives nationales d'outre mer, FR CAOM 30Fi103/40.)

The Vietnamese particularities dovetail into two larger, historic discussions that could potentially frame contemporary development of the delta. A number of Asian settlement concepts developed through the writings of Western Sino-experts Karl Wittfogel 'Hydraulic civilizations' 1956 and Terry McGee ('*Desakota*,' 1967, 1991). They interrogate Asiatic-specific modes of urbanization and their interdependent relationships to landscape. Hydraulic civilization refers to the *raison d'être* of societies, which required substantial and centralized works of water control and, in turn, was reflected in political power and leadership. The concept of an urban countryside/rural metropolis—the mix of consumptive and productive dispersed landscapes—is behind the Indonesian word-play of *desakota*.

Wittfogel, the Frankfurt School historian and sinologist, referred to classical economists, "who with varying consistency recognized that the higher agrarian civilizations of the 'orient' and their urban rural conditions followed a pattern of development decidedly unlike that of the West" (Wittfogel, 1956:152). For him, Asia boasted a juxtaposition of urban and rural, a specific type of social formation founded upon centralized state authority with its own forces and relations of production emerging out of water engineering and control. He set forth a thesis that "hydraulic societies" and despotism were functionally connected. Specifically, he argued that in naturally arid regions, only an absolutely obedient, virtually enslaved regime could possibly have mobilized the concentrations of labor needed to man and maintain the irrigation canals and dikes on which intensive agriculture depended. "Where agriculture required substantial and centralized works of water control, the representatives of government monopolized political power and societal leadership, and they dominated their country's economy. By preventing the growth of strong competitive forces, such as a feudal knighthood, an autonomous church, or self-governing guild cities, they were able to make themselves the sole masters of their society. It is this combination of hydraulic agriculture, a hydraulic government, and a single-centred society that constitutes the institutional essence of hydraulic civilization" (Wittfogel, 1956:153). Wittfogel's nomenclature admittedly stresses more human action rather than geography and, as he argues, the notion of a hydraulic civilization having developed "a *bureaucratic* land-lordism, a *bureaucratic* capitalism and a *bureaucratic* gentry … and a cumulative tendency of unchecked power …" (Wittfogel 1957:4, italics his).

The comprehensive system of the "hydraulic civilization," which employed extensive *corvée* labor and placed a constant upward pressure on the level of population growth, not only created productive water works (for irrigation and drainage) and protective water works (for flood control), but also provided drinking water and communication conduits. Digging, dredging, and damming led to the significant remolding of the ground plain. Wittfogel's thesis is predominately concerned with naturally arid regions, but also comments upon regions of wet-rice cultivation which requires a relatively equitable distribution of water and necessitates a system of canals, dikes, irrigation canals, terraces, and locks to regulate water levels. The notion of a hydraulic civilization—beyond its ties to politics—holds invaluable lessons and inspiration for contemporary deltas. Wittfogel's hypothesis of a "hydraulic civilization" clarifies a man/nature, water catchment/settlement relationship that, leaving the despotism argument aside, needs to once again become fundamentally understood, interpreted, and (re)designed.

As in so many deltas, the fertility of the Mekong Delta's land attracts farmers and generates massive development. And, as in most fertile deltas, development is as intense as it is dispersed. Development opportunities are equally spread all over the territory and are also so excessive that hierarchy is senseless. In opposition to the central place theory of geographer Walter Christaller (Christaller, 1933), settlements are rather equal and above all multiple. If they are differentiated, then it is rather by specialization (such as port city, coastal city, educational center, etc.) than by hierarchy, the exception perhaps being Cantho, the largest city in the delta, centrally located, and at the confluence of the Hau (the lower branch of the Mekong) and Cantho rivers. During the first wave of the delta's urbanization, with the Nguyen lords' 1739 "march to the south," settlers were allowed to freely occupy land and primarily settled in the alluvial plain around Cantho. Land reclamation and settlements were later established through decentralized agricultural colonies (*dinh dien*) and military colonies (*don dien*). When the delta fell under French control in 1876, it was considered the "granary" of Saigon, and massive infrastructure building and densification occurred to exploit commercial agriculture. During the Vietnam War, the Vietnamese "agroville" program was succeeded by the American "strategic hamlet" program, both self-contained, camp-like communities which failed miserably on all accounts.

Today, the Mekong Delta remains a predominantly agricultural area, not to say a rice region. It is densely occupied and generates a very specific settlement pattern, namely McGee's *desakota,* from Bahasa Indonesian *desa* for village and *kota* for town or city. *Desakota* is a hybrid *in situ* urbanism which consists of the overlapping of functionally independent entities into already densely populated agricultural regions, materializing in traditional agriculture existing alongside industry (capital intensive and cottage industry), entertainment (including theme parks and golf courses), retail (malls and strip shopping), and housing (from squatter housing to gated communities).

Theorist Stephen Cairns has recognized three noteworthy repercussions of *desakota*: "First, *desakota* regions are so vast and difficult to navigate that they produce administrative 'blind spots' in which planning regulations are not enforceable in any uniform way. Second, *desakota* regions feature a high mobility of goods and services and a transient pattern of settlement. Third, and closely related, *desakota* regions resist being taken-up into a more formal system of inter-connected, functionally specialized zones" (Cairns, 2002:118). Importantly, there is potential in the concept that to date has remained useful in terms of describing the context. The recognition and abstraction of the synergy between the urban and rural, the consumptive and productive landscapes, can become potent in the reconceptualization of the delta's "field conditions" to create a more differentiated landscape mosaic, which generates neither hierarchy nor concentration, but rather accentuates diffusion of organized dispersal.

Contemporary Challenges

It is clear that the water-society nexus has had tremendous influence on the Mekong Delta's political formation, socio-spatial order, and economic growth. Numerous scholars have developed lengthy studies on these topics (Taylor, 2001; Shannon, 2004; Mollinga, 2008; Molle et al., 2009; Biggs, 2010; Benedikter, 2014). The contemporary paradigm of land and water management is currently undergoing a fundamental shift that climate change will inescapably further articulate. Living with the floods (in opposition to defensive living behind dikes) and the interconnected canal system is no longer as evident. As previously mentioned, the entire hydrological regime of the Mekong River[1] has been and continues to be reengineered by Vietnam's riparian neighbors, in particular with dam construction by China and Laos. Water pressure in the Vietnamese Mekong is consequently decreasing. The delta is under pressure from overexploitation of basin resources while upstream damming has fundamental consequences downstream, including the alteration of the flow and nature of the river, disruption of fish migration and significant impact to inland fisheries and food security, threats to aquatic biodiversity and terrestrial system changes due to the natural flood cycles crucial to ecosystems, and, importantly, disruption of recession agriculture—the practice of planting crops in floodplains watered by receding floods (Chellaney, 2011).

The delta today is already bearing witness to the impacts of climate change. Sea level rise, saline intrusion, massive inundation, and periods of intensive drought and rising temperatures have begun to affect the region and are revealing unavoidable consequences on the productive landscapes, rural settlements, as well as for urban areas. The manipulated water pressure from upstream dam building, coupled with sea level rise, makes saline intrusion a tangible threat to enormous territories, especially in the remote southern Ca Mau peninsula and the other coastal regions. The construction of large, protective sea dikes is now on the way in exactly the same region that has always thrived solely from natural processes of erosion and sedimentation (De Meulder and Shannon, 2013). The sea dikes would safeguard coastal security and prevent saline intrusion. Other parts of the landscape, particularly the low-lying Plain of Reeds (where the Mekong enters Vietnam) will be faced with more dramatic and systematic inundation. In short, a new geography is in the making.

The existing Mekong Delta Region Plan (approved in 2009) was solidly based in conventional planning concepts. As if it were illustrating a textbook of ideal abstract planning principles, it superimposes a strong hierarchical system of cities and rigorously applies traditional, sectorial land use planning concepts, which is radically in opposition with the natural condition of the territory, as well as the intense urban culture that mixes urban and rural, consumptive and productive land, formal and informal economies in an inimitable and thriving manner. The plan that resonates rigid, central economic planning dogmas also foresees a conventional transport model with an almost equal distribution of roads and highways over the territory. This dense infrastructural grid would literally equalize accessibility and be expected to catalyze internal, as well as external exchange in the wake of which trade and industrial development would emerge. It was seen as an urgent necessity to absorb the very fast-growing labor force that exceeds the needs of agriculture. Until today, however, a large part of the surplus labor force is, as it traditionally always has been, attracted to Ho Chi Minh City and many of the sometimes very large scale "Industrial Parks" (IP) and "Economic Zones" (EZ) in the Mekong Delta. The gap between planning based on projection targets and political proclamations in the traditions of central plan economics and current realities on the ground could not be more caricaturized. Processing of local agri- and aquacultural produce is instead a main and growing industrial activity. The Mekong Delta is "losing" population. The question, of course, is how incorrect it is that the Mekong Delta only retains the population that it can effectively support as a predominantly agri- and aquacultural region. Government policies stimulate generic and large-scale industrialization in order to provide local jobs to the fast-growing population that frequently and naturally opts for migration.

Presently, the Mekong Delta Region Plan has no agency, no mandate on the ground. The spatial development of the region is effectively steered—as much as development can be guided in wild ad-hoc development—by the master plans of the region's 12 provinces and one first-class city (Cantho).[2] Each plan has been prepared with the socio-economic objectives of the individual provinces/city in mind and is obviously not coordinated with any of the others. (Is there any place on earth where plans on different levels form a coherent whole?) The provincial master plans are an ambitious exercise in the spatial translation of socio-economic projections (and hence attempt to substantiate larger objectives, meaning political and financial claims of the central state) that are too abstract and often alien to the dynamics, characteristics, and realities of the specific geographies concerned. As the objectives and proposed measures are the same for the whole of Vietnam, the existing planning and policy mainly lead to an undifferentiated generic development.

In addition to climate change impacts, the almost systemic mismatch between planning and realities on the ground was a primary motivation for a revised Mekong Delta Region Plan. Another motivation was the turbulent economic and urban development of the last two decades. Massive growth of the middle classes led to explosive building activity. This might not be uncommon in the Asian context, but it remains remarkable nonetheless. In the Mekong Delta region, a generic Vietnamese urbanization is blanketing the territory. At a very rapid pace, the existing indigenous and distinctive urban architectures and morphological particularities are being wiped out—including the predominant importance of the water structures that previously determined the spatial structures of cities and towns. The majority of the settlements that characterize the region are generic, young, and not necessarily aging very well. They are without noticeable variation and are fueled by the generous provision of stereotypical, road-based new allotments and other service building sites, with generally oversized road infrastructure (and a remarkable lack of parking space). One can only pity such a government-driven unleashing of speculation. In the best cases—as can be witnessed in cities such as Ca Mau—intelligent cut-and-fill principles were applied, which resulted in lakes around which new urban quarters were developed.

A New Delta Plan: Return to the Waters

The revision of the Mekong Delta Region Plan[3] seizes the opportunity of climate change (and the way it reconfigures the geography of the territory) to realign the plan (and consequent development) with the (evolving) characteristics of the territory. The main question of the new proposed regional plan was how to organize a constructive interplay between unavoidable landscape dynamics and the social, economic, and cultural dynamics of the region. The main asset of the delta (surely in view of the expected worldwide food shortage) remains its enormous agri- and aquacultural potential. Such a territorial identity is largely the outcome of the interplay between land and water, or more precisely topography (including bathymetry), soil qualities (most broadly alluvia, saline, and acid sulphate) and conditions, and salt and freshwater parameters (volume; quality, including carried sedimentation; height and extent; tides and seasonal variations). Recognition and accentuation of the delta's underlying geography, which has been classified into six broadly agro-ecological sub-regions[4] (excluding the seven hills and mountainous areas near the Cambodian border), can reestablish its core identity, counterbalance the relative homogeneity of the region's urbanism and rearticulate the productive landscapes of the delta's dynamic ecological territory (Figure 18.4).

The Vietnamese predicted consequences of climate change by 2050,[5] which in general will create more extreme conditions and inevitably strengthen the differentiation between the six sub-regions (Figure 18.5). The regions include three coastal sub-regions along the East Sea and Gulf of Thailand where saline intrusion will drastically change vegetation and urban, as well as agri- and aquaculture development potentials: the Eastern Coastal Zone, Ca Mau Peninsula at the tip of the delta, and the Long Xuyen Quadrangle along the Gulf of Thailand and bordering Cambodia. The Dong Thap Muoi floodplain (500,000 ha), popularly known as the Plain of Reeds, is the delta's major catchment basin. It is a large natural depression and lowest elevation of the delta (0.5 m below mean sea level) where profound inundation will extend. Finally, there are two centrally located regions: the freshwater alluvial (along the two branches of the Mekong River itself) and Trans-Bassac zones. These regions will remain relatively "safe" from both saline intrusion and inundation and are historically and presently where most of the delta's development has occurred.

These six sub-regions define significant differentiation in the identity of the Mekong Delta. Each of them has its own characteristics due primarily to its soils and hydrology,

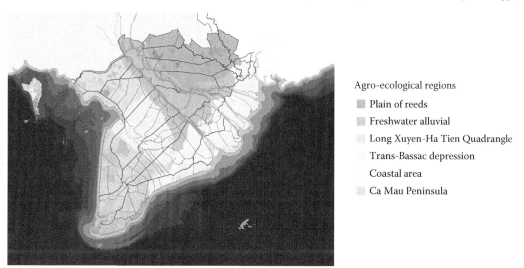

Agro-ecological regions
- Plain of reeds
- Freshwater alluvial
- Long Xuyen-Ha Tien Quadrangle
- Trans-Bassac depression
- Coastal area
- Ca Mau Peninsula

FIGURE 18.4
Agro-ecological regions of Vietnam's Mekong Delta. The identity of the Mekong Delta is grounded in its landscape and its six main ecological/geographic regions. (Image credit: RUA, 2016.)

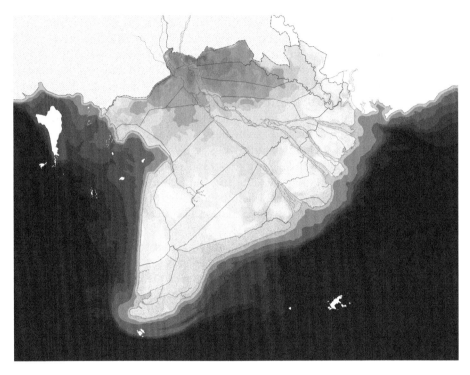

FIGURE 18.5
Vietnam's official predicted inundation and saline intrusion in 2050. Inundation is expected to severely affect the Plain of Reeds in the northwest of the delta, while saline intrusion will be extreme along the coastal area, in the Ca Mau Peninsula, and in part of the Long Xuyen-Ha Tien Quadrangle. (Image credit: RUA, 2016.)

and the revision of the plan advocates the anchoring of development scenarios on these distinctions which, as specific and natural locational assets, can be ultimately turned into unique competitive advantages that complement each sub-region (Figures 18.6 through 18.8). The provinces each could define a further level of differentiation, but until now have only functioned at the national level to implement general policies. Differentiation on this level rather seems to have to do with the dynamism of the administration and its ability to acquire national funds, than with specification of needs or character. Finally, the water management units (120 in total, on average 20 by sub-region, or nine in each province) are of strategic importance, since water regulation (controlling quality and quantity of water allocation over time) defines the territory through compartmentalization. At an even lower level, the drainage structure and parcel system define the mesh of the mosaic landscape, but it is clear that it is the 120 water management units that define, the most important scale of differentiation in the landscape, in addition to the six sub-regions . The Mekong Delta Region Plan explicitly chooses to anchor future economic potential in these six regions and to create a regional center for think tanks and innovation centers for new productive research in each one. In many instances, these coincide with provincial capitals, but not always. Also, the provincial capitals of the MDR will be given socio-economic boosts to support the renewed productivity of each specialty of the sub-region.

Aside from the six agro-ecological sub-regions, the proposed Mekong Delta Region Plan defines three major urban and rural morphologies as a system of wet and dry areas derived from soil types, sedimentation, erosion, and subsidence, as well as the 2050 predictions of inundation and saline intrusion. In all three typologies, the hierarchy of

FIGURE 18.6

Plain of Reeds: Tram Chim flooded forest. The protected area covers 7,588 hectares and is a national park, specifically created to protect several rare birds whose habitat is the inland submerged wetland of inundated grasslands, open swamps, and melaleuca. There is an opportunity to expand the Ramsar-protected landscape. (Image credit: K. Shannon, 2014.)

FIGURE 18.7

Vast rice fields of alluvial plains. The Mekong Delta is the rice basket of Vietnam. Vast waving fields of green cover most of the territory, providing not only food security for the nation, but also an export economy. The proposed plan argues that the Mekong Delta should continue to capitalize and improve on its agricultural legacy. (Image credit: K. Shannon, 2014.)

FIGURE 18.8
Mudflats and mangroves of the East Sea at Bac Lieu. The tidal mudflats and mangrove forests of Bac Lieu along
the East Sea are important dynamic ecosystems that are reclaimed for aquaculture and industrial development.
There is a challenge to keep ecologic and economic forces in check. (Image credit: K. Shannon, 2014.)

topography, infrastructure, and settlement is reconfigured in relation to the site-specific
realities and future projections (Figure 18.9). A "coastal areas" type would be developed
for the Eastern Coastal Zone, Ca Mau Peninsula, and the coastal part of the Long Xuyen
Quadrangle to form part of a more dynamic and robust coastline. Eventually, sea dikes
would be reconceptualized as just one possible element in the complex landscape that
organizes the transition between land and sea. Massive mangrove forests would be the
more general and natural component of this landscape zone. They would simultaneously
allow for a better mix of saline and freshwater systems, providing protection against
the increasing threat of storm surges and allowing for a more sustainable aquaculture
economy. This zone would also create an economy embedded within an ecology rather
than a vulnerable monoculture that replaces an ecology. Mangrove forests can accelerate
the gaining and consolidation of land by natural sedimentation processes at the coast
of the East Sea and the Gulf of Thailand. Where natural conditions are suitable, dune
formation, land gains, and their consolidation can be enhanced with simple "sand or
sediment catching" by way of vegetation screens, breakwaters, mangrove forests, and
so on. In this respect, the Mekong River's yearly discharge of 68–145 million tonnes of
sediment is noteworthy. Large amounts of these sediments are carried back and forth
along the southern and eastern coasts, depending on the monsoon seasons. Settlements
in these "coastal areas" would be developed to work with the East Coast's parallel and
alternating rows of sand ridges and back-swamps, successive peat layers, and subsiding
lands of the Ca Mau Peninsula. Directly along the coast, settlements would comprise
clustered, floating buildings that center on collective spaces in clearings of mangrove
forests. Aquaculture (fish and shrimp farming) would be interwoven into a rich mosaic
able to withstand the tidal pulses from the East Sea and Gulf of Thailand. The *desakota*
principles would adapt to this new situation; organized dispersal of floating settlements
would respond to tidal pulses and sea level rise. Densification is not a main issue in
coastal areas where expected population growth is to be marginal, however, the security

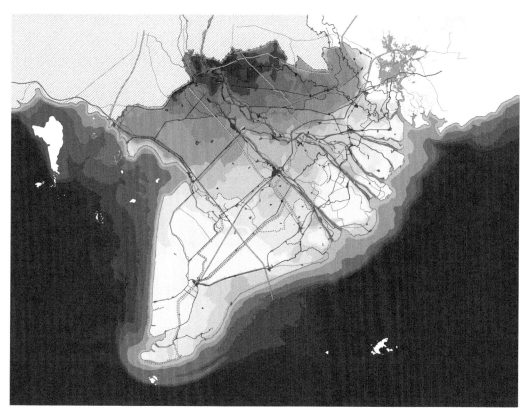

FIGURE 18.9

New urban morphologies and infrastructure to respond to climate change. Three distinct morphologies are developed: coastal areas that take into account sea level rise and saline intrusion and are protected by massive new mangrove afforestation; a freshwater alluvial area that needs qualitative upgrading; and the extended floodplain near the Cambodian border which will have to deal with high degrees of inundation. (Image credit: RUA, 2016.)

and the re-articulation of the relationship of settlements with evolving environmental conditions and their economic exploitation is. Continued dispersal seems indicated, despite the national centralization policy in rural areas. Within this setting of new mangroves and water bodies, a new and complementary aquacultural economy could be built with a larger emphasis on harvesting naturally renewable resources (there is great potential for wind and tidal energy) and high-end specialized products, such as rare medical plants. As such, these additions, which would succeed what until now was a largely mono-cultural economy, will strengthen the robustness of the economy and the sustainability of the development of the Mekong region, while simultaneously rebuilding a biologically diverse environment (Figure 18.10).

An "extended Plain of Reeds" type would be developed for Dong Thap Muoi and part of the Long Xuyen Quadrangle. The area is expected to be significantly inundated as time marches forward, and thus the region could once again thrive as an area of controlled flooding and water storage. Buildings would be on stilts and the central spine of clustered areas of urbanization would be a dike, at a minimum safety height to avoid inundation in the predicted worst-case flood scenario. Large monsoon collection basins would be inscribed into the existing canal and irrigation system and would be able to release freshwater into the delta system during the dry season, as well as serve as aquaculture

FIGURE 18.10
(a,b) Coastal morphologies. In the coastal areas, the new settlement strategy would comprise clustered, floating buildings that center on floating collective spaces in clearings of mangrove forests. Aquaculture (fish and shrimp farming) and renewable energy would be interwoven into a rich mosaic that is able to withstand the tidal pulses from the East Sea and Gulf of Thailand. (Image credit: RUA, 2016.)

landscapes. Nowhere here, other than in one center of excellence/think-tank for productive landscape experimentation, is densification on the agenda. Fish farming, indigo farming, and other activities that can flourish in such extreme environmental conditions would be strengthened (Figure 18.11).

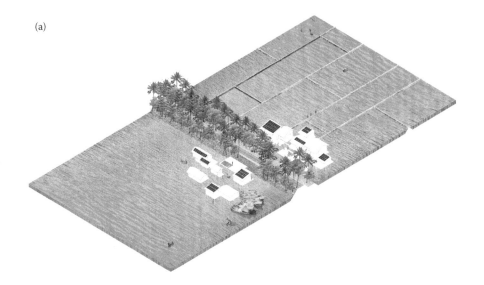

FIGURE 18.11

(a,b) Upstream flooded morphologies. Upstream flooded morphologies are premised on the prediction that massive inundation will occur in the coming years and progressively increase over the decades. Buildings would be on stilts and the central spine of clustered areas of urbanization would be a dike, at a minimum safety height to avoid inundation in the predicted worst-case scenario. Monsoon collection basins would be able to release freshwater into the delta system during the dry season, as well as serve as aquaculture landscapes. (Image credit: RUA, 2016.)

A "freshwater alluvial" type would be developed for the Trans Bassac Depression and the Freshwater Alluvial zones where a requalified system of urbanism on relatively high land would work together with a diversified productive landscape of rice, orchards, and vegetables. The existing settlement types would be modernized, densified, and clustered.

This area, the relative "highland" of the delta, is the most fertile and the most productive, and will be the region of the delta least affected by climate change. Paradoxically, it will gather the largest concentration of people in the near future as a consequence of climate change and the restructuring of the agriculture in the other areas of the Mekong Delta. Therefore, extension of settlements will be severely restricted and needs to be cleverly redirected toward densification for the diffuse *desakota* urbanization patterns that safeguard agricultural land, the main asset of the region (Figure 18.12).

The infrastructure for the new Mekong Delta Region Plan is reconceived in light of climate change and vulnerability of the territory. The deliberate choice is not to build new, extremely expensive infrastructure where there will be inundation anyway. A number of planned highways by the Ministry of Transportation could not be considered optimal investments, as they were going to be flooded by 2050. Instead, the plan proposes to create connections, not as equal access or a homogenous grid across the territory, but via a strategic selection of highways, comprising a main highway collector, a central spine located on guaranteed dry land in the center of the delta, and feeders or linkages to important socio-economic destinations, including Ho Chi Minh City, ports, and transversals to Vietnam's border gates. The TransAsian Highway is proposed to be perpendicular and complement the spine. The further from the spine, the more modest the profile of the feeders, and probably the more natural structures such as mangroves and water elements would take over as structuring devices of the territory. The feeders recuperate existing "national roads" that are upgraded and modernized where necessary, an operation that should go hand-in-hand with the upgrading and modernization of their parallel canals. The canals and embankments, the national roads, their respective interstices and adjacent lands literally superimpose a man-made framework that unites the territory. As such, the systematically planted section of this recurrent sequence is of the utmost importance. It not only organizes the water and road mobility flows, but simultaneously articulates the water/land relation (including the management, regulation, and protection of floodwaters) and hosts construction, as well as vegetation of all kinds, such as the majestic tree rows that are instrumental for creating a livable micro-climate. Planted segments alternate spontaneous vegetation and old construction with brand new investments.

In this infrastructural section that Lewis Mumford undoubtedly would have qualified as polytechnic (Mumford, 1934), all is condensed and nothing is monocultural. The polytechnicity is evident from the manner in which the delta's infrastructural sections merge the technical with civic and economic with ecological performances. Throughout the Mekong Delta, this is done in quite an urban way, balancing order and spontaneity, regularity and messiness, fixity and openness, regulation and freedom. While modernizing and upgrading the infrastructural section, the new regional plan cherishes and intends to enhance the polytechnicity. The intention is to extend the planned new highways, which inevitably will have to become adapted to the dynamic tropical climate. The plan also very importantly stresses and rebalances road transport with water transport, for passengers and cargo. It remains a simple fact that canal construction and maintenance is many times more economic than road construction, given the particular, unstable soil composition within the Mekong Delta.

Finally, there would be a major coastal flagship project located in the southeastern coastal province of Bac Lieu, home to nearly 900,000 people and south of all the Mekong branch outlets into the East Sea. The project would drastically reformulate Vietnam's current infrastructural practices. It would assemble a number of the Vietnamese government's on-going projects, including tidal and wind energy parks, a central international airport

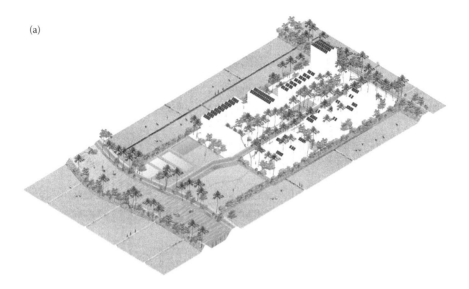

FIGURE 18.12
(a,b) Alluvial plain morphologies. In the alluvial plain, the existing settlement types would be modernized, densified, and clustered: it is the most productive and will be the region of the delta least affected by climate change. The productive landscape will be modernized, diversified, and combined with renewable energy. (Image credit: RUA, 2016.)

for the Mekong Delta, an international deep seaport (an ambitious role comparable to Thailand's Kra Canal—which would shorten shipping routes through the Malacca Straits), the start of the TransAsian highway, and the building of a sea dike (which is the first defense structure of this nature in the Mekong where dikes are historically not used). The project would create an interplay with the natural processes existing in the delta's dynamics to

FIGURE 18.13
Coastal mangrove necklace. Mangrove forests are proposed to protect against storm surges and to create a healthier brackish water environment. Integrated shrimp and fish farming, in addition to renewable energy production in mangroves would be encouraged, as opposed to exclusively shrimp farming. Restoration of coastal mangrove belts would include the planting of numerous native species, including many plants with uses and applications in medicine. (Image credit: RUA, 2016.)

generate synergies, increase coastal protection, and simultaneously create a new type of land reclamation program. The Mekong Delta Regional Plan proposes to enhance a controlled and increased sedimentation along the East Sea in order to drastically scale up the natural land-gaining process. It capitalizes on the flows of discharge to build land. The systematic implantation of obstacles (breakwaters) on the sub-aqueous delta platform would break the existing streams' deposits. As such, they are efficient coastal protection devices. Moreover, they slow down the water current and halt it locally. This strongly enhances sedimentation and consequently induces an accelerated formation of sand banks that ultimately results in a land-gaining process, further consolidated by the systematic planting of mangroves. This will not only strengthen the ecological structure of the Mekong Delta, but also increase its resilience in case of storms (Figure 18.13).

The proposed process can be compared with the well-known techniques of dune formation and reconstruction through the placing of windbreaks and consolidation through succession and adapted plants. The main technical issue of the land-gaining program defines the design of the "obstacles" (breakwaters), the distribution pattern that would be tailored in relation to dominant water current directions and their variations, and the indicated openings and distances between the obstacles. The optimization of these parameters would steer the process that can be adapted to the dynamics of the sea and desired land formation and targeted interesting alterations between water and land, deep and shallow water, and so on. These variations would allow the creation of optimal conditions for a variety of habitats (shrimp, oysters, and other seafood, fish, birds, plants, wood). Differentiations in the shape and depth of the breakwaters, distribution pattern and orientation, and distances in between the breakwaters will allow the generation of a variety of landscapes and could include new waterways of various characters. The technique can be applied both in areas along the East Sea that already gains land naturally (and speed up the process there), as well as areas nowadays suffering from erosion by intensifying density and dimension of breakwaters as coastal defense systems.

This principle of almost natural, but accelerated land creation might appear farfetched, but in the end, it merely capitalizes on the natural processes of sedimentation (and erosion) that exist throughout the entire Mekong Delta region. Ultimately, the Mekong Delta is one gigantic, unstable mud plain that oscillates between being ground and water, land and sea. Hence, there is actually no real or precise shoreline, but an entire territory that could be considered as one extended shoreline. The sea dike that was planned before is replaced, in this plan, by a larger territory with a substantial (ever-growing) depth. The heroic "hold the line" paradigm of the dike is exchanged for a majestic landscape. This new induced landscape reproduces a natural, self-renewing Mekong Delta and works as an infrastructure. It protects (from the sea) and organizes transitions between salt, brackish, and freshwater, while supporting a wide range of activities. A mosaic of appropriations anchor themselves on the natural conditions, ranging from new ecological shrimp farms to tidal energy installations to fishing villages and other sustainable coastal settlements, from fish farms to nurseries of medicinal plants and oyster banks and other seafood nurseries, to bamboo and melaleuca plantations, farms and new state-of-the-art fishery ports.

As previously mentioned, the landscape will host a number of the flagship projects of the Vietnamese government: the region's international airport, wind farms, and a deep seaport. These are consciously built offshore. This way, they neither require (land expropriation nor cause the loss of valuable productive land (which remains the main asset of the Mekong Delta). They will, on the contrary work as mega-breakwaters, form important stepping stones, and are consciously designed and positioned to accelerate the very land reclamation that, in the long run, will absorb them. As the landscape absorbs these areas, they simultaneously become the infrastructure that gives the necessary support to ecological development (coastal protective mangrove forests) and the various water and land variations that will generate the mosaic of appropriations enumerated earlier.

The revision of the Mekong Delta Region Plan is a first step to reengage with the context-embedded dynamics that nature imposes on the territory. Renewed logics of the "hydraulic civilization" and organized dispersal of *desakota* are inherently part-and-parcel of the contemporary plan, which is to strengthen its agricultural importance in light of the delta's intrinsic value and Vietnam's need of national food security. The landscape of the delta is the result of a highly managed and choreographed interplay between land and water in all the volumes, measures, and gradients one can imagine. The main operation of the revised Mekong Delta Regional Plan is to predict the optimum water-land interplays in light of climate change. The basic unit of the plan's 120 water compartments (grouped in five water sub-basins), six agro-ecological sub-regions, three broad landscape and settlement types, the selection and layout of infrastructures all have to do with these new water conditions. They re-articulate the register. The proposal is to graft the further development of the Mekong Delta on this sophisticated water register. The conventional infrastructure and the new landscape induced here work as warp and woof, but roles are inverted. Landscape becomes infrastructure, resistance becomes source, bypasses act as parallel avenues to other features, rather than the over-mechanized and centralized—should we say run-down—territory of the post-industrial era. This is an infrastructure for the twenty-first century, one that purposefully reengages with the dynamics of nature and the notion of a larger deltaic coast; a new and self-renewing ecology that, from the onset, instantly supports a multitude of activities and simple technical interventions which can induce a majestic and varied world (Figure 18.14).

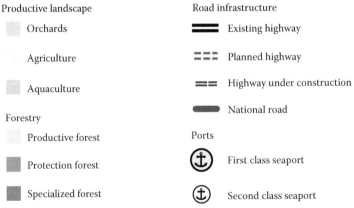

FIGURE 18.14
Mekong Delta Region Plan 2030 Vision to 2050. Infrastructure and development are realigned to take advantage of climate change and the vulnerabilities of the landscape. The delta is optimized as a productive landscape a new state-of-the-art polysynthetic reclamation is conceived on the East Sea. (Image credit: RUA, 2016.)

Endnotes

1. The Mekong River is the largest river in the tropical Western Pacific in terms of length (4,750 km)—the world's 12th longest—and freshwater/sediment inputs. It originates high

up in the Tibetan Plateau, flows through six countries: China, Myanmar, Thailand, Laos, Cambodia, and enters the East Sea through Vietnam's Mekong Delta region. The basin measures 795,000 km², making it the 21st largest river basin in the world. Its delta plain is the third largest in the world. The Vietnamese component of the delta shares 300 km of border with Cambodia, has more than 700 km of coastline, and includes approximately 360,000 km² of territorial sea.

2. The Vietnamese planning system classifies cities based on their population numbers, as well as the cities' importance in the overall national socio-economic structure. Cantho, whose borders remain the size of a province, is the only first-class city of the Mekong Delta region, meaning its planning and economic policies are directly overseen by the national government in Hanoi.

3. Revision of the Mekong Delta Region Plan 2030–2050 is an ongoing work that began in November 2014. RUA, Research Urbanism Architecture, is the international consultant to SISP (Southern Institute for Strategic Planning). RUA and SISP were commissioned by the Ministry of Construction in Hanoi, Vietnam. The RUA team is led by Bruno De Meulder and consists of Kelly Shannon, Claudia Rojas Bernal, Christina Hood, Donielle Kaufman, Michael Waibel, Tracy Collier and Eric Heikkila. From January to March 2016 the RUA team was assisted by M. Fancler and Superjacent (K. Majewski, A. Paradowski, C. Torres).

4. The six agro-ecological sub-regions have long been recognized by geographers in Vietnam; their precise boundaries vary depending on authors. See for example David Biggs. 2010. *Quagmire: Nation-building and Nature in the Mekong Delta*. Seattle, WA: University of Washington Press, and Nguyen, Duy Can, Le, Tanh Duong, Nguyen, Van Sanh and Fiona Miller. 2007. "Livelihoods and resource use strategies in the Mekong delta" (Chapter 2), in *Literature Analysis, Challenges to Sustainable Development in the Mekong Delta: Regional and National Policy Issues and Research Needs*, Tran T. B., Bach T. S., and F. Miller, eds. Bangkok: The Sustainable Mekong Research Network (Sumernet). 69–98.

5. SIWRP (Southern Institute of Water and Resource Planning). 2012. Master Plan in the Mekong Delta for Impact of Climate Change—Sea Level Rise, Ho Chi Minh City.

References

Benedikter, S. 2014. *The Vietnamese Hydrocracy and the Mekong Delta: Water Resources Development from State Socialism to Bureaucratic Capitalism*. Berlin: LIT Verlag, Dr. W. Hopf.

Biggs, D. 2010. *Quagmire: Nation-Building and Nature in the Mekong Delta*. Seattle, WA: University of Washington Press.

Brocheux, P. and D. Hémery. 2009. *Indochina: An Ambiguous Colonization, 1858–1954*. Berkeley, CA: University of California Press.

Cairns, S. 2002. "Troubling real-estate: Reflecting on urban form in southeast Asia." In: Bunnell, T., Drummond, L., and Ho, K. C. (eds) *Critical Reflections on Cities in Southeast Asia*. Singapore: Times Academic Press, 101–123.

Chellaney, B. 2011. *Water: Asia's New Battleground*. Washington, DC: Georgetown University Press.

Christaller, W. 1933. *Die zentralen Orte in Süddeutschland*. Jena: Gustav Fischer. (Translated in part, by Charlisle W. Baskin, 1966, as *Central Places in Southern Germany*. Upper Saddle River, NJ: Prentice Hall).

De Meulder, B. and K. Shannon. 2013. "Mangroving Ca Mau, Vietnam: Water and forest as development frames." In: De Meulder, B. and Shannon, K. *Water Urbanisms East UFO3: Explorations in Urbanism*. Zurich: Park Books, 118–137.

GSO (General Statistics Office). 2015. *Statistical Handbook of Vietnam 2014*. Hanoi: Nhà xuất bản Thống kê (Statistical Publishing House). https://www.gso.gov.vn/default_en.aspx?tabid=774 (accessed October 15, 2016).

Kummu, M., Lu, X. X., Wang, J. J., and O. Varis. 2010. "Basin-wide sediment trapping efficiency of emerging reservoirs along the Mekong." *Geomorphology* 119:1181–1970.

Le, A. T., Chu, T. H., Miller, F., and T. S. Bach. 2007. "Floods and salinity management in the Mekong delta, Vietnam" (Chapter 1). In: Tran, T. B., Bach, T. S., and Miller, F. (eds) *Literature Analysis, Challenges to Sustainable Development in the Mekong Delta: Regional and National Policy Issues and Research Needs*. Bangkok: The Sustainable Mekong Research Network (Sumernet), 15–68.

McGee, T. 1967. *The Southeast Asian City: A Social Geography of the Primate Cities of Southeast Asia*. London: G. Bell and Sons, London.

McGee, T. 1991. "The emergence of *desakota* regions in Asia: Expanding a hypothesis." In: Ginsburg, N., Koppel, B. and T.G. McGee. *The Extended Metropolis: Settlement Transition in Asia*. Honolulu, HI: University of Hawaii Press, 3–25.

Molle, F., Mollinga, P. P. and P. Wester. 2009. "Hydraulic bureaucracies and the hydraulic mission: Flows of water, flows of power." *Water Alternatives* 2(3):328–349.

Mollinga, P. P. 2008. "Water politics and development: Framing a political sociology of water resources management." *Water Alternatives* 1(1):7–23.

Mumford, L. 1934. *Technics and Civilization*. New York, NY: Harcourt, Brace & Company, Inc.

Nguyen, D. C., Le, T. D., Nguyen, V. S., and F. Miller. 2007. "Livelihoods and resource use strategies in the Mekong delta" (Chapter 2). In: Tran, T. B., Bach, T. S., and F. Miller (eds) *Literature Analysis, Challenges to Sustainable Development in the Mekong Delta: Regional and National Policy Issues and Research Needs*. Bangkok: The Sustainable Mekong Research Network (Sumernet), 69–98.

Nguyen, H. N. 2016. "Tidal regime deformation by sea level rise along the coast of the Mekong delta." *Estuarine, Coastal and Shelf Science* 183:382–391.

Nguyen, T. A. 2002. Mekong delta water quality and sustainable aquaculture development. *Proceedings of the workshop held in Tra Vinh*, Vietnam, March 2002: Shrimp Farming Sustainability in the Mekong Delta, Environmental and Technical Approaches. France: IFREMER.

Osborne, M. 2000. *The Mekong: Turbulent Past, Uncertain Future*. London: Allen and Unwin.

Phan, L. K., van Thiel de Vries, J. S. M., and Stive, M. J. F. 2015. "Coastal mangrove squeeze in the Mekong Delta." *Journal of Coastal Research* 31(2):233–243.

Scott, J. C. 1998. *Seeing like a State: How Certain Schemes to Improve the Human Condition Have Failed*. New Haven, CT: Yale.

Shannon, K. 2004. *Rhetorics & Realities. Addressing Landscape Urbanism. Three Cities in Vietnam*. Leuven: unpublished doctoral dissertation.

SIWRP (Southern Institute of Water and Resource Planning). 2012. Ho Chi Minh City: Master Plan in the Mekong Delta for Impact of Climate Change - Sea Level Rise.

Taylor, P. 2001. *Fragments of the Present: Searching for Modernity in Vietnam's South*. Honolulu, HI: Allen & Unwin and University of Hawai'i Press.

USDA. 2014. Table 24: World rice trade (milled basis): Exports and imports of selected countries or regions. United States Department of Agriculture Economic Research Service: Rice Yearbook 2016. http://www.ers.usda.gov/data-products/rice-yearbook/rice-yearbook/#World%20Supply,%20Utilization,%20and%20Trade (accessed October 15, 2016).

Wittfogel, K. 1956. "The Hydraulic Civilizations." In: Thomas W. L. (ed) *Man's Role in Changing the Face of the Earth*. Chicago, IL: University of Chicago Press, 152–164.

Wittfogel, K. 1957. *Oriental Despotism: A Comparative Study of Total Power*. New Haven, CT: Yale University Press.

Xue, Z., Liu, J. P., DeMaster, D., Leithold, E. L., Wan, S., Qian Ge, Q., Nguyen, V. L., and T. K. O. Ta. 2014. "Sedimentary processes on the Mekong subaqueous delta: Clay mineral and geochemical analysis."! *Journal of Asian Earth Sciences* 79:520–528.

Section IV

Sites and Structures

Specific design solutions for sites of flooding and coastal change.

19

The New Hondsbossche Dunes

Adriaan Geuze and Edzo Bindels with Riette Bosch

CONTENTS

A New Delta Committee

A large proportion of the Netherlands lies below or around sea level. This land has to be protected from flooding by the sea and, along the northwestern coastline, natural dunes form the first line of defense. The nine kilometers (km) of coast between Petten and Camperduin had an exceptional appearance of a dike clad with stone, heightened over and over again. Climate change is a new reality that cannot be ignored. Predicted sea level rise has compelled the Dutch Government to look far into the future, to widen its scope, and to anticipate developments further ahead. For that reason, in 2007 the Cabinet appointed the Sustainable Coastal Development Committee, a "new" Delta Committee mandated to formulate a vision on the long-term protection of the Dutch coast and its hinterland. This mandate is broader than that of the first Delta Committee which was primarily concerned with "hydraulic engineering works to counter an acute threat."

For the second Delta Committee, the threat is not acute but the mandate remains urgent. The challenge to the Netherlands in the coming centuries is not primarily a threat; rather it offers new prospects and opportunities to reconsider a future approach. Critically, it offers the chance to investigate powerful synergies with other societal functions, such as living and working, agriculture, nature, recreation, infrastructure, and energy. Changing the way that the Netherlands manages its growing environmental concerns creates new options for a country fundamentally vulnerable to the power of water. Working *with* water, as opposed to against it, may improve the quality and robustness of the Netherlands's coastal environment and enable innovative ideas and solutions and the sort of resilience necessary to protect against future environmental crises. Where there is water, new forms of nature can arise. So, "building with nature" was introduced; in the case of the Hondsbossche en Pettemer Zeewering it meant that the dike would not remain as the solution.

For the North Sea coast (Holland, the Zeeland headlands, and the Wadden Islands), the emphasis lies on maintaining coastal safety by continuing the practice of beach nourishment that will offer permanent safety far into the next century. The Committee advises that nourishment must be carried out in such a way that the coast can grow to meet the needs of society in the next century. This vision should allow the "weak links" to be dealt with. A growing coast creates extra space for nature and recreation (including seaside resorts).[1] Research from the last three decades has provided sound methods to quantify dune strength for coastal safety assessments. This has subsequently led to the development of soft and dynamic dune engineering methods, capable of creating a robust coastal defense of a predetermined strength. Furthermore, research in the 1980s yielded quantitative approaches to prevent structural coastal erosion. This resulted in the implementation of a national policy to counter structural coastal erosion by sand nourishment.

The New Hondsbossche Dunes are therefore a valuable case study for illustrating the success of sand nourishment as a coastal engineering measure in the Netherlands and its potential application in other coastal areas. Sand-based coastal management strategies are clearly effective, sustainable, and flexible. Research and experience have shown that sandy coasts, although soft and dynamic, can in fact be engineered to provide predefined safety levels while supporting other functions as well.[2,3,4]

In 2012, West 8, an urban design and landscape architecture firm, was invited to join the consortium of Van Oord Dredging and Marine Contractors and Boskalis, for the North Holland Water Board to support their team, which included: Svašek Hydraulics (hydrology and dredging), Witteveen + Bos (engineer), and Altenburg & Wymenga ecologisch onderzoek (ecology). West 8 was responsible for the environmental design of the new coastal defense initiative at Hondsbossche Pettemer seawall, the weak link in the coastal line of defense, now known as the New Hondsbossche Dunes.

In the New Hondsbossche Dunes, the principle of "building with nature" is fully exploited. Working closely with the consultant team, West 8 designed a natural dunal landscape as an alternative to a sandy dike. Intended to strengthen the Dutch coastline, the landscape intervention also strengthens the identity of the region, enhances and embraces the natural environment, improves accessibility, and bolsters recreational activities in the area.

By connecting the Schoorl and Pettemer dunes, one of the longest uninterrupted dunescapes in the Netherlands is created. Designed to complement the existing dunes, the new landscape features varied topography, native vegetation, young drifting dunes, an enclosed ecologically significant dune valley, a sculptural memorial to the "drowned villages" of Petten, and a south facing lagoon at Camperduin which features an open connection to the sea. The new 10 km beach varies in width and culminates with a 25-meter (m) high dune, with panoramic lookout across Petten. The highly technical, carefully engineered solution for the new dune volumes efficiently uses sand and minimizes dune erosion to meet the dual goals of coastal protection and environmental quality. The intervention begins a new chapter in the Dutch art of land forming, one that unites a technical solution with a poetic narrative.

Project Brief

The North Holland District of the Dutch Water Board is responsible for managing the North Holland dunes and protecting the surrounding communities from rising waters and flood events. For some time, the Hondsbossch and Petten seawall (HPZ) and the adjacent dunes of Petten have been the "weak link" in North Holland's primary seawall. The client sought to strengthen this area, using only sand. In this context, the competition brief was to design, construct, and maintain a secure sandy coastal defense between Camperduin and Petten that would provide sustainable protection for the next 30 years.

Driven by a mandate to improve the safety and resiliency of this "weak link" in the North Holland line of coastal defense, the Consortium van Oord and Boskalis' winning proposal uses natural processes to safeguard the community, while simultaneously creating new opportunities for nature, recreation, and further developing the identities of Petten and Camperduin as popular seaside resorts. The result is an artificial natural landscape that sees safety, nature, and recreation existing hand in hand.

The winning, and now realized, 140 million Euro proposal—the New Hondsbossche Dunes—is a technically complex man-made beach/dune volume that is spatially integrated into the existing landscape. It encompasses recreational beaches at key access points in the north and south and strengthens the ecology of the area through the creation of a plethora of habitats, such as an enclosed dune valley located centrally within the new landscape that has become a paradise for birds.

Historical Context

For centuries coastal erosion and land reclamation have fought for supremacy in the area around Hondsbossch and Petten. Built in 1880, the Hondsbossch and Petten seawall is a national monument, representing hundreds of years of coastal defense. Despite this, however, the seawall has unfortunately often turned out to be largely ineffective, mostly due to the need for annual supplementation in several places. The new "sandy" approach is more robust and allows nature to do the work by itself for a longer period. The surplus of sand that is added to the weak nodes will spread slowly over the entire coastline and therefore strengthen it.

The inefficiency of the HPZ was further compounded by the draining of the area's former dune islands, salt marshes, and mudflats. The sea gates were closed, creeks dammed, and marshes drained. Rising sea levels would further erode the sandy shore, leaving the village of Petten at the mercy of the sea, at risk of flooding, and constantly needing to be built anew, in a safer place. The time had come to embark on the next step in the "tradition" of land formation and fortify HPZ with a long sandy defense.

The construction of the New Hondsbossche Dunes adds a contemporary twist to the rich history of coastal protection and water management in the Netherlands. Dikes and barriers are icons in Dutch history, but never before has the coast been strengthened with artificial dunescapes of this size. From an ecological perspective, the project is of considerable import. The construction of this impressive new dunescape is sandwiched between several Natura 2000[5] sites. Therefore, special attention was given to the spatial and ecological integration, as well as the area's existing natural and landscape qualities. The introduction of a new dune bridge between the existing nature reserves literally creates a safe space for people and nature to coexist.

Design Vision: Landscaping

The process of land forming has to do with the dynamics of wind, currents, and waves and the (hydrological) combination of low/high salt, freshwater, and wet-dry conditions.

The design intervention establishes the necessary abiotic environmental conditions and essential habitats (conditional control) to build new land. Over time morphological and ecological processes will determine the dynamic evolution of the different habitats and their species composition. However, this organic process takes place within a secure coastal framework that forms the core of West 8's design approach.

West 8's design vision is based on the premise that it is possible to construct a safe and sustainable coastal landscape by utilizing the powerful, dynamic processes that dominate this coastal environment. These include natural erosion and sedimentation (above and below sea level) and the stabilizing influence of native vegetation. A systems approach is adopted, which focuses on "shoring with nature." Because the design puts natural processes to work, it cultivates a robust ecosystem that is to a large extent maintained by those same dynamics. This new buffer is designed so that it need not be replenished for the next decade, giving nature plenty of time to recover and develop both above and below sea level.

On land, the new dune volume utilizes strategically planted vegetation in order to change dune heights by either releasing or retaining sand. The planting of 640,000 square meters (m^2) of European beachgrass (*Ammophila arenaria*) and the creation of new open spaces are the first steps in developing vital dune communities, encouraging small-scale habitat variation. By allowing natural processes to work, future maintenance is limited to the subtle redirection of these processes. The open spaces will erode with the play of the wind and the sand will accumulate at the first line of beachgrass.

Technical Resolution: Utilizing a Large Sand Buffer System

The design of the New Hondsbossche Dunes is underpinned by three design principles:

1. Safety is the principle objective for the construction of a sandy coastal defense system.
2. A systems approach must be utilized that follows the principles of "building with nature."
3. To address the principles of safety, ecology, and recreation, the design is zoned according to the length of the coast and the transverse direction (Figure 19.1).

To create a sandy coastal defense and reduce future maintenance requirements, a large sand buffer is constructed underwater (along the beach) and also on land against the HPZ. This helps protect against sand erosion, and compensates for the loss of sand caused by hydraulic transport, wind, and human activity. The extra break in the cross-section leaves room for the wind to recreate the dunescape. Accumulated sand supports the growth of vital inland dunes, which will be carefully monitored to ensure the sand does not encroach too far inland. The robust, man-made dunescape between HPZ and the sea provides maximum safety for the nearby coastal communities and the hinterland beyond.

The new dunes and natural areas are literally supported by the existing HPZ. On the landward side, a smooth transition from coast to polder is created and planted with buckthorn and wild roses, which will evolve into a buckthorn valley over time. On the

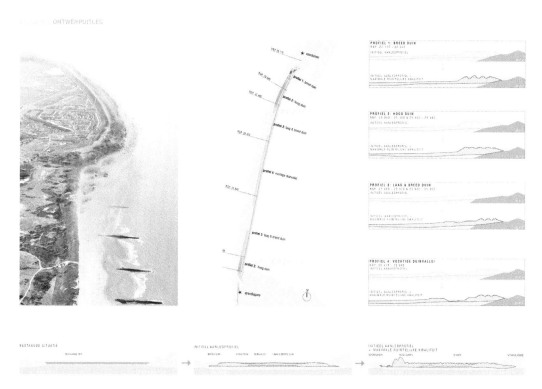

FIGURE 19.1

(Left to right, top to bottom): Visualization showing the new stretch of high dunes at Petten connected by a 10 km stretch of sandy coastline. Plan showing location of different longitudinal profiles. Coastal section of different longitudinal profiles: showing the new technical intervention, its relationship with the existing dike, and the refinement of the design enhanced landscape quality. Diagram showing (left to right) technical solution, optimized technical solution using sand, and soft design engineered solution that creates a multifunctional ecological landscape. (Images courtesy of West 8.)

seaward side, 9 km of new beach is framed by two uninterrupted dune volumes that are separated at the mid-point by a natural dune valley. Where the new dunes meet the existing dunes at the villages of Petten in the north and Camperduin in the south, a number of new landscape elements are added. At Petten, a stretch of dunes is realized, climaxing in a 25 m high dune with sweeping 360-degree panoramas. In Camperduin, a south-facing salt water lagoon that is open to the sea is created, providing amenities for visitors, fostering biodiversity, and establishing an iconic landmark in this Dutch coastal landscape (Figure 19.2).

Extensive sand calculations and coastal modeling led to the addition of 35.6 million cubic meters (m^3) of sand (enough to fill 400 football fields), placed between January 2014 and January 2016, using trailing suction hopper dredgers and spraying boats. These boats feature large, powerful pumps and engines that enable them to dredge sand, clay, sludge, and even gravel from sea or river beds. The vessel stores the dredged material in its own hopper and discharges the leftover water overboard. The hopper can be emptied in a variety of ways, including by depositing the material through doors or valves on the bottom of the vessel; using pumps and jets to make the material fluid again and pump it ashore through

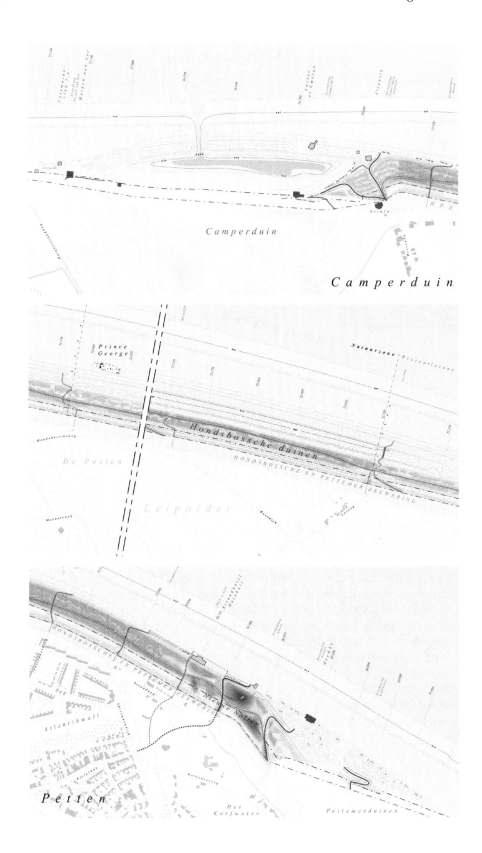

a pipeline; or spraying, where the hopper contents are pumped directly from the vessel to the desired site.

Design optimization significantly reduced the amount of sand needed. This meant fewer trips between the project location and the sand extraction site, resulting in lower fuel consumption and lower carbon emissions. In addition, better coordination between the trailing hopper suction dredgers during project execution meant shorter waiting times and subsequently lower fuel consumption while waiting. As a result of these coastal protection efforts, the central natural zone can remain untouched for the 10 years following the project's completion, supporting the development of nature in this area. Following this initial decade, a 20-year maintenance period will see an extra 4.2 million m^3 of sand added to the coastline over the course of three consecutive sand replenishment initiatives.

To determine the level of sand required, and the sediment deposits after the initial construction period, a FINEL2D system was used to simulate the landscape's evolution over the next three decades and predict the effects of various construction and maintenance regimes on the dunescape. These variations helped to determine the most appropriate form, location, and timing for replenishment. The natural area will not require any sand replenishment until 2026. While the north and south recreation zones are scheduled to undergo two successive sand replenishment initiatives, the first in 2023 and the second in 2030, the later addition of a sand buffer means no replenishment will be required for two to eight years after the official maintenance period has ended in 2035.

Detailed calculations on water management and water quality show that the design and construction methods combined with independent factors, such as climate change and sea level rise, do not affect the presence of brackish vegetation in the Abtskolk and De Putten regions, situated on the landside of the HPZ. Since the water level in the surrounding polders is below sea level, the salinity of ground- and surface water is guaranteed to remain constant. After an initial period of sweetening, and the depositing of windblown sand, the level of the new dune valley will rise. The influence of sand deposits, plus the reestablishment of sweet water, will rebalance the groundwater to create a sweet water habit that is very unique, particularly in such close proximity to the Dutch coastline. This will in turn allow the colonization of vegetation specific to this coastal ecosystem to establish and improve the biodiversity of the region (Figures 19.3 through 19.6).

The introduction of this elongated sand dune forms a natural bridge between the dunes of Schoorl and Petten. In the longitudinal direction, increasing the acreage of dune habitat contributes to the natural quality of the regional network of dunes, coastline, and nature reserves. In the perpendicular direction, the beach reserve, inland nature reserve, and sweet water dune valley preserve enhance the habitats of local birdlife (such as Dwarf Geese) and create a connected ecological corridor for birds and other fauna. This is a condition rarely found along the Dutch coastline. Thus, the New Hondsbossche Dunes present a remarkable step forward in the development of sustainable coastal defense systems with the added benefit of the creation of a new beach and dunescape each with tremendous natural and recreational qualities.

FIGURE 19.2
Plan view of three new coastal zones (top to bottom): Southern leisure zone encompassing the coastal lagoon at Camperduin; central natural zone and wetland dune valley in the Hondsbossche Dunes; and stretch of dunes at Petten. (Images courtesy of West 8.)

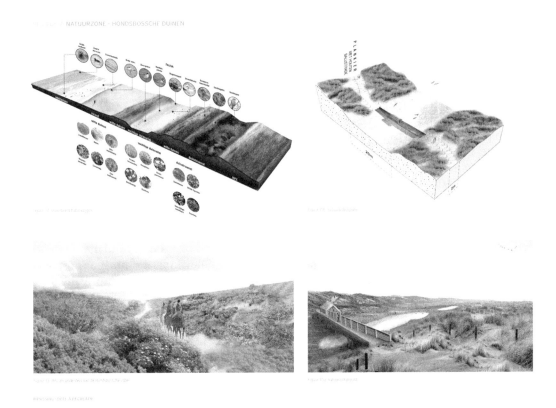

FIGURE 19.3
Left to right: Diagram showing different habitat types across dune cross-section; HPZ as part of the Hondsbossche valley, and overlaid with new recreational programs, such as horse riding; birdwatcher's cabin situated within the central nature zone. (Images courtesy of West 8.)

The Layered Importance of the Dune Landscape within the Netherlands

Flexibility is paramount to the success of the New Hondsbossche Dunes. The inherent elasticity within the design allows visitors' needs to be addressed, an attractive coastal landscape to be formed, the area's accessibility to be improved, and the technical and functional requirements of the brief to be met with genuine consideration for nature. The programming of the large variety of landscapes is underpinned by five key principles.

A. *"Building with Nature" Should Add Value*: An integral part of building *with* nature is the notion of building *on* nature. Nature must be as obvious as possible. The team chose not to create a fixed dune design, but rather to initiate a solution that approximates the natural situation from its inception. Nature is enhanced along the full length of the new dunes, not only in just the dedicated central nature zone. Indeed, the character of the new dunes varies longitudinally from north to south and also along the perpendicular direction. The project presented an opportunity to create the longest contiguous dunes of the Dutch coast (and even northwest Europe) (Figure 19.1). This will provide major economic stimulus for the coastal

FIGURE 19.4
Dune stone showing historic layers of the landscape, located next to the highest dune in Petten. Seasonal beach houses in embryonic dunes at Petten. (Images courtesy of West 8.)

villages of Petten and Camperduin, through an expanded recreational program, new facilities, and improved landscape quality, while helping to reinforce the respective identities of these two beachside communities.

B. *Varying Dunes*: A rigid dune, resembling a dike placed in front of the monumental Hondsbossche and Petten seawall, does not do justice to this centuries-old icon and national monument. The brief stipulated that in order to protect the historic sight lines of the seawall, the new dune volume should not be visible from the hinterland. The team's solution adds a new chapter to the rich epic of the Netherlands' struggle against the sea. The design respects the cultural and historical significance of the HPZ and surrounding context, and the naturalized dunes play a critical role as an ally in building a more resilient future. From a historical perspective, the solution may seem bold and technically challenging, but its extensive engineering and modeling guarantee that the principal safety objective is met.

C. *Connecting the Site with Its Context*: This proposal employs big picture thinking, and significant consideration is given to both the natural environment and anthropological factors (Figure 19.7). Incorporating the dunescape into the physical environment necessitated a seamless and natural transition between the artificial and existing dunes around Camperduin and Petten that pay homage to the contrast between the coastal landscape and the polders. The design introduces

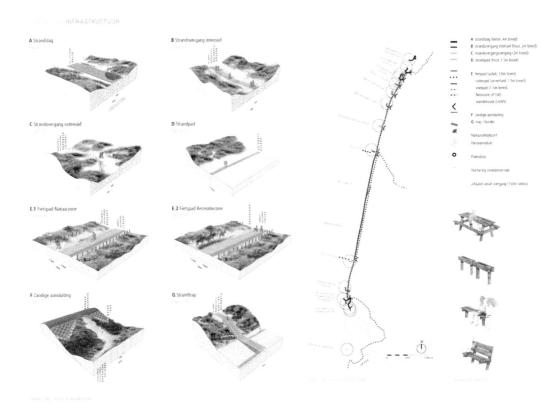

FIGURE 19.5
Dunescape infrastructure designed in different scale and context including also public furniture and recreational facilities. (Images courtesy of West 8.)

eight pedestrian linkages between the hinterland and the new beach, as well as a continuous sea view cycling paths and horse tracks which form part of the coastal cycling route, the LF1, that runs from Den Helder to Belgium. At Patten, a stretch of high dunes is created. Atop of the highest dune is a viewing point that looks out toward the old polder landscape and across the new sandy defense. A "dunal rock" shows the different layers of the landscape over time. In the heart of the seaward landscape sits a sweet water dune valley which will develop in time as an undisturbed habitat for rare and endangered flora and fauna. Parallel to the monumental dike a dry dune valley will slowly be filled with sand and overgrown with wild roses and buckthorn. At Camperduin the dunes open up toward the south, embracing a lagoon which has an open connection with the sea.

D. *Not All Recreationists Are Sun Worshippers*: The beach does not just attract sunbathers. The new dunescape caters to a diverse range of interests and offers a plethora of activities for many (potential) visitors. The new landscape is accessible to the young and old; the able-bodied and disabled; hikers, cyclists, horse riders, bird watchers, sunbathers, swimmers, athletes, parents, and children—this is a place for everyone. The newly created landscape typologies dictate the proposed usages. For instance, the beach near Camperduin and Petten is more easily accessible and therefore more intensively used, while the nature reserves located several kilometers from human

(a)

(b)

FIGURE 19.6

(a) At Camperduin the dunes open up toward the south, embracing a lagoon which has an open connection with the sea. (b) The new beach, dune landscape, and artificial lagoon replicates the predominantly coastal ecology found along the Dutch coastline. (Image courtesy of Jeroen Musch.) *(Continued)*

(c)

FIGURE 19.6 (Continued)
(c) Looking across the young drifting dunes toward the new 25-m-high superdunes at Petten. (Images courtesy of Jeroen Musch.) (Image courtesy of Jeroen Musch.)

FIGURE 19.7
Before and after diagram showing extent of dune replenishment between Petten and Camperduin that forms the new lines of coastal defense. (Images courtesy of West 8.)

activity are sparsely inhabited. In this way, a natural flow of people is created through the landscape that supports recreation and protects nature.

The difference in context triggers the different usage of the site. Kids can run down the superdune at Petten and plunge into the waves. North of the dune family is a quintessential Dutch tradition; a row of simple beach cabins is positioned within a living embryonic dune landscape. While the wet dune valley is hard to reach, a secluded bird cabin offers nature lovers a wonderful view of the birds. South of the superdune is an extra wide beach access that can accommodate large numbers of people. The diversity of programming in this area can be extended over time as recreational facilities establish themselves and bring new businesses with them. At Camperduin the warm south dune slope is inviting in the early spring for sun worshippers. Surfers will be a bit further south and the less daring a few hundred meters north at the beach club.

E. *Do More with More*: Coastal interventions will add greater value if they adopt an integrated approach. In the case of the New Hondsbossche Dunes, the design of a soft rather than a hard landscape intervention that successfully meets the safety and environmental objectives of the project brief represents the next step in the ongoing art of Dutch engineering. Future generations can continue to refine and develop this new soft approach as yet another way to protect the Dutch hinterland from the sea.

National Recognition

Testament to the successful realization of the project, the New Hondsbossche Dunes received the 8th edition of the prestigious Arie Keppler Prize in the planning category. The professional jury praised the project's

considerable investment in opportunities to broaden the recreational possibilities of the landscape, using water, through the new path network, improved beach accessibility, viewpoints and historical references. The suite of new furnishings within the landscape are elegantly understated and carefully implemented. The jury appreciates the allusions to area's history, with the old seawall on the landward side remaining intact as an archaeological relic behind and beneath the new dunes. At that time of its construction the original dike changed the landscape dramatically. Once again this is the case. The construction of these new dunes will keep people's feet in North Holland dry for years to come. Here is a virtue of necessity created according to the law of nature, through the medium of an impressive piece of art in engineering.

Endnotes

1. Working together with water: A living land builds for its future, Findings of the Deltacommissie 2008|summary and conclusions, Deltacommissie.
2. Coastal morphologist, Delft University of Technology, Department of Civil Engineering, WL|Delft Hydraulics and Netherlands Centre for Coastal Research, Delft, The Netherlands.

3. Coastal engineer, Ministerie van Verkeer en Waterstaat, Rijkswaterstaat, North Sea Directorate, Rijswijk, The Netherlands.
4. Coastal morphologist, Ministerie van Verkeer en Waterstaat, Rijkswaterstaat, National Institute for Coastal and Marine Management, RIKZ, University of Twente and Netherlands Centre for Coastal Research, Den Haag, The Netherlands.
5. Natura 2000 is the largest coordinated network of protected areas in the world. Designated sites offer a haven to Europe's most valuable and threatened species and habitats. In May 1992, the governments of the European Communities adopted legislation designed to protect the most seriously threatened habitats and species across Europe. Today, Natura 2000 protects around 18% of land in the EU countries (787,767 km²).

20

Adaptive Landscapes for Coastal Restoration and Resilience in Contemporary China

Mary G. Padua and Stanley Lung Wai Cham

CONTENTS

Introduction

The work of China's first private landscape architecture practice, Turenscape, frames the exploration of a professional journey that parallels China's post-reform urban projects with the search for resilience in the synergy of architecture and the environment. A potent mix of population growth, rapid modernization and urbanization, economic prosperity, and extensive social transformation, has led to an unprecedented scale of environmental degradation. Combined with an era of domestic and foreign investment, China now faces a critical opportunity to transform its approach to urban planning and design; to experiment with ways to bring development back in sync with the natural world framework in an emergent green revolution. Historical factors that have formed the physical background of contemporary urban China are driving a new consciousness, an understanding of the importance of educating urban planners, landscape architects, and bureaucrats in how adaptive urbanism and resilient design solutions can integrate with planning and construction to address the distinct environmental challenges China now faces.

Contemporary Urban China

Over the last 30 years, the world witnessed the People's Republic of China (China) transforming itself, as the nation shifted to a market-based economy where capitalism, entrepreneurial activities, and consumerism took hold. This 30-year process of modernization, nation-building, and hyper-rapid urbanization was initiated by Deng

Xiaoping, China's paramount leader, who succeeded the communist revolutionary leader and founder of the People's Republic of China, Chairman Mao Zedong. Deng's late 1970s reforms and open-door policy called for radical and revolutionary change across several dimensions (Padua, 2007; Campanella, 2008). Deng's Four Modernizations and open-door policy (*gaige kaifang*) involved the sectors of agriculture, industry, science and technology, and national defense (Hsü, 1990). Deng recognized that China's success as a nation-state would need to break from the isolation, famine, and poverty created by Mao's failed policies and the Cultural Revolution, when urban populations were sent to the countryside for moral reform and schools were closed for ten years, circa 1966–1976. The subsequent post-Mao and post-reform period of hyper-rapid urbanization could not have occurred without China's system of governance. Framing this system is critical to understanding the nation's modernization process, urban development, and tensions around conflicting priorities for the protection of natural resources and management of arable land. It also illustrates the capacity to implement large-scale coastal and urban projects on a time scale that is often not possible in other places.

China's political complex of governance currently functions through a single-party (the Communist Party of China, CPC) system of centralized government with four branches: executive, legislative, judicial, and military.[1] The nation largely operates as a hierarchy with three core levels: central; provincial-level jurisdictions; and cities or counties. Nation-building policies and priorities are set by a national-level development planning instrument known as the "Five Year Plan" (FYP), imported from the Soviet Union during the Mao period. However, the FYP process has evolved immensely from its initial static use with a shift occurring in the 9th (1996–2000) FYP when it became adaptive from one FYP to the next (Heilmann and Melton, 2013; Lockett, 2016).

Formulating and implementing the FYP is a complex continuous cycle of information-gathering, policy analysis and formulation, evaluation, and review (Heilmann and Melton, 2013). Generally, the FYP has three key active-oriented features: (1) it sets out policy directives for local governments to follow through with management according to objectives; (2) it connects accomplishments with local achievements and assessments for career advancement; and (3) as a central planning policy apparatus, it spurs market activities (Heilmann and Melton, 2013; Hu, 2014; Bell, 2015). Each FYP is initiated two years prior to its launch following the mid-term review of the FYP in place. FYP priorities and programs are subsequently carried out in a series of sub-plans through an inter-locking hierarchical political web of China's three levels of government. The combination of the FYP flurry of activities, various other government operations, and CPC leadership has created the crucible for China's tremendous 30-year transformation. It is within this light that China has functioned as a living laboratory for the world's largest urban experiment where ecological planning and modern landscape architecture has emerged.

China has transformed itself from a largely rural and impoverished nation with only 18% of China's population living in cities in the late 1970s (Zhang 2001; Goodkind and West 2002; Zhu, 2007). Fast forward to 2015 when China's National Bureau of Statistics stated that over 56% of its population lived in urban areas. In just over 30 years, China's urban population has more than doubled, with more than 767 million people out of a population of nearly 1.4 billion living in cities, including an emergent consumer-oriented middle class. Scholars and architectural critics have noted that historical cycles that took a century or more to unfold in Europe or North America can be compressed into less than a decade in China today (Friedman, 2005; Campanella, 2008; Ouroussoff, 2008; Padua, 2014). In effect, for the first time in the Anthropocene era, the world has witnessed China's vast transformation into a largely urban society and global economic powerhouse.

China's historical development, human settlement patterns, and geographic locations of its towns and villages are linked to its physical landscape and are a result of thousands of years of political territorialism, urbanization, and periods of disunity, war, and peace (Skinner, 1977; Spence, 1991; Yu, 1992; Elvin, 2004; Padua, 2014). Its complex physical landscape is defined by large plateaus, steep mountain ranges, fertile valleys, and alluvial plains drained by a network of rivers that emerge into its long coastline. The first cities in China were located in the fertile basins, along rivers and in valleys with further urbanization concentrated along the coastline following the influence of foreign trade. While China's hyper-rapid urbanization in the post-reform era has been enormous, the current urban landscape is the continuation of a process that has ancient roots (Padua, 2014).

In terms of China's natural environment, until the advent of foreign trade, circa eighteenth century, Chinese society's human footprint and relationship to nature largely coexisted in harmony. This changed dramatically when Mao's mid-twentieth century modernization efforts, including land reforms and his version of rural-urban industrial and agricultural production activities, caused significant environmental degradation (Shapiro, 2001; Ho, 2003; Economy, 2004; Zhao and Woudstra, 2007). Mao's staunch revolutionary leadership believed in the power of science to dominate nature, whereby the natural environment would be transformed to meet the needs of the people (Shapiro, 2001; Ho, 2003; Economy, 2004). However, while Mao's steel-making factories along rivers and agricultural experiments in rural areas, along with his failed economic policies, negatively impacted China's environment, the destruction was not at the scale that the nation has experienced over the last several decades (Kram et al., 2012).

During China's post-reform era, rural areas have been amalgamated into the urban districts of new cities, as well as subsumed in the expansion of older cities and towns (Zhu, 2004). Major cities like Beijing and Shanghai have grown and expanded due to market pressures, the advent of suburbanization, and the rise of China's middle class, with their enormous appetite to own automobiles. At the same time, the Pearl River delta region in the nation's south evolved into the "workshop of the world,"[2] peaking at the turn of the twenty-first century (Cantin and Taylor, 2008). In turn, massive transportation infrastructure-building occurred involving deep water ports, highway networks for private cars and trucks, as well as an expanded railway system with high-speed passenger trains (Weng, 2007; Zhang, 2015). In parallel, tremendous degradation of China's environment, natural resources, and ecosystems have occurred (Wang et al., 2003). One metric is the amount of cement consumed in China between 2011 and 2013: approximately 6.6 gigatons (one gigaton is the equivalent of one billion tons and one U.S. ton is the equivalent of two thousand pounds); this represents 140% of the amount that the United States consumed (4.5 gigatons) during the twentieth century (Smil, 2013).

China's conventional city-building practices during both the Maoist period and the post-reform era of hyper-rapid urbanization, along with increased frequencies of extreme global weather patterns, have contributed to China's current environmental crisis. China's natural woodlands, grasslands, lakes, and wetlands, the physical environment that would normally have allowed natural hydrological processes, have been disappearing rapidly. Additionally, the nation exhibits water scarcity, extremely poor air quality, low utilization of rainwater resources, over-exploitation of groundwater, serious pollution of urban water bodies, overall shrinking of rivers, lakes and wetlands, as well as soil erosion (Piao et al., 2010; Qiu, 2010; UN, 2015).

Similar to traditional civil engineering practices in Western countries,[3] China's modern city-building practices involved the realignment and channelization of rivers to optimize land for development. Most of the nation's 87,000 dams were hastily built between the 1950s and 1970s and the quality of construction was low. During the post-reform era, natural watersheds were drastically transformed. The levelling of natural topography for urban development, and the construction of new built fabric with its dominance by impermeable surfaces along with channelized river courses, has radically accelerated the velocity and volume of surface water during rainy seasons. This has also contributed to hazardous flood events. These practices and poor management of water resources have converted natural water systems into a network of concrete open channels and underground pipes, designed to efficiently move water away from cities and rural areas. Often, raw sewage passes through this water conveyance system. This conventional practice has been sorely ineffective and a three-fold complex has emerged: (1) the system's capacity is unable to accommodate the speed and scale of hyper-urbanization and has resulted in catastrophic flash floods in urban areas; (2) water has been drained away from rural and urban areas, depleting the nation's water resources along with municipal groundwater levels falling by one to two meters each year throughout the country; and (3) the destruction of riparian ecosystems has disconnected people from their natural environments. In 2013, half of Chinese cities failed to reach national standards for flood prevention with more than 230 cities affected by major flood events (Duan et al., 2016). Additionally, the use of fertilizers and pesticides has negatively affected water quality, and this in turn affects the availability of domestic drinking water. The combination of the growing impact of climate change and impact of hyper-rapid urbanization on China's water and natural resources are proving to be serious threats for China's future survival.

The scale of China's hyper-rapid urbanization and the causal link to environmental degradation has created serious national and international concerns (Liu and Diamond, 2005; Kram et al., 2012). Invariably, China's action-oriented FYP and subsequent implementation at the local municipal and provincial government levels, fostered a competitive market for domestic and foreign investments. This created an entrepreneurial milieu and enabled the more enlightened local government officials to experiment with remaking the identity of their cities. Simultaneously, progress on activities directly connected to FYP national priorities served as metrics for evaluating government officials and contributed to their career advancement. This has created a favorable climate for the implementation of large public projects that both create public amenity and open space, as well as serving the needs of environmental restoration and management.

Turenscape and the Emergent Green Revolution

In 1998, Kongjian Yu established China's first private landscape architecture firm, Turenscape, having returned from study and practice in the United States in 1997 (Yu, 2010). Yu has also been an influential educator, as a professor at Peking University's Department of Urban and Environmental Sciences where he established Peking University's research center for landscape architecture in early 1997 (Ibid.). Over time, Yu developed this center and successfully established the College of Architecture and Landscape Architecture in 2010. The late 1990s was the period when China's city-making movement and urban fever began its sweep across the nation and this provided momentum for the work of the firm Yu and Padua (2007). Yu is committed to the view that contemporary landscape architecture praxis is a vehicle for battling China's environmental crisis. His view on landscape architecture in China is that the old guard was intent on a regressive view of the profession characterized by traditional Chinese gardens and the neo-classical Beaux Arts tradition for parks and gardens as the only acceptable spatial forms (Ibid.), and that contemporary practice must engage with issues of performance in the creation of resilient places. For example, in Zhongshan Shipyard Park's design concept, Yu emphasized the importance of the site's natural ecology, temporal dynamics of the river's daily tidal fluctuations of over one meter, as well as Mao's industrial heritage introducing a new genre of urban parks to the Chinese people (Padua, 2003; Yu and Padua, 2007). This pioneering project commenced a professional journey for Yu and Turenscape that revealed China's post-reform urban experiment. Their projects continue to explore the intersection of the dynamics of water (natural and urban), local culture and natural ecology, ancient agrarian practices, mitigation of degraded natural resources, as well as local community needs (Saunders, 2012; Padua, 2003, 2006, 2008, 2013a,b). Turenscape's numerous built works in China represent resilient design solutions that utilize adaptive strategies and ecological design principles in the context of climate change.

This impact has been underpinned by Yu's political engagement as a dedicated member of the CPC influencing China's various levels of government officials and CPC party leadership. He has regularly lectured at the Mayors' Forum held in Beijing two to three times a year (Yu and Li, 2003; Yu, 2010). He has also been interviewed numerous times, appearing on national and international television. His completed projects are regularly visited by government officials as part of their professional development. Some of his writings are required reading for government officials and school children (Yu, 2010; Steinitz, 2012). The two projects presented here, were commissioned directly by the mayors from Jinhua and Qinhuangdao after they engaged with Yu at the Mayors' Forum in Beijing (Padua, 2013).

Yu regularly consults [4] for the Ministry of Land and Resources (MLR), a vital government branch in the State Council that oversees the planning, management, and conservation of land nationwide (Ren, 2013). The MLR holds authority over land use control, laws, and policies in China and is considered one of the most powerful ministries. On the other hand, the Ministry of Housing and Urban-Rural Development has oversight of development control and related laws and policies for city, town, and village planning and construction, the building industry, and municipal works. However, given MLR's challenges of managing land use control nationwide, monitoring and regulating land development at the local level remains problematic, particularly given the national priorities for economic growth and nation-building. It is important to note that the national-level cabinet of ministries execute their own action plans in alignment with the FYP's mentioned earlier in the chapter.

A brief look at national priorities stated in the 9th through 13th FYP's and China's leaders reveals China's growing concerns about environmental degradation. As stated earlier, the FYP is China's major national policy making tool for economic and social development that is implemented at the provincial and local levels. The following FYP summary highlights the political context within which Yu was operating. It also implies ways Yu's vision may have influenced the national narrative to halt environmental degradation and slow China's rampant post-reform hyper-rapid urbanization.

The 9th (1996–2000) FYP incorporated aspects of *China's Agenda 21*, a white paper on population, environment, and development in the twenty-first century with specific language on sustainable development (Lin, 1998; OECD, 2007). Priorities in the 10th (2001–2005) FYP were geared toward economic development through advances in science and information technology. Additionally, for the first time, language in China's FYP raises environmental sustainability to the level of a national priority with directives on environmental mitigation and critical improvements needed for polluted river basins and lakes, and air quality (Hellmann and Melton; ADB). Other policy action items included improving forest coverage and urban greening, as well as improving environmental awareness. During the 9th and 10th FYP periods, China was focused on raising its international status with world-branding activities including World Trade Organization (WTO) membership, advocating for Beijing as the host for the 2008 Summer Olympics, as well as for Shanghai as host for the 2010 World Expo. China was successful in all of these efforts.

At the mid-term assessment of the 10th FYP and as preparations for the 11th FYP were underway, analysts were concerned that no real progress was achieved in terms of environmental protection in the 10th FYP period (Heilmann and Melton, 2013). At the same time, urban scholars have indicated that the 10th FYP was one of the periods when tremendous urbanization and so-called urban fever was taking hold with no real checks and balances, lack of inter-agency coordination, or regulations on urbanization practices and their impact on natural resources (Friedman, 2005; Yu and Padua, 2007). Hence, language in the 11th FYP (2006–2010) shifted from the basis of GDP economic growth to President Hu Jintao's notion of the "harmonious society" (*hexie shehui*) and "scientific concept of development" (*kexue fazhan guan*): people-centered development that is coordinated and sustainable, for the promotion of overall harmonious development of economy, well-being, and society (Fewsmith, 2004; Naughton, 2005). The 11th FYP was built from deficiencies and achievements from activities arising out of the 10th FYP. President Hu Jintao at the 17th National Congress of the CPC held mid-October 2007 further articulated the scientific concept of development and harmonious society by proposing that China resituate itself through the lens of an "ecological civilization" (*shengtai wenming*). Hu's goal was for a comprehensive and future-based civilization where human society and nature lived in harmony (Fewsmith, 2005; Gare, 2012; Hu, 2017).

Five years later, President Xi Jinping elevated the concept of ecological civilization to a national-level policy at the 18th National Congress of the CPC (Zhu, 2016; Pan, 2014). By then the mid-term assessment of the 11th FYP found environmental protection activities were not meeting CPC expectations. Hence, the 12th (2011–2015) FYP builds from both strengths and deficiencies in the 11th FYP, and environmental protection measures were further emphasized as a national priority. Along with information technology and biotechnology, environmental protection was designated a "pillar industry" (Hilton, 2011) The 12th FYP has been noted as accelerating China's green revolution[5] by setting several environmental targets including: low carbon and energy efficiency, as well as improvements to natural resources like air and water quality (Casey and Koeleski, 2011). The 13th (2016–2020) FYP accelerates China's green revolution even further. For the first time, the FYP includes a section titled "Ecosystems and the Environment;" and it emphasizes China's commitment to intensifying efforts to restore the degraded environment, conserve natural resources, improve mechanisms for ecological security and global climate change (CPC, 2016; Beijing Review, 2017; Koleski, 2017). While the 13th FYP is in its early stages, programs for creating ecological security and establishing conservation zones for environmentally sensitive areas were recently launched, along with the establishment of new national parks and a reformed national park system (Bai et al., 2016; McLaughlin, 2016).

The FYP process is tremendously complex, and politics in China is equally if not more complex. This national development policy has clearly demonstrated that China has finally made conservation and restoration of their disappearing natural resources and highly degraded environment a national priority. The FYP process and China's pace of change has enabled Yu and Turenscape to demonstrate the significance of landscape architecture and ecological planning as a major profession to take seriously.

Much of the built work by Yu and Turenscape demonstrates ways to tackle China's environmental degradation in urban areas at the site scale. Yu and Turenscape have been able to elevate the science side of the discipline of landscape architecture, especially in expanding knowledge on urban hydrology, regenerative processes, and resilient design. However, in the last 10 years or so, it appears that their work on green infrastructure and national ecological security planning, has also advanced CPC thinking and the national agenda for sustainable development. A case in point is Yu's green sponge and low impact development practices. This green sponge strategy was elevated to China's national stage in December 2013 at the first Central Urbanization Work Conference of the CPC Central Committee held in Beijing. There, President Xi Jinping spoke about the need to develop sponge-like cities that naturally accumulate, filter, and purify rainwater (Shepard, 2016). Within a year, the Sponge City (*haimian chengshi*) national program (SCP) was launched in alignment with the environmental targets laid out in the 12th FYP.

According to the English translation of the State Council's literature for the construction of sponge cities, the program initiative "promotes water resilient, low impact development integrated with urban planning and construction, allowing cities to function much like a sponge by absorbing rainwater that mitigates flooding while also storing and purifying water to meet future needs" (Yong et al., 2017). A sponge city in this regard refers to sustainable urban development including flood control, water conservation, water quality improvement, and natural ecosystems protection. The intention is that the retro-fitted water system for the city would absorb, store, filter, and purify stormwater, and release it for reuse when needed (Li et al., 2017).

Three national-level Ministries (Housing and Rural-Urban Development; Finance; and Water Resources) are currently responsible for directing and guiding the SCP initiative. In the guidelines for the SCP application, Turenscape's Yanweizhou Park was listed as a prototype. Hundreds of applications were vetted and in April 2015, 16 cities were announced for the initial pilot; and one year later in April 2016, 14 cities were added (Ibid.). In China's hyper-rapid fashion, the designated cities for the SCP have been adapting their urban fabric to accommodate green infrastructure. A recent survey and assessment of the SCP initiative reveals several challenges and ways the concept may or may not succeed (Ibid.). Critical to advancing the SPC and its success is whether or not the central government can recognize green infrastructure as one of several contributing factors for restoring the natural environment, improving water quality, controlling floods, and dealing with coastal resilience and extreme weather due to climate change.

The launch of the SCP program initiative suggests the primacy of the profession of landscape architecture in China, a major world player, and Yu's successful advocacy. It is indicative of Yu's tremendous influence on China's twenty-first century national environmental agenda. The hope is that Yu and Turenscape's work will continue to flourish and advance thinking on green infrastructure and low impact development. For example, a preliminary assessment of the SPC suggests a scale shift is needed.

Typically, when dealing with natural hydrological processes, a landscape architect would consider the watershed scale or the regional scale of a river basin and water catchment area. While the SCP was being launched, central government announced the National New-type Urbanization Plan (NUP) 2014–2020 (*guojia xinxing chengzhenhua guihua*); it was released in March 2014, a few months prior to the SCP initiative (Wu, 2016). In alignment with the 12th FYP, it incorporates sustainability (Bai et al., 2014). The NUP established a framework of city-region clusters or urban agglomerations (*chengshiqun*) in this national-level plan (Taylor, 2015). It builds on current mega-regions and designates 20 of these spatial clusters at three levels: (1) 5 national levels; (2) 9 regional levels and (3) 6 local levels (Fang, 2005).

It also draws on the national spatial concept of *"liang heng san zong,"* two horizontal and three vertical axes (Fang and Yu, 2016). What is most compelling is the NUP locates these urban clusters and mega-regions within the nation's six river basins (Fang, 2005). Through the lens of resilience, ecological planning, and green infrastructure, this spatial organization of urban agglomerations nested within river basins provides an incredible opportunity to shift the SPC pilot to the river basin scale.

There is no doubt that Yu and Turenscape will continue with their hydrological green infrastructure experiments and derive resilient design solutions, especially given their ability to navigate the hyper-rapid pace of change in China. Turenscape's award-winning riverfront project, Houtan Park, on a linear brownfield site, was ne of these experiments, a green sponge investigation that was featured at the 2010 Shanghai World Expo. It focused on environmental education and demonstrated ways that polluted river water could be cleansed systematically through the use of green infrastructure. While Yu has expressed concerns about the lack of building standards, codes, and a modern regulatory framework, he and Turenscape have had their work implemented (Yu, 2010). Their Houtan Park resulted in eight design patents with the majority related to green infrastructure components. Turenscape has been able to redeploy these design components in projects built since then, as well as refine their ideas. The SCP and its refinements in the coming years is just the tipping point for Yu and Turenscape's drive for design excellence and restoring the natural environment. The potential to shift the green sponge to the river basin scale would be optimal.

The challenges facing Yu and Turenscape are the conflicting efforts currently underway and the political complexity of the political landscape surrounding water issues. A case in point is the South-North Water Diversion (SNWD) project, a vast infrastructure project that commenced in 2002 (Tran et al., 2016; Webber et al., 2017). According to Barnett et al. (2015), it is one of the largest water transfer schemes in the world; it connects four major river basins, three megacities, six provinces and hundreds of millions of people. The intention was to divert freshwater from the Yangtze River in south China to the drier regions in the north. At least US$80 billion has been spent on reservoir and canal construction and millions of people relocated in the process. It is not difficult to imagine the negative environmental impacts and damage to local ecosystems. Apparently, once the system is fully operational, the current water shortage will still not be alleviated (Ibid.).

China's heritage of large-scale hydro-engineering is generally known, for example, the Grand Canal circa 605 BCE, and Dujiangyan Irrigation System, a UNESCO world heritage site built circa 200 BCE. Additionally, China has an ancient urban morphology of flood adaptive human settlement patterns (Yu et al., 2008). However, with current concerns for sustainable development, green urbanism, and green technology, the SNWD is clearly the binary opposite. The argument for deploying tactics of the green sponge to capture and store water in north China would potentially offer a responsible and resilient alternate solution, as well as contributing to the nation's water security.

The hope is that the SCP initiative will succeed and central government will have the scientific evidence to consider different scenarios for applying the green sponge. This could include the jump up in scale to the river basin along with connectivity among and between the various towns and cities located there. Another scenario might consider certain green infrastructure components, for example, water harvesting, retention, and storage to meet water supply needs in water scarce northern areas. Deploying a variety of green infrastructure tactics, including cleansing, capture, and storage could contribute to increasing water supplies for two-thirds of China's cities that exhibit water scarcity (Li et al., 2017). Furthermore, with the anticipated population growth and implementation of the NUP and related city-region clusters, more potable water will be needed.

Two projects, Yanweizhou Park and Qinhuangdao Beach Restoration, illustrate a body of work developed by Turenscape, exploring landscape strategies with the potential to mitigate coastal change and make cities more resilient in the face of flooding. Yanweizhou Park is located along the confluence of two rivers in Jinhua, a city with a population of more than one million in Zhejiang province. The second project, Qinhuangdao Beach Restoration, is located in a popular coastal resort city due east of Beijing in Hebei Province.

Resilience at the Confluence of Two Rivers

In 2010, the city of Jinhua's municipal government designated Yanweizhou Park as a pilot project to rebrand their city's identity. Kongjian Yu had been developing ideas of how to use urban open space as a "green sponge," a form of green infrastructure with performance goals that deal with both flood control and water conservation, and Turenscape were commissioned to put this into practice in the new park.

The objective for Jinhua Park's green infrastructure was to restore the park's natural ecology, enhance habitat, demonstrate environmental responsibility, and provide opportunities for the local citizens to have a nature-based experience near the city's center. The park's new design would also provide environmental education and teach the local citizens about wetlands and habitat restoration. It would introduce the community to the concept of resilience and climate change, and ways that the green infrastructure could deal with seasonal flooding and storm surge events, as well as demonstrate ways to improve water quality through cleansing. It is important to note that Yu's concept of the green sponge as a significant strategy for Chinese urbanism became highly influential when the idea of sponge cities was raised in China circa 2012 and was later emphasized by President Xi Jinping at the 2013 Central Government Working Conference on Urbanization held in Beijing in December (O'Meara, 2015; Sidner, 2017).

The site of Jinhua Park covers 26 hectares and is located at the confluence of two rivers, Wuyi and Yiwu. Across the river along one edge is a residential district and a traditional riverfront park is across the other river. Existing conditions at the park site included a bird habitat in wetland remnants; a sand quarry previously used for construction activities; and an area set aside for a performing arts building. Like many riverfront cities with more than a million people, the city's river system was realigned and channelized to facilitate efficient urban development. The natural ecology of the riparian corridor was significantly degraded; and the concrete walls along the realigned riverbanks, combined with the built form of the city, exacerbated flooding and storm surges during the monsoon season (Figure 20.1).

The English translation for Yanweizhou is "sparrow's tail" and aptly describes the site's narrow landmass situated at the confluence of the Wuyi and Yiwu rivers. In plan, the site resembles a sparrow's tail and its isolation from people has allowed egrets and other native bird species to thrive in Jinhua's last existing wetland habitat. Turenscape was challenged by the site's bird habitat and lack of physical access due to the rivers' 100-meter width, and the sand quarry, as well as the spatial requirements for a large outdoor assembly area adjacent to the performing arts building.

The design solution deployed resilient tactics that took into account the site's seasonal rainfall, storm surge, flooding, and rising water levels from the river system, and the provision of physical access for the city's residents (Figure 20.2). The park's design also allowed the community to learn about their natural environment and experience the

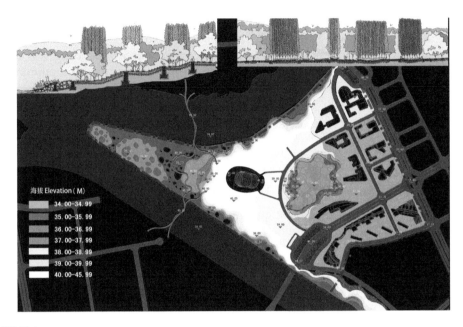

FIGURE 20.1
Grading Plan and Section: Concrete walls were removed and replaced with vegetated terraces, and designed using a balanced cut and fill strategy. The site's resilient design intention is intended to treat urban water, manage seasonal floods and storm events and is 100% permeable.

FIGURE 20.2
Aerial photo taken during the wet monsoon season and 20-year flood event illustrates the park's porosity and resilience, as well as its location at the confluence of two rivers (May 2016).

wetland habitat, including bird-watching. Direct physical and visual access to the river's dynamic waterfront edge and riparian environment also provided educational, as well as recreational opportunities. Using green infrastructure and low-tech approaches, Turenscape transformed the lifeless landscape into a resilient green sponge.

Turenscape's green sponge concept involved a combination of native vegetation plantings, constructed wetlands, landform manipulation and naturalized river embankments. In their design intervention, the vegetated terraces allow polluted river water from storms or floods to be cleansed and absorbed onsite. The vegetated terraced landform was borrowed from China's historic agrarian practice of rice paddy cultivation. In this practice, farmers capture and retain water to use for irrigation during the dry season. Adapting this age-old practice into their design solution, Turenscape created a resilient and self-sustaining landscape where the form of the stepped terraces with native wetland plantings slowed water and absorbed silt generated during the rainy monsoon season and flood events (Figure 20.3). In their implementation, artificial irrigation and fertilizers were not utilized. Additionally, Turenscape incorporated stonework into the design of the vegetated terraces. This masonry work also created a path system that provided visitors visual and physical access to the riverfront edge and wetland area.

As part of the green infrastructure system, the remnant sand quarry area was replaced with a large retention pond. Turenscape's design intention was that it would absorb and cleanse floodwater before it entered the local groundwater table, as well as recharging the watershed's aquifer. The pond's water originated from the river system and Turenscape's design filtered this water through sand-gravel layers. This system was designed so that the pond water would meet water quality safety standards for recreational swimming. Like the vegetated terraces, the pond was a design component of Turenscape's low impact development strategy. A pedestrian path system, incorporating pavilions at key vista points, was integrated into the planting terraces and designed to withstand floods and stormwater surges generated during the monsoon season. For safety purposes, the city

FIGURE 20.3
Aerial photo taken during the dry season illustrates the vegetated riverfront terraced edge with lush wetland grasses and natural silting from the river.

FIGURE 20.4
The Bayong Qiao bridge rebrands the city's identity and provides an iconic destination in addition to its function as a physical connection between the park and the city's residents and tourists (2014).

closes the park for public use during flood events. Additionally, the city is able to save money on costly conventional wastewater treatment facilities. This is due to the park's capacity to treat 984 cubic meters (260,000 gallons) of water a day; and is the equivalent to a wastewater treatment system that would cost US$1 million, according to Yu. Organic pollutants are also harvested and used as nutrients for the establishment of vegetation in a wetland garden. Another major design move was the elevated 700-meter-long Bayong Qiao Bridge (Figure 20.4). For the first time, park visitors from across the two rivers were able to gain physical access to the park site, creating direct connections for residents in the adjacent communities. This monumental bridge is structurally designed to withstand a 200-year flood. Ramps from the bridge to the park are designed to be submerged and withstand 20-year floods. The selection of the bridge's red and yellow colors was inspired by local folk tradition and dragon dance festival activities that take place during the Chinese New Year celebrations each year. These colors are incorporated into the bridge's handrails and at night are accentuated with artificial lighting. The night lighting was intended to reinforce the park's image as the city's icon.

In keeping with the green sponge theme of resiliency and landscape performance, the ground plane for the outdoor assembly area adjacent to the performing arts building (an opera house), was composed of permeable materials: recycled gravel, permeable pavers, and circular bio swales. These surface materials were intended to capture and absorb surface water. The area's concentric pattern and curvilinear forms resemble ripples in a lake formed by raindrops. Curvilinear tree planters with yellow fiberglass benches and taller vegetation were also incorporated and created smaller intimate areas for visitors.

In 2014, shortly after the park's construction was complete, a 20-year flood occurred during the annual monsoon season. After the majority of the park's wetland area was submerged in the rising waters for two days, it reemerged a week later filled with natural life. The fact that Yanweizhou Park survived this flood event is a testimony to the success of the resilient design solution. Since its opening, over 40,000 visitors have crossed the park's bridge, another measure of the park's success. Yanweizhou Park serves as a resilient prototype in China and represents a crucial adaptive strategy for mayors to consider when redeveloping their waterfronts to survive large-scale floods.

Tourism and Resilience

As stated earlier in the chapter, China's relentless hyper-urbanization caused the rise of a consumer-oriented culture with changing societal needs. With shifts in the everyday work schedule for China's growing middle class, leisure time increased (Davis and Kraus, 1995). Furthermore, China's new urban population exhibited a tremendous appetite for owning cars. Combining these various factors, the demand for recreational facilities intensified. In turn, this created tremendous pressures for domestic tourism.

This growth took a toll on Qinhuangdao, a popular coastal resort area due east of Beijing in Hebei province, and now part of the "Jing-Jin-Ji" urban cluster (*chengshiqun*) or mega-region (Fang, 2005; Johnson, 2015). The large-scale new construction of a high-speed railway and road network with major elevated expressways increased the number of visitors to the coast from the Beijing area. This new transportation infrastructure and greater number of tourists negatively affected the coastal environment causing erosion and degradation of the area's natural ecology. The surge of tourists, traffic, and road congestion during the summer months and other major national holidays has multiplied pressures on this coastal city, the location for Turenscape's Qinhuangdao Beach Restoration.

The area's popularity is due to its rugged cliffs and a broad stretch of sandy beach located about 290 kilometers from Beijing. Qinhuangdao's cultural heritage has appeal to the Chinese people. It's a place where Emperor Qin and Chairman Mao visited; and it also developed as a beach resort destination for foreign expatriates during China's nineteenth century colonial period. The CPC leadership regularly frequents the area, and this resort location is comparable to the USA's Camp David where political leaders travel for a retreat, have political discussions, sometimes with visiting world leaders (Padua, 2013). The area lies within a regionally significant bird sanctuary, the Qinhuangdao Wetland Reserve. The area is also geographically interior to an estuarine zone of several rivers that open into the Bohai Sea, the innermost gulf of the Yellow Sea. Its location at the confluence of migratory paths for birds flying between northeast Asia, China's southern region, Indochina, Australia, and East Africa has made Qinhuangdao famous for bird-watching during the spring and fall seasons. With the rise of environmental consciousness, the area has increased in its popularity for bird-watching by both amateurs and professional ornithologists.

Essentially, Turenscape's tri-partite design for Qinhuangdao Beach Restoration consists of a coastal shoreline, intertidal zone, and wetlands (Figure 20.5). Like Yanweizhou Park, this nearly 10-year-old park serves as a design precedent for coastal resilience, adaptive strategies, and ecological design. It demonstrates regenerative design and ways to mitigate and restore impaired environments to governmental officials. It also demonstrates different design solutions for environmental protection: habitat protection, restoration and enhancement, as well as ways that constructed wetlands function in green infrastructure and mimic natural systems. Turenscape's innovative low impact design rationalizes coastal access to an existing critical bird sanctuary, the regional Qinhuangdao Wetland Reserve.

FIGURE 20.5
Turenscape restored the degraded coastal wetland in the project's intertidal zone.

The major design interventions included: pedestrian access with a boardwalk traversing through the shoreline habitat and intertidal zone; restored and constructed wetlands surrounding a new Wetland Museum on a former theme park site in the intertidal zone; and replacing a concrete embankment with more resilient materials (Figure 20.6). The transformed shoreline acts as an edge for a constructed lake. Newly created artificial

FIGURE 20.6
Turenscape replaced the shoreline's concrete embankment with riprap, native plants, and a pedestrian boardwalk.

FIGURE 20.7
Nine small vegetated islands were created to provide a new habitat and enhance the area's diversity.

islands in the new lake body were intended to enhance the existing bird habitat and coastal area's biodiversity (Figure 20.7).

Turenscape was challenged by the highly sensitive physical environments for the bird sanctuary and sandy beach, a critical primary dune habitat. Adding to these challenges was the project's compressed timeline with only six months from commission to opening. Kongjian Yu recommended phased construction according to the physical areas and their level of ecological impairment. Yu also realized that the ecologically sensitive sandy beach zone and compressed schedule would not allow for the typical technique of excavating for poured-in-place concrete foundations needed to construct a 4-mile long pedestrian boardwalk made of timber. Yu quickly innovated with a novel design solution for dealing with the foundations for the boardwalk's post and beam wood frame construction: a portable prefabricated lightweight fiberglass box. These units were fabricated off-site, transported onsite, and manually filled with sand for structural stability. Essentially, this novel solution reduced the environmental impact created by conventional concrete foundation construction with the related noisy heavy truck traffic traversing the critically sensitive primary dune habitat (Figure 20.8). Additionally, the prefab technology allowed completion of the boardwalk construction within the short time frame.

Through the deployment of various regenerative design techniques, Turenscape was able to accomplish their goals: to rehabilitate the severely damaged natural environment and restore and reveal to tourists and local residents the site's natural beauty (Figure 20.9). Simultaneously, their resilient design solution transformed an environmentally degraded beach and intertidal zone into an ecologically healthy and aesthetically attractive landscape.

The two case study projects presented in this chapter are significant precedents for adaptive urbanism and resilient design solutions for coast and waterfront. From the perspective of governmental officials, these two parks have been transformative and educational. In China, Yu and Turenscape have demonstrated ways landscape architecture praxis can advance design knowledge on climate change and resilience within that nation's vast urban experiment. From the world view, Yu and Turenscape have created two projects that are environmentally and socially responsible, as well as beautiful solutions for dealing with urban floods, climate change, coastal erosion, habitat restoration, and promoting healthy communities.

FIGURE 20.8
The four-mile boardwalk traverses through coastal plant communities and shady groves of trees with views of the sea.

FIGURE 20.9
The Wetland Museum is situated in the re-vegetated intertidal wetland zone.

Endnotes

1. For a clear description and organizational chart of China's system of governance, see How China Is Ruled, BBC News website. http://news.bbc.co.uk/2/shared/spl/hi/in_depth/china_politics/government/html/1.stm.

2. In the period spanning the late 1990s through the first decade of the twenty-first century, China developed the reputation as the workshop of the world. Generally, this defined China as the contemporary equivalent of nineteenth century Manchester, England. See Parker (2015) for description of England's reputation as workshop of the world during the industrial revolution. China's workshop of the world identity was a result of the export-quality goods manufactured by low cost labor working in numerous factories concentrated in the Pearl River Delta (PRD) region. The PRD is geographically located in southern China and primarily in the Guangdong province. Guangzhou (formerly Canton), the provincial capital and China's first foreign port, is located at the northern end of the delta with Macao, Hong Kong, and their respective sister cities, Zhuhai and Shenzhen designated as Special Economic Zones, at the southern end and mouth of the Pearl River. The factories (some independently owned and others foreign-owned) manufactured a range of goods including computer mother boards, refrigerators, televisions, and shoes using very low-cost labor—all within easy access to deep water container ports. See Gao (2012) for a detailed description of the PRD geography, factory town distribution, infrastructure, overseas Chinese influence, foreign investment and globalization. See Smart and Lin (2007) and Chan and Zhang (1999) for a socio-cultural examination of the migrant population, illegal work force and China's Hukou policy. See Johnson and Woon (1997) for an understanding of PRD's spatial characteristics of the area in the 1980s and 1990s before rapid urbanization took hold.

3. The Los Angeles River in southern California, the original settlement for the city of the same name, was channelized in the first decade of the twentieth century for flood control and management. Later in the century, a grassroots effort activated the need to remake and reclaim the river's riparian ecosystem (transformed into a linear system of public parks) for social equity and healthy communities, as well as for purposes of economic revitalization. See Price (2008) and Gumprecht (1999) for the river's environmental history and the website www.lariver.org.

4. Confirmed by Yu in various meetings with Dr. Mary G. Padua at Turenscape's headquarters in Beijing in the past several years. At one of these meetings, Yu noted that it is customary in China to consult for only one cabinet-level ministry, given the politics and competition among the various ministry leaders.

5. China's green revolution is believed to have been formalized in the 12th (2011–2016) FYP. Scholars tend to frame modern China and its development within a revolutionary praxis. For example, Sun Yat-sen's revolution and the rise of the Republic of China as a modern nation divorcing itself from the Qing dynasty court and imperial past; Mao's communist revolution and the establishment of the People's Republic of China; and Deng's economic revolution and various reforms. The 12th FYP reframed China's economic development in terms of green technology and innovation for renewable energy resources with environmental protection as one of the pillars of industry. See Hilton (2011) and Hu (2014).

References

Bai, X., Shi P., and Y. Liu. 2014. "Society: Realizing China's urban dream." *Nature* 509:158–160.

Bai, Y., Jiang, B., Wang, M., Li, H., Alatalo, J., and S. Huang. 2016. "New ecological redline policy (ERP) to secure ecosystem services in China." *Land Use Policy* 55:348–351.

Barnett, J., Rogers, S., Webber, M., Finlayson, B., and M. Wang. 2015. "Transfer project cannot meet China's water needs." *Nature* 527(7578):295–297.

Bell, D. A. 2015. *The China Model: Political Meritocracy and the Limits of Democracy.* Princeton, NJ: Princeton University Press.

Cantin, E. and M. Taylor. 2008. "Making the 'workshop of the world': China and the transformation of the international division of labour." In *Global Economy Contested: Power and Conflict across the International Division of Labour,* ed. M. Taylor. New York, NY: Routledge, 51–77.

Campanella, T. 2008. *The Concrete Dragon.* Princeton Architectural Press, New York.

Casey, J. and K. Koeleski. 2011. *Backgrounder: China's 12th Five Year Plan.* U.S.-China Economic & Security Review Commission, Washington, DC.

Central Committee of the Communist Party of China. 2016. *The 13th Five-Year Plan for Economic and Social Development of the People's Republic of China 2016–2020.* Beijing: Central Compilation and Translation Press.

Chan, K. W. and L. Zhang. 1999. "The Hukou System and rural-urban migration in China: Processes and changes." *The China Quarterly* 160:818–855.

Davis, D. and R. Kraus. 1995. *Urban Spaces in Contemporary China: The Potential for Autonomy and Community in Post-Mao China.* Washington, DC: Cambridge University: Woodrow Wilson Center Press.

Duan, W., He, B., Nover, D., Fan, J., Yang, G., Chen, W., Meng, H., and C. Liu. 2016. "Floods and associated socioeconomic damages in China over the last century." *Natural Hazards* 82(1):401–413.

Economy, E. 2004. *The River Runs Black: The Environmental Challenge to China's Future.* New York, NY: Cornell University Press.

Elvin, M. 2004. *The Retreat of the Elephants: An Environmental History of China.* New Haven, CT and London: Yale University Press.

Friedman, J. 2005. *China's Urban Transition.* Minneapolis, MN: University of Minnesota Press.

Fang, C. 2005. "Important progress and future direction of studies on China's urban agglomerations." *Journal of Geographical Sciences* 25(8):1003–1024.

Fang, C. and D. Yu. 2016. *China's New Urbanization: Developmental Paths, Blueprints and Patterns.* Heidelberg, New York, Dordrecht, and London: Springer Verlag.

Fewsmith, J. 2004. Promoting Scientific Development Concept, China Leadership. *Monitor No. 11, Hoover Institution, Stanford University.*

Fewsmith, J. 2005. "The Changing Methodology of Beijingology." *Presented at seminar on "Behind the Bamboo Curtain: Chinese Leadership, Politics, and Policy,"* Carnegie Endowment for International Peace, Washington, DC, November 2. pp. 1–9.

Gao, Y. 2012. *China as the Workshop of the World: An Analysis at the National and Industry Level of China in the International Division of Labor.* New York, NY: Routledge.

Gare, A. 2012. "China and the struggle for ecological civilization." *Capitalism Nature Socialism* 23(4):10–26.

Goodkind, D. and L. West. 2002. "China's floating population: Definitions, data and recent findings." *Urban Studies* 39(12):2237–2250.

Gumprecht, B. 1999. *The Los Angeles River: Its Life, Death, and Possible Rebirth.* Baltimore, MD: The Johns Hopkins University Press.

Heilmann, S. and O. Melton. 2013. "The reinvention of development planning in China 1993–2012." *Modern China* 39(6):580–628.

Hilton, I. 2011. "Introduction: The evolving blue print." In *Creative Commons' Attribution-NonCommercial-NoDerivs 2.0 England and Wales License and 2.5 China License.* e-book, pp. 4–10.

Ho, P. 2003. "Mao's war against nature? The environmental impact of the grain-first campaign in China." *The China Journal* 50:37–59.

Hsü, I. C. 1990. *China Without Mao: The Search for a New Order.* Oxford: Oxford University Press.

Hu, A. 2014. *China: Innovative Green Development.* Heidelberg, New York, Dordrecht, London: Springer Verlag.

Hu, A. 2017. "Theory of green development." In *China: Innovative Green Development.* Heidelberg, New York, Dordrecht, London: Springer Verlag, pp. 15–58.

Johnson, G. and Y. Woon. 1997. "Rural development patterns in post-reform China: The pearl river delta region in the 1990s." *Development and Change* 28(4):731–751.

Johnson, I. 2015. "As Beijing becomes a supercity, the rapid growth brings pains." *New York Times.*

Koleski, K. 2017. *The 13th Five-Year Plan. U.S. China Economic and Security Review Commission Staff Research Report.*

Kram, M., Bedford, C., Durnin, M., Luo, Y., Rokpelnis, K., Roth, B., Smith, N. et al., [Smith, N. ed.] 2012. *Protecting China's Biodiversity: A Guide to Land Use, Land Tenure, and Land Protection Tools.* Beijing: The Nature Conservancy.

Li, H., Ding, L., Ren, M., Li, Ch., and H. Wang. 2017. "Sponge city construction in China: A survey of the challenges and opportunities." *Water* 9(594):1–17.

Lin, G. 1998. "Implementing China's Agenda 21: From national strategy to local action." *Impact Assessment and Project Appraisal* 16(4):277–287.

Liu, J. and J. Diamond. 2005. "China's environment in a globalizing world." *Nature* 435:1179–1186.

Lockett, H. 2016. "Adaptive anachronism: China's latest five-year plan could be its last." *China Economic Review.* http://www.chinaeconomicreview.com/china%E2%80%99s-latest-five-year-plan-could-be-its-last (accessed August 20, 2017).

McLaughlin, K. 2016. "Can a new park save China's big cats?" *Science Magazine.* http://www.sciencemag.org/news/2016/08/can-new-park-save-chinas-big-cats.

Naughton, B. 2005. The New Common Economic Program: China's Eleventh Five Year Plan and What It Means. China Leadership Monitor 16.

O'Meara, S. 2015. "China wants to build sponge cities." *The Atlantic.* https://www.theatlantic.com/international/archive/2015/11/china-sponge-cities-floods/417498/.

Organization for Economic Cooperation and Development. 2007. *Environmental Performance Reviews China.* Paris, France. www.oecd.org.

Ouroussoff, N. 2008. "Lost in the New Beijing: The old neighborhood." *New York Times.*

Padua, M. 2003. "Industrial strength." *Landscape Architecture* 93(6):76–85, 105–107.

Padua, M. 2006. "Touching the earth." *Landscape Architecture* 96(12):100–109.

Padua, M. 2007. "Hybrid modernity: Framing the park in post-Mao China." Council of Educators in Landscape Architecture (CELA). *Negotiating the Landscapes Proceedings Penn State*, State College, PA, 65–80.

Padua, M. 2008. "A fine red line: Design tests the boundary between art and ecology." *Landscape Architecture* 98(1):90–99.

Padua, M. 2013a. "Bridge to somewhere else." *Landscape Architecture* 103(2):80–87.

Padua, M. 2013b. "Triptych by the sea." *Landscape Architecture* 103(2):98–107.

Padua, M. 2014. "China: New cultures and changing urban cultures." In *New Cultural Landscapes*, eds. M. Roe and K. Taylor. USA and Canada: Routledge, pp. 202–222.

Pan, J. 2014. *China's Environmental Governing and Ecological Civilization.* Heidelberg, New York, Dordrecht, London: Springer Verlag.

Parker, S. 2015. *Urban Theory and the Urban Experience: Encountering the City.* New York, NY: Routledge.

Piao, S., Ciais, P., Huang, Y., Shen, Z., Peng, S., Li, J., Zhou, L. et al., 2010. "The impacts of climate change on water resources and agriculture in China." *Nature* 467:43–51.

Price, J. 2008. "Remaking American environmentalism: On the banks of the L.A. River." *Environmental History* 13(3):536–555.

Qiu, J. 2010. "China faces up to groundwater crisis." *Nature* 466:308.

Ren, X. 2013. *Urban China.* Cambridge, UK, Malden, MA: Polity Press.

Saunders, W. 2012. *Designed Ecologies: The Landscape Architecture of Kongjian Yu.* Basel: Birkhäuser.

Shapiro, J. 2001. *Mao's War against Nature: Politics and the Environment in Revolutionary China.* Cambridge, CT: Cambridge University Press.

Shepard, W. 2016. "Can 'sponge cities' solve China's urban flooding problem?" www.citiscope.org.

Sidner, L. 2017. Sponge City: Solutions for China's Thirsty and Flooded Cities. *New Security Beat, the blog of the Woodrow Wilson Center's Environmental Change and Security Program.* https://www.newsecuritybeat.org/2017/07/sponge-city-solutions-chinas-thirsty-flooded-cities/.

Skinner, G. W. 1977. "Regional urbanization in nineteenth century China." In *The City in Late Imperial China*, ed. G. W. Skinner. Stanford, CA: Stanford University Press, pp. 211–249.

Smart, A. and G. C. S. Lin. 2007. "Local capitalisms, local citizenship and translocality: Rescaling from below in the Pearl River Delta Region, China." *International Journal of Urban and Regional Research* 31(2):280–302.

Smil, V. 2013. *Making the Modern World: Materials and Dematerialization*. John Wiley & Sons, NJ, USA.

Spence, J. D. 1991. *The Search for Modern China*. New York, NY: W. W. Norton & Company.

Steiner, F. 2012. "The Activist Educator." In *Designed Ecologies: The Landscape Architecture of Kongjian Yu*, ed. W. Saunders. Basel: Birkhäuser, 106–115.

Taylor, J. R. 2015. "The China dream is an urban dream: Assessing the CPC's National new-type urbanization plan." *Journal of Political Science* 20(2):107–120.

Thompson, B. and K. Sorvig. 2000. *Sustainable Landscape Construction*. Washington, DC: Island Press.

Tran, M., Konkagul, E., and R. Connor. 2016. *The United Nations World Water Report: Water and Jobs*. UN World Water Assessment Program.

United Nations. 2015. *The Human Cost of Weather-Related Disasters 1995–2015*. UN Office for Disaster Risk Reduction, Geneva, Switzerland.

Wang, Y., Morgan, R. K., and M. Cashmore. 2003. "Environmental impact assessment of projects in the People's Republic of China: New law, old problems." *Environmental Impact Assessment Review* 23:543–579.

Webber, M., Crow-Miller, B., and S. Rogers. 2017. "The South–North Water Transfer project: Remaking the geography of China." *Regional Studies* 51(3):370–382.

Weng, Q. 2007. "A historical perspective of river basin management in the Pearl River delta of China." *Journal of Environmental Management* 85(4):1048–1062.

Wu, F. 2016. "China's emergent city-region governance: A new form of state spatial selectivity through state-orchestrated rescaling." *International Journal of Urban and Regional Research* 40:1134–1151.

Yong, J., Zevenbergen, C., and D. Fu. 2017. "Can "sponge cities" mitigate China's increased occurrences of urban flooding?" *Aquademia: Water, Environment and Technology* 1:1–5.

Yu, K. and D. Li. 2003. *The Road to Urban Landscape: A Dialogue with Mayors*. Beijing: China Architecture & Building Press.

Yu, K. and M. Padua. 2007. "China's Cosmetic Cities: Urban fever and superficiality." *Landscape Research* 32(2):225–249.

Yu, K. 1992. "Experience of basin landscapes in Chinese agriculture has led to ecologically prudent engineering." In *Human Responsibility and Global Change: Proceedings of the International Conference on Human Ecology*, eds. L. O. Hansson and B. Jungen. Sweden: University of Gothenburg, 289–299.

Yu, K. 2010. Kongjian Yu: Turenscape landscape architecture, urban design, architecture, Beijing. *Harvard Design Magazine, No. 33 / Design Practices Now, Vol. II*.

Yu, K. and M. Padua, eds. 2007. *The Art of Survival: Recovering Landscape Architecture*. Victoria, Australia: Mulgrave Publishing.

Yu, K., Zhang, L. and D. Li. 2008. "Living with water: Flood adaptive landscapes in the Yellow River basin of China." *Journal of Landscape Architecture* 3(2):6–17.

Zhang, L. 2001. *Strangers in the City: Reconfigurations of Space, Power, and Social Networks within China's Floating Population*. Stanford, CA: Stanford University Press.

Zhang, X. 2015. "Globalization and the mega-region: investigating the evolution of the Pearl River delta in a historical perspective." In *Megaregions: Globalization's New Urban Form?* eds. J. Harrison and M. Hoyler. Northhampton, MA: Edgar Publishing Ltd, 75–200.

Zhao, J. and J. Woudstra. 2007. "In agriculture, learn from Dazhai, Mao Zedong's revolutionary model village and the battle against nature." *Landscape Research* 32(2):171–205.

Zhu, G. 2016. Ecological Civilization: a national strategy for innovative, concerted, green, open and inclusive development. *UN Environment Programme*.

Zhu, J. 2004. "Local development state and order in China's urban development during transition." *International Journal of Urban and Regional Research* 28(2):424–447.

Zhu, Y. 2007. "China's floating population and their settlement intention in the cities: Beyond the *Hukou* reform." *Habitat International* 31(1):65–76.

21

Going with the Flow: Building Resilience in Southeast Queensland

James Davidson and Samuel Bowstead

CONTENTS

Context

This essay provides an overview of the body of work of James Davidson Architect (JDA), an architectural practice based in Brisbane, Australia. Since leading Emergency Architects Australia's Queensland Flood relief effort in 2011, the practice has embarked on a series of experimental flood-resilient projects that serve as case studies for this essay. Ranging from built residential work to more radical "amphibious" designs, these projects allow an analysis of flood-resilient design principles and how these interact with local building codes and planning policy. This essay will discuss each case study before drawing out flood appropriate design principles with reflections on the lessons they provide to policy frameworks and design practice more broadly.

In January 2011, an *inland tsunami* flowing up to eight meters high hit the Lockyer Valley (see Figure 21.1) region west of Brisbane in Southeast Queensland and caught the world's attention. Up to 160 millimeters (mm) of rain fell in a period of 36 hours, creating widespread

FIGURE 21.1
Flooding running through Toowoomba outside the Lockyer Valley. (Image credit: Winson, 2011.)

flash flooding that swept away the town of Grantham and several local farms, resulting in the deaths of 12 people (Insurance Council of Australia, 2011). Hailed as "Australia's most expensive natural disaster" (Queensland Reconstruction Authority, 2013), the floods across the state affected 200,000 homes across approximately 1 million square kilometers, a landmass equivalent to the size of France and Germany combined. Flood damage was estimated to be AUD\$5 Billion (Queensland Government, 2012:32).

In addition to the flash flooding from the Lockyer Valley, Wivenhoe Dam, Brisbane's main water supply and only major piece of flood infrastructure, reached capacity and came within 600 mm of overtopping (Insurance Council of Australia 2011:i). Emergency flow releases from the dam resulted in an unprecedented level of water entering the lower Brisbane River, which combined with Lockyer Creek and the Bremer River, inundating 28,000 homes and businesses, including low lying areas of the Brisbane CBD (Ibid.).

Emergency Architects Australia's Queensland Flood Response

In response to these devastating floods, Davidson, then director of the not-for-profit Emergency Architects Australia (EAA), and Bowstead, a young university graduate, coordinated a volunteer team of 60 architects, 10 engineers, and 100 students to provide 230 pro bono professional house inspections in those suburbs inundated by floodwaters. These building assessments were designed to give practical advice and guidance in post-flood reconstruction to largely uninsured homeowners. In relying on a thorough technical report, assessments were geared toward providing a sense of direction for occupants in wading through not only rebuilding, but also the planning process associated with reconstruction, something which many households had never dealt with before.

Building assessments lasted between one and two hours and were conducted as an informal conversation to give homeowners the time to fully understand the overall extent of damage to their home and the processes moving forward. Upon completion, these technical reports were then packaged with photographs and location-specific notes and returned to individual homeowners as a full record of the items discussed. If deemed necessary by the assessing architect, a follow-up pro bono structural engineering inspection was arranged to give further reconstruction advice. At the conclusion of the program in mid-2011, EAA volunteers had assisted over 1,000 families across 34 suburbs in Ipswich and Brisbane.

Affordability and Reactionary Planning

Given the high risk of repeat events of this kind in Brisbane and Ipswich, and to reduce costs and undue hardship on those least able to bear it, EAA submitted the following recommendations to the Queensland Floods Commission of Inquiry (2012):

1. That the extent of "no-build" flood liable areas be redefined and/or that minimum floor levels and suitable construction techniques and materials be mandated where prohibition of development is not desirable or feasible. For example: plasterboard, particleboard, cavity insulation, and electrical circuits cannot stand inundation and should not be permitted below well-established likely flood levels.

2. That flood risk levels be clearly identified in development control instruments and codes and mandated on real estate documents.

3. That as the population most greatly impacted by the flood was not insured and was the least able to afford insurance, government should provide insurance in areas where it was and is now clear that the land at risk was released by government for development. The ambiguities in private insurance policies relating to flooding cover should also be clarified by mandate.

The findings set out in the EAA report (Davidson, 2011) have serious implications for future planning and development control policies and mechanisms in coastal and riverine flood-prone areas. While it can be argued that no further construction should be permitted in these areas, this may not always be desirable or feasible. It became very clear from these inspections that, for a variety of easy-to-understand reasons, different building designs and materials performed very differently when it came to effecting repairs—with considerable cost implications. Common contemporary materials such as plasterboard, particleboard, and cavity, slab on ground construction, cannot stand inundation or contact with water. The use of suspended floors (at present not favored by home builders who prefer brick veneer and concrete slab on ground for ease of construction and cost reasons) and other features of traditional Queensland building design would have considerably reduced the need to replace the interior finishes and cabinetry. This would have benefited affected houses minimizing reconstruction costs, as well as reducing the loss of personal possessions.

The Queensland building tradition emerged when the former British penal colony was opened to free settlement in 1838, causing a boom in construction of the inner-city timber homes that continue to define the region today. The ubiquitous elevated "Queenslander" style emerged due to the region's steep topography, sub-tropical climate, and the abundant availability of hardwood timber. The vernacular is typified by an elevated timber structure on piers, with single skin hardwood paneling and a steeply pitched metal roof. Buildings were designed with a passive approach to the climate in mind, where wide openings and undercroft spaces allowed for cross ventilation, as well as the obvious benefits of escaping water on the ground. The Queenslander remains an example of flood-resilient design in comparison to its contemporary counterparts.

The final EAA report was at odds with several of the local government planning responses whose measures were aimed at getting people back into homes as quickly as possible. These temporary local planning instruments (TLPIs) included relaxations on building heights and setbacks (Brisbane City, 2014), and resulted in houses being permitted to be raised many meters off the ground to meet revised flood planning levels (see Figure 21.8). However, the issue with this approach is the lack of certainty that these levels will result in complete immunity from future flooding. The expense of carrying out this work seems counterintuitive given the uncertainty surrounding rainfall extents and flood heights. Thus, it is proposed to *go with the flow*—a design approach that accommodates flood risk using flood-resilient design principles; working *with* water, rather than avoiding it altogether.

Flood-Resilient Residential Design

Based on practice experience and design advocacy work over the years since the Brisbane floods, Davidson and Bowstead have defined flood-resilient residential design as:

> The use of materials, construction systems, and design typologies that can withstand substantial and multiple inundations causing little or no material damage, including measures to enable home owners to safely store belongings prior to an inundation event, and easily clean out and move back in quickly after such an event.

JDA's practice has had a great deal of success in utilizing flood-resilient design that enables home owners to wash out their homes after an inundation using the concept of wet-proofing. Wet flood-proofing allows floodwaters to enter and leave a building with minimum damage, whereas the alternative approach "dry flood-proofing" prevents floodwaters from entering a building by either permanent or temporary barriers. Due to the differentials in hydrostatic forces acting on a building, "dry flood-proofing" is not considered suitable for most buildings in Australia (HNFMSC, 2007:21).

The concept of wet-proofing directly relates to planning terms that restrict the type of use of a particular space in a building. The term *habitable* space describes spaces such as living rooms, bedrooms, and kitchens where minimum standards for safety and amenity must be met. *Non-habitable*, on the other hand, includes service spaces such as bathrooms, laundries, garages, and storerooms that are more flexible in their requirements. Because wet-proofing involves the design of spaces which can withstand floodwaters, these commonly become non-habitable spaces within flood-resilient residential design.

In the wake of the 2011 floods, a range of innovative flood-resilient housing systems have been implemented that incorporate a range of strategies for flood resilience relevant to coastal and riverine communities around Australia and the world (Davidson, 2012).

The first three case studies are the work of James Davidson Architect, and the following examples are from Bud Brannigan Architect and Ziegler Build.

Building on their experience with EAA, James Davidson Architect has continued to explore architectural approaches to flood design, having undertaken a range of research projects, pro bono community work, and built commissions since the Queensland floods. Their approach to flood design focuses on allowing water to flow through the non-habitable, service spaces on the ground floor of a building with little damage. The practice has designed and built several innovative flood-resilient housing projects. Stemming from pro bono services, much of the practice's flood work focuses on maintaining affordability and high-quality design outcomes, while at the same time being mindful of local planning restrictions. Using wet-proofing principles, the aim is to create flexible non-habitable spaces, which ameliorate aesthetic (streetscape) concerns and maximize value to both the homeowner and wider community.

Sharp Street House

This post-war house in Fairfield, Brisbane was flooded throughout its ground floor spaces. The clients had limited insurance and difficult personal circumstances, which meant the post-flood construction needed to occur affordably, quickly, and to Code. They desired not only to reposition the existing house higher, but to rebuild it underneath with flood-proof materials. Given their limited finances, they chose not to renovate the upper level, which was left for a later date. This approach resulted in the combination of both experimental and conventional strategies in all aspects of the house's design and construction. Downstairs, single skin designer block masonry was used with polished concrete floors and sunken sill details to allow easy washing out of future floodwaters.

Removable marine-plywood cabinets were constructed of flood- resilient materials, with self-supporting stone benchtops standing on their own powder-coated steel frame. This would allow for ease of removal prior to an event and effective cleaning afterward. These techniques will both substantially lower repair costs in a future flood and reduce insurance premiums. The combination of flood-friendly materials with strategic planning of uses between the non-habitable ground and habitable upper floor allowed a contemporary, natural aesthetic while providing spatial flexibility for the clients.

The ground floor was divided into several small, adjoining spaces including a kitchen, living room, study, and laundry. There was no internal opening per se, with each space flowing into the other. This spatial strategy has cross-ventilation advantages, as well as easy access for effective cleaning after a flood event. All rooms have two access points to allow floodwaters to flow through unimpeded and assist in post-flood cleaning which may include the use of a high-pressure hose. A double-height void space over the kitchen connects the upper and lower living areas and enables easy transfer of ground level furniture prior to a flood event. The void also becomes an important element in escape planning during a flood. A steel stair is cantilevered off the eastern exterior wall. This stair is wrapped with a hardwood timber frame and translucent polycarbonate sheet wall for light transference to lower undercroft spaces. Both materials perform well during flooding as they clean easily. All sills on the ground floor are set down using a decorative block insert. Importantly, the ceiling remains exposed. This leaves no risk for complications in times of high flood but does pose acoustic and thermal insulation issues between floors, as normally the ceiling cavity is where insulation would be placed.

Graceville Avenue House

Only a few hundred meters from the Brisbane River and backing on to flood-prone parkland, this post-war property at Graceville flooded up to the level of the roof eaves (Figure 21.2). The clients were required to raise their house more than 5 meters to comply with new temporary planning instruments, which at the time required habitable floor levels at 500 mm above the 2011 flood datum. The interior of the original house was gutted, and the owner's lack of insurance meant sticking to their tight budget was paramount.

Prominent features of this building include the use of water-resilient affordable building materials for the lower floor which was raised approximately 3 meters, namely concrete block work with a render finish. The sloping ground level was elevated to create a plinth that addresses issues of height and scale, particularly on the street front, as well as assisting to maintain ease of access to the backyard. Removable flood-proof cabinetry like that in the Fairfield house described previously was installed on the ground level. Hardwood timber was used for permeable screens and the stair, which is more resistant to floodwater while providing cross-ventilation and passive cooling in normal conditions.

FIGURE 21.2
Street elevation of Graceville Avenue, Graceville. (Photo by James Davidson Architect.)

The ground floor features two large open plan spaces separated by a service core. The rear space features a kitchen and dining area, which opens out onto an elevated courtyard which the practice calls a flood plinth. This plinth, which is 1.5 m off the ground, serves to raise the backyard, reducing the over-scaling effect of the house being so far off the ground and provides an opportunity for the household to easily access their backyard. Large openings feature along the short width of the plan, with timber sliding doors and a roller garage door. Upstairs comprises three bedrooms, a living room, and a bathroom. Both floors are connected via a large screened walkway on the northern side, which serves as the main address to the street. Steel framing sits within rendered single skin concrete block walls that provide ample bracing and ease of post-flood cleaning. Due to the poor quality of the alluvial soils in this location, concrete piers, up to 12 m below ground level, were required on site (Figure 21.3).

Polished concrete was used through the ground floor, and all openings are hardwood timber, as is the central island benchtop. Other joinery uses treated marine plywood. The upper floor has been treated equally in terms of materials. The kitchen joinery uses a self-supporting stone benchtop hung to the wall on internal steel brackets. Separate joinery modules below the timber benchtop are removable, with a similar detail to the Fairfield house described earlier.

The plan of Graceville Avenue has wide openings to allow for the cross flow of water between the park behind and river in front. Graceville Avenue continues the use of heavy materials with standard block work rendered with a waterproof mix. The decision to use these materials was also one of cost and convenience, rather than in Sharp Street where the architect could more playfully experiment with decorative block work and flood-resilient construction.

(a) (b)

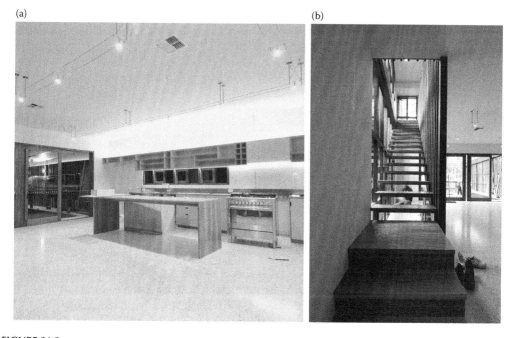

FIGURE 21.3
(a) and (b) Graceville Avenue house interior. (Photos by James Davidson Architect.)

Orleigh Street House

Sited directly opposite the West End banks of the Brisbane River, the flooding risk facing this property is visually evident. As a traditional Queenslander with hardwood flooring and wall finishes, the building itself escaped relatively unscathed. Using the opportunity to maximize the location of the property, the clients wished to build a flood-appropriate livable spaces under the existing building. Utilizing similar finishes and details to Davidson's other flood houses, the building serves as an excellent example of a typological combination of a traditional Queenslander renovated with new flood-resilient materials.

As a previously elevated Queenslander, the renovation of the West End house focused on the creation of a new ground floor and restoration of the original first floor. The scope of work for the project was extensive. The ground floor features a large open-plan living room and kitchen area, separated by joinery modules. An upstairs bathroom straddles the stair, which is hung on the exterior of the house. The living and kitchen area downstairs make use of large sliding door openings to outdoor spaces at the front and rear of the property, including a timber deck, which creates a covered basement at the front of the house. Again, like the two previous houses, these large doors facilitate the ease of floodwaters passing through during an event and post-flood washing out. Upstairs comprises three bedrooms and a bathroom. The upper floor was raised approximately 2 meters during the renovation and was refurbished throughout.

Single skin standard rendered concrete block tied into a conventional concrete slab is used for ease of cleaning. Polished concrete has been used through the ground floor with all block work being treated with a waterproof render tanking internally and externally. A suspended ceiling hides the upper floor structure. All openings are hardwood timber with cabinetry utilizing treated marine plywood for long-term water resistance. Decorative hardwood battening is used in external areas to allow floodwaters to pass. The kitchen cabinetry again uses a self-supporting composite stone benchtop hung on internal steel brackets. Separate joinery modules are hung to the same structure and are removable, creating a seamless ground floor (see Figure 21.4).

Ryan Street House

This Queenslander worker's cottage was originally built in the early 1900s and was under renovation by Bud Brannigan Architect when it was extensively damaged during the floods of 2011. Waters rose to 500 mm above the upper floor of the house, approximately 3 meters above ground level. Given that the house was under construction at the time, the builder's home warranty insurance covered the damage and clients proceeded immediately with a redesign of the house (Bowstead, 2015). The 95-square-meter (m^2) renovation was completed in August 2011.

The ground floor comprises an open plan incorporating a kitchen, dining, and living areas, as well as a multi-purpose room. Large bi-fold doors open to a courtyard garden and backyard to take advantage of the relationship to the ground. The use of an atrium/ void space above the kitchen forms the center point of the house, taking advantage of the sheer height the upper floor was raised to. An openable window to the rear amplifies this space.

FIGURE 21.4
Orleigh Street house interior. (Photo by James Davidson Architect.)

The upstairs has three bedrooms and a bathroom and was raised approximately 1.5 m after the initial renovation. Brannigan used a system of single skin construction for the lower floor, with translucent fiberglass and polycarbonate sheeting finish on the exterior. The original house is supported by a modified steel frame on a conventional concrete slab, much like the previous case studies. Polished concrete was used through the ground floor, apart from the kitchen area where there is a hardwood timber inlay. Spotted gum has similarly been used in all lower floor studwork, as well as in stair and joinery details. Translucent polycarbonate sheeting compliments horizontal timber cladding on the upper floor. Marine plywood has been used on the ceiling.

The kitchen has been designed using a plastic MDF composite "Uniboard" with a 60 mm concrete benchtop, designed to be removable. Cabinetry has been minimized using a movable island bench on a stainless-steel frame and significant appliances such as the oven have a decoupling switch. Vertical battens over several openings form security screens while still allowing ventilation and view through. The spotted gum stud frame has been doubled in several locations to allow a small gap for services. A removable plate on the interior covers this gap between double studs. The stair builds on a similar aesthetic. The marine ply ceiling panels are individually attached with a shadow line, allowing for easy removal (Figure 21.5).

Joara Street House

Designed by owner-builder Todd Ziegler of Ziegler Build, the Joara Street House involved the demolition of an existing post-war house extensively damaged after the 2011 flood (Figure 21.6). The owners were significantly under-insured at the time of the flood and were forced to look for alternative affordable construction systems to rebuild their house

FIGURE 21.5
Ryan Street interior. (Photo by James Davidson Architect.)

(Fremantle Media, 2014). The primary method they settled on was the use of 30 shipping containers that were craned into place and retrofitted *in situ* to create a bespoke, yet affordable flood-resilient home. The 300 m² Joara Street house was completed in December 2013. Having flooded up to 2.5 m above ground level, Zeigler Build took advantage of the revised permissible height limit of 9.5 m to design a three-story house with a sacrificial non-habitable ground floor and two-story habitable levels above the defined flood planning level.

Each floor is based on the tectonic shape of shipping containers arranged in various formations to produce space. The ground floor features a garage, laundry/bathroom, studio space, and rumpus room, united by a central walkway from the front yard to the rear. The first floor incorporates the main living areas of the house and three bedrooms. The top

FIGURE 21.6
Elevation of Joara Street, Graceville. (Photo by James Davidson Architect.)

floor contains the main bedroom and a large deck covering the lower floor. The first floor was built to sit 500 mm above the defined flood line and uses a double height space over the dining area. Decorative screening mitigates some of the scale between levels externally, while large openings, supported by deep steel reinforcing reveals, act similarly internally (Figure 21.7).

Construction relies on the individual structural integrity of the stacked shipping containers, which are reinforced where necessary, for example over openings. The house sits on a series of 30 × 300 mm "feet" supported by deep micro-piers, instead of traditional piers. Marine plywood flooring has been used throughout while the original shipping container lining has been left exposed in all areas. Upper floors use a mixture of exposed container lining and cavity stud walls with perforated plywood finish. Cabinetry is generally conventional throughout and could pose an issue during the next flood event. Recycled plywood materials have been used for the bedroom doors and stair balustrade, while other elements, such as the hardwood garage door, also incorporate a reuse of sturdy materials.

FIGURE 21.7
Joara Street house between containers. (Photo by James Davidson Architect.)

Flood-Resilient Residential Design Principles and Strategies

While Davidson's practice has experience in providing design services for flood-resilient housing, they do not support residents remaining in their homes during an actual flood event. The flood principles outlined in the previous case studies have been developed to allow safe egress and storage of valuable property during a flood event, as well as to enable rapid and easy clean-up post-flood with minimal damage to assist in reestablishing normality as soon as possible. Davidson and Bowstead propose the following elements as guiding principles when planning for flood resilience:

Scale building envelopes to preserve and build on existing streetscape.

Materials should follow the principles of wet-proofing and be water resistant and/or waterproof and encourage permeability for ease of washing out.

Character of streetscape can be used as a guide for the implementation of resilient materials.

Form should complement the existing buildings in the area.

Program safe zones where property can be protected, no matter the scope of renovation, following the principles of dry flood-proofing (elevation).

Access between levels needs to allow visual and physical connection between habitable and non-habitable zones.

In building on the flood principles evident in these case studies, the following architectural design strategies that form the basis of flood-resilient housing design in Southeast Queensland have been identified. These strategies not only challenge contemporary perceptions regarding the effects of flood events, but also show new architectural forms are possible when adapting to climate variability and change.

Resilient and Affordable Materials

Resilient material usage was a common theme across all five case studies. Despite their obvious difference in choice of materials, there is evidence of an effort to avoid potential damage caused by floodwaters. All case studies use single skin construction in some form to avoid the use of cavities that are problematic to clean and can retain toxic substances found in floodwaters. The use of polished concrete is also sensible considering this type of construction is relatively inexpensive, as opposed to other non-slab-on-ground options. Although resilient materials are generally more expensive than conventional materials, it is important to consider the benefits not only in terms of cost-saving from replacing materials such as plasterboard in the event of a future flood, but also as one of increased property value for the quality of finish and improved performance.

Equally, all case studies demonstrate the use of resilient materials with openings on the ground floor to enable floodwaters to pass through and for ease of post-flood cleaning. Door thresholds, such as Sharp Street's recessed sills show how important construction detail is with this strategy. Such a strategy that enables permeability of the ground floor challenges many of the conceptual ideas about flood risk and program, as well as providing a technical challenge. Such permeability is only permitted with a material palette of resilient details and finishes. More important, however, is the use of these strategies in creating useful, valuable spaces in areas rendered "noninhabitable" by the local authority.

Layering Resilient Materials

Layering resilient materials should complement streetscape, scale, and character of the location. Tiered block work, masonry, steel, fiber cement, and hardwood timbers in a hierarchy can correspond to the local material condition. There should be no cavities in the building envelope or between floors if possible. The layering of materials is particularly evident in Orleigh Street, where the hierarchy of rendered block work and hardwood balustrading break up the elevation with a significant vertical presence on the street.

FIGURE 21.8
When scale goes wrong. Neighboring house to the Graceville Avenue case study under construction in 2015. (Photos by James Davidson Architect.)

Scale: Raising and Programming Appropriately

Across the five cases there is a common move to raise habitable upper floors to a permissible height above a defined flood planning level. While it is a given that each design did so to comply with Brisbane City Council policy (Brisbane City Council, 2014), the scope of the move remains a point of difference. The programmatic arrangement of habitable spaces in each house serves as a significant lesson using the flood-resilient principle of escape (see Figure 21.8). Evidently, non-habitable spaces on the ground have been designed intentionally to be useful beyond merely a service role, allowing maximum use of the ground floor and the exterior. The potential with this position is it offers a way to reduce the proliferation of large under-utilized ground floor spaces when homeowners have raised their homes meters above the ground.

Form: Terraces and Flood Plinths

The use of terracing and flood plinths to elevate ground levels in those areas where elevation is required can achieve defined flood levels. Flood plinths assist in reducing scale and formal mass while maintaining a functional connection between ground levels and backyards. This strategy is particularly evident in Graceville Avenue, where a 1.5-m plinth allows maximum connection to the exterior, while avoiding potential issues of scale and form.

Program: Elevated Storage and Refuge

Easy access to elevated habitable spaces above a defined flood level should also include spaces in roof cavities and upper levels where property will be protected and people can escape in the event of flash flooding.

This aligns with a common strategy seen in all case studies of modularity where details, particularly in the cabinetry, are designed to be removable. Brannigan's electrical and ceiling details are modular and removable and provide for the decoupling of electrical appliances, particularly on the ground floor, which allows a potential use of the upper floor during two to three days of flooding, or at least easy repair after an event. Similarly, parts of the lower floor joinery in Sharp Street and Graceville Avenue are designed to be removable. Modularity in this case is a detailed compliment to earlier resilient material strategies.

Access: Voids

Voids are an important means of gaining light between floor levels, as well as establishing a direct connection between habitable and non-habitable areas in the house. Voids are also a means of safe egress for people to evacuate to higher ground and assist in storing valuables upstairs above the flood line.

Building Controls and Flood Insurance

The idea of floodable spaces is not a new one. However, its use as an architectural strategy reflects the severity of the conditions within Brisbane; one where the sheer value of land creates the necessity of an approach beyond simple elevation or retreat, coupled with the ecological reality of being a city built on a floodplain. Using strategies of permeability on the ground floor, in conjunction with resilience, allows the doubling of usable floor areas which is particularly significant given the extremely high median house price in these Brisbane suburbs. While floodable spaces as a design strategy could be technically at odds with local planning codes and insurance risk profiling, one major national insurance company has recently reduced flood premiums at the Graceville house by 40% due to flood resilient design. James Davidson Architect's recent work with the Queensland Reconstruction Authority (QRA) has highlighted Building Codes Queensland's willingness to consider amending local building controls in order to improve residential flood resilience. The QRA engaged James Davidson Architect to study the effects of continual flooding on properties in Queensland and produce a guideline for flood resilience that can be used to improve the effectiveness of such controls (Davidson 2016). The report generated several recommendations including:

1. Provisions in the Australian National Construction Code (NCC) and Queensland Development Code (QDC) can be improved to encourage greater use of water resistant materials and flood-resilient construction systems.

2. The divide between planning and building controls when evaluating flood-resilient building design systems creates confusion for the wider community and local authorities and it needs to be simplified.

3. Develop a flood-resilience matrix similar to the Bushfire Attack Level (BAL) system for bushfire design that gives the wider community and local authorities greater certainty in designing homes and planning communities.

4. Incorporating flood-resilient systems into housing design could see low-risk uses located under the defined flood level, resulting in significant reconstruction cost savings, reduced insurance premiums, and a greater uptake of flood insurance.

The findings from the report have been utilized by the QRA to engage with Queensland Government Housing and Public Works, Building Codes Queensland, and the Queensland Building and Construction Commission to develop a material selections guideline to be used in conjunction with building certification in order to encourage the implementation of flood-resilient design principles in the renovation of existing building stock and new-build housing in Queensland. In parallel, JDA is also working with the Brisbane City Council and Suncorp Insurance to develop a resilient-material and flood-appropriate construction methodology rating system to encourage homeowners to consider flood resilience when either renovating or building anew. By connecting local government and the insurance industry the approach is set up to reward homeowners through premium insurance relief commensurate to the level of material risk. Therefore, from an insurance point of view, the above design strategies would have a direct and positive impact on the cost of insuring properties in comparison to conventional designs.

Conclusion: Going with the Flow

Based on Davidson's professional experience, the strategy of *going with the flow* is vital in maintaining a middle ground between flood risk and economic reality in flood prone areas. The case studies presented in this chapter illustrate the range of innovative design strategies arising out of a lived experience of extreme flooding in Southeast Queensland. Using resilient materials, permeable ground floors were created allowing the doubling of usable space in a two-story home. These corresponded with the local character of the streetscape, meaning the often-problematic outcomes of elevation in flood zones on streetscape were avoided. Programmatically, all case studies used elevation to comply with local regulations, while using terraces and plinths to address scale and form. Elevated storage and removable components complimented this, enabled by access voids between upper and lower levels. Interestingly, these strategies are more in line with traditional Queenslander forms, where a combination of elevation on stumps, hardwood material resilience, and permeability allowed buildings to respond to floodwaters with much less consequence. This was backed up by EAA's findings, which discerned superficial damage in contemporary cavity, slab-on-ground homes was significantly higher than those in older areas with traditional homes.

Despite being from three different designers, each of these built case studies is within a 3-km radius, all with similar flood conditions, and all demonstrating the use of these design strategies. This is particularly significant in furthering the argument that permeable, resilient ground floor spaces are important in addressing all concerns of safety, amenity, and affordability. These findings show the importance of more flexible planning policies, where spaces in the flood zone are resilient and usable during the 20 or so years between flood events. New forms of architecture were created in response to working with natural process and accommodating floodwaters, rather than the simplistic and costly approach of elevating houses above a fictional flood line without any certainty of long term immunity to disaster.

References

Bowstead, S. L. 2015. Amphibious Brisbane: A critical comparison of five post-flood houses in Brisbane City. *Masters dissertation*. St Lucia: University of Queensland.

Brisbane City Council. 2014. *Brisbane City Plan*. Available via http://www.brisbane.qld.gov.au/ (accessed May 15, 2015).

Davidson, J. S. 2011. *Building Assessments Report to the Queensland Floods Commission of Inquiry*. Brisbane: Emergency Architects Australia.

Davidson, J. S. 2012. *Accommodating Water: Adaptive Architectures, Reactionary Planning and Designed Resilience in the USA, Netherlands and UK*. Brisbane: Final Report for the Winston Churchill Fellowship Trust.

Davidson, J. S. 2016. *Building Controls for Flood Resilience*. Brisbane: Discussion Paper for the Queensland Reconstruction Authority.

Fremantle Media. 2014. *Container Home, on Grand Designs, Season 5, Episode 15*.

Hawkesbury-Nepean Floodplain Management Steering Committee. 2007. *Reducing Vulnerability of Buildings to Flood Damage: Guidance on Building in Flood Prone Areas*.

Insurance Council of Australia. 2011. *Flooding in the Brisbane River Catchment: January 2011*. Vol 1 (accessed August 20, 2015).

Queensland Government. 2012. *Queensland Flood Commission of Enquiry Final Report*. http://www.floodcommission.qld.gov.au (accessed August 20, 2015).

Queensland Reconstruction Authority. 2013. *Flood Snapshot 2010/2011*. http://qldreconstruction.org.au/u/lib/cms/>floodplains---section-2.pdf (accessed August 20, 2015).

profession, especially in a post-Paris Accord political environment, architecture should work to innovate, develop, and revise our practices to build new knowledge about how we accommodate, repel, or efficiently use water in conditions ranging from land to sea, as posed by Pliny. We must consider new meaning and generative opportunities for the study and design of the dynamic aqueous environments in which much of the world's population dwells.

The cultural and technological narrative of water and flooding manifests in architectural practices throughout history and into the present day. From our conceptualization of original shelter to grand urban schemes, architecture has a long history of relating to water through site interventions and building technology. However, unlike the narrative of the flood found in myriad cultures and traditions, we do not commonly believe that a Creator is sending a deluge to wipe away all sin; but sea level is rising and there is no really big boat to carry us safely through. With few exceptions, such as the Netherlands, our lack of planning and infrastructure makes this a design problem that will challenge architects and related fields to rethink traditional models of practice and methodology for decades to come.

How do we simultaneously resist the destructive capacity of water and yet invite it into our communities and buildings through controlled apertures and channels? How can humankind build and rebuild safely, proportionally, and poetically on a ground that often seems more like the sea? What if designers were to take on the challenge of building with a consideration of water beyond drainage and potable supply lines? Suppose we resist a reliance on engineering to offer brittle defenses to a perceived problem and embrace the beauty *and* the danger of aqueous landscapes. We must begin with an understanding of how architects have approached the concept of ground and site historically and technically. Following on that, this essay identifies three temporal building conditions and examples of architectural endeavors within the coastal and increasingly flood-prone environment: (1) on and within water (constant inundation); (2) on land/water edges (the most dynamic); and (3) floodplains (dynamic inundation). Within that framework of land to water conditions, architects have explored a variety of tactical approaches from which we can extrapolate and imagine a future where a variety of elevated structures and communities continue to prosper in the new coastal environment.

Defining Site: The Coast and the Flood

Water is essential for virtually all known life forms, covering 71% of the earth's surface. While fundamental to the survival of the physical world, it also has wide-ranging spiritual and cultural significance. Our relationship to water shaped the beginnings of civilization when human settlements were sited primarily near continental edges and inland waterways. Water as a cultural symbol originates in narrative form across almost every known religion. The flood story remains one of the most pervasive and foundational aquatic narratives. Parables and myths such as Noah and the Flood are morality plays that teach us as much about today as they do about the past. As we continue to devise thoughtful and effective long-term responses to sea level rise and recent events related to tsunamis, typhoons, and hurricanes, we may begin to appreciate lessons yet to be learned from these stories. If nothing else, Noah's preparation, met with scorn by his contemporaries, demonstrates that intelligent and well-informed planning predicates a greater chance of survival in times of grave and sudden change.

Pliny the Elder, when describing in *The Natural History* the conditions of what was to become of the Netherlands, could have been discussing many coastal settlements. In this landscape, as flooding and sea level rise continue to push ocean levels to new heights, threatening people and structures along them, we might still ask ourselves the questions of whether we are living on the land or the sea. It may be helpful to view the coast through the framework of three distinct temporal conditions:

1. *Building on the water*—Throughout time various cultures have committed to a life on the sea, not just in naval vessels, but in houses and communities that are built on the water. In this condition, ground for pedestrian movement and occupation must be artificially constructed.

2. *Building on the edge*—Many buildings and cities are constructed right to the edge of large bodies of water. The edge demands a very particular type of structure, typically heavily reinforced against the tension of the moment where land meets the sea, but occasionally embracing that condition.

3. *Building in the floodplain*—In the increasingly fluid global dynamic, this condition has become ubiquitous, even in previously waterless environments. Sea level rise and subsequent changes in gravity drained land have created flood-prone conditions affecting vast swathes of settled landscapes.

As architects grapple with the impacts of climate-change-driven sea level rise and how it affects our collective human capacity to occupy our continent's littoral territories, it is imperative that we address the issue of flooding through adaptation and innovative ways of building that look toward the future. It is clear that today's environmental conditions in coastal environments are increasingly impacting the human activities that can occur there, and affecting our ability to safely inhabit cities and communities that provide people's livelihoods. An obvious solution, proposed by many following disasters of all types, is to simply move away from the threat. However, the prospect of displacement for the millions of coastal inhabitants is unfathomable and in fact, is considered one of the major future sources of stress for massive civil unrest around the globe. The U.S. Department of Defense has declared climate change displacement as a key disruptor for peace nationally and internationally. This essay champions the prospect that there are a number of strategies for not only living safely, but also living poetically, in concert with the beauty and sublime poignancy of water.

Defining Site: Architectural Strategies

Architecturally, we might theorize that our primal response to water is one of the originating determinants of building form and technique. In fact, our relationship to water forms our most basic understanding of the purpose of shelter, from keeping water out to controlling ways that water can pass through. As architects, we tend to learn about water as a problem requiring resolution; it is something to be shed by our roofs or hydrostatic pressure in need of release and rerouting away from foundations and basements. It is generally accepted that uncontrolled water in a building is one of the most destructive forces imaginable, and resisting the infiltration of water has been the driver of technological innovations in building systems over the millennia and particularly in the last 200 years. In short, resisting water is a technical matter for which architects and engineers have developed increasingly sophisticated building systems. However, this technically focused approach has reduced our thinking to the point that we regard water more as H_2O than as water, as Ivan Illich describes in his seminal study: *H_2O and the Waters of Forgetfulness*. During the last two decades, as the effects of climate change have been registered in meteorological, societal, and cultural shifts, it is becoming more widely held that deluges, the scale of which was once thought mythic in substance and scale, may become relatively quotidian events for humankind. Our insistence on treating water as a problem has limited our ability to engage with water as a source of joy, beauty, and cultural significance. But resistance is becoming increasingly futile, and it is time for architects to engage with water in a more exposed and thoughtful way.

To investigate what it means to build in the watery world of the coast, one must begin with a basic understanding of what architecture is. The architect Gottfried Semper (1803–1879) most eloquently speaks of our relationship to the earth and architecture when he writes of the *mound,* the *hearth,* the *enclosure,* and the *roof* in the *Four Elements of Architecture.* The noted architectural theorist and historian Kenneth Frampton and others have extended Semper's discussion of the four elements explaining it as a dialogue between the stereotomic and the tectonic, with the stereotomic referring to construction that is heavy and dense, like the earth or actually on the earth, as compared to the relatively light and delicate skin and roof (Figure 22.1). Through their elocution on the subject, many architects have taken on the challenge of an architecture that seems drawn out of the ground rather than placed on the ground. By extension, this "grounded" architecture seems to belong to a place, rather than anyplace. In the contemporary interpretation, as differentiated from Semper, who posited that the hearth was the primary architectural act, we would posit that in today's

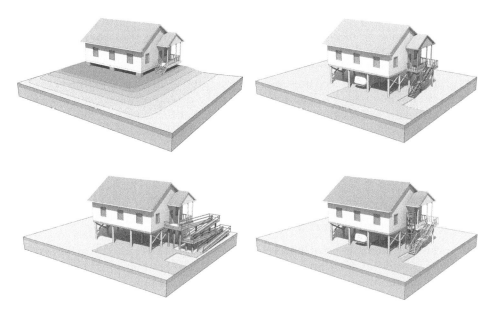

FIGURE 22.1
Diagram of an elevated structure on fill, elevated structure on piles with stairs, elevated structure on piles with ramp, and elevated structure on piles with elevator. (Image by J. Erdman.)

world, constructing the mound is the primary architectural act. It remains consistent with Semper's view that the purpose and function of architecture is to shelter humans from the weather and provide a symbolic claim upon site.

Further, Frampton has introduced the idea of architecture as a practice of topos, typos, and tectonic. In his explanation, topos refers to architecture as earthwork or mound as a place. The topos theory relates to the moving body through space and time in relation to place and ground. In this sense, one of the most important connections of building to site is through the threshold and movement from outside to inside. In the historical and theoretical underpinnings of topos, it is the sense that architecture grows from a hard surface or rock form of the earth, and so our primary understanding of architectural foundation means that we build from a stone-like material. The primary move of any design is to physically connect to the ground in order to occupy the land, but in the watery landscape of coastal architecture, a hardened ground cannot be assumed, and this direct continuous occupation of a landscape is not a given. Architecture is no ship, and we must now consider architectural endeavor in the context of a demanding, complex, and vulnerable coastal environment.

The strategy of elevating shelter above the reach of water, relates directly to our understanding of the mound. The land on which coastal architecture is built is often soft, dynamic, and subject to inundation. Psychologically, socially, and culturally, we are generally conditioned to expect a close relationship to the land, and the history of architecture demonstrates this repeatedly but it has largely been ignored in architectural practices of the twenty-first century. The elevational strategy, along with building material and assembly strategies, can be explored tactically depending on the temporal conditions of the site; whether it be in direct and continuous contact with water, the edge between land and water, or the most dynamic condition of the floodplain. Balancing adaptive and innovative building methods with resilient decisions about land use should inform architectural practice in the future.

Returning to the original question, what if our ground is not of a material from which one can construct, such as water, or mud, or shifting sands? It is not part of the typical human condition to imagine ourselves living on boats, without some fixed horizontal plane. Our biological evolution has not equipped us for a purely aquatic environment, and so, from our inner ears to our pedestrian feet, we humans are better suited for a stable and consistent ground, constructed or found. The first move in this landscape is most often not to scrape the surface of the earth, but to penetrate the surface with a series of point loaded piers which are connected in the horizontal plane, at a distance above the transitory and changing land. In this operation, the connection to the earth becomes more distant and tenuous. The Modern era introduced the concept of the plinth as a part of architectural discourse, but rarely is it imagined that the ground over which it hovered would actually be water. However, creating a plinth, or artificial ground, is a useful operation that can be seen in vernacular architecture along coasts around the world, from the shrimping villages of Vietnam to the Cajun cottages of Louisiana.

Over millennia we have countless examples of human's efforts to shape the ground, to build above the surrounding landscape, to shape earth and stone in order to create order out of the wilderness. But in today's world we have more people and less purpose in our constructions than before; we are not driven by a purpose to raise towers to gods, but rather towers to commerce. And the American Dream of one house, one family proliferates but doesn't speak easily to the need for simple and poetic ways to move from an aqueous, fluid ground to the stability and safety of a home above.

In the world of architecture with a capital A, less a grammatical distinction here than a semantic one, there are examples in the industrialized context bearing useful and poetic building practices, particularly from the Venetian architect and designer Carlo Scarpa, and the national approach of the Dutch. Both practice in the aqueous landscape of coast, and both offer reason for optimism about our ability to continue to live along our coastal shores.

The prevailing architectural strategy at the end of the Industrial Age has been a reliance on engineering to respond to the conditions of land and sea. This passive stance has left human inhabitation in the hands of instrumentalist theories, seeing water only as a problem, rather than an opportunity to engage with the human soul and our native inclinations to live in harmony with nature. For example, infrastructure built more than 50 years ago and without anticipating increased flow load is no longer able to handle the drainage needed in many coastal communities around the globe. Residents who may have previously anticipated flood events in 10-year cycles are now subjected to inundation multiple times in a single year, living and working in buildings that are ill-equipped to handle this condition. Residents in these areas are now forced to adapt their buildings to ongoing flooding or relocate.

Engineers and other construction experts have developed different tactics for elevating inhabitable structures beyond the reach of ground level inundation, including earthworks such as mounding, and architectonic elements such as walls and piers. The gold standards of coastal construction for the global community seem to be provided by the U.S. Federal Emergency Management Agency (FEMA), which regularly documents best practices for building in many environments, including the coast and other flood-prone environments. In the formation of these documents they rely heavily on existing building codes and technical review by construction contractors and engineers. A review of the most up-to-date manuals reveals that architects and designers have no role in the creation or conceptualization of these strategies, leaving them devoid of any response to place or culture (Figure 22.1).

In line with Illich's notion of reducing water to H_2O (a problem), the best information about building on the coast provides no guidance in terms of how we might begin to engage the watery condition in meaningful ways. As a profession, architects have remained curiously mute on the subject.

Architectural Tactics: Building on the Water

Building practices evolve from practical and cultural distinctions of geomorphology, access to technological developments, and climate. Secondary factors relate to population growth, density, and activities. Vernacular architecture provides some clues about how community and individual building practices can respond to the conditions of a water-based site. The island inhabitants of southeast Asia have developed interesting techniques for maintaining long term occupation of such a landscape. Studies of the Bajau Laut, a fishing and seafaring society in Indonesia reveal detailing and aggregation methods adapted over the past century to climate changes and sea level rise without losing their long-standing connections to place and economic viability.

The Bajau Laut, also known simply as the Bajau (Crabbe, 2006), have built a community over the last century without a land-based architecture or site. Their simple wood plank dwellings are constructed on piles driven through shallow water into a sandy substrate (Figure 22.2).

FIGURE 22.2
Bajau Laut community living above the water. (Haziz San, CC by 2.0, https://www.flickr.com/photos/hazize/.)

Typically, the earthwork of their shelter is by nature not a found earthwork, but rather a constructed pile and plank shelf upon which dwellings are constructed. However, a few communities have been using coral blocks harvested from local reefs to create a perimeter wall with earth infill and building upon this makeshift mound. The mounded lots can be seen to sustain trees and vegetable gardens that feed the community. As an example of how some attempts to protect can actually do more harm than good, the practice of cutting blocks from the coral reefs has significantly damaged the protective reef. Coastal scientists are working to encourage Bajau communities to use concrete block rather than coral in their mound building.

As a group, the Bajau Laut build their communities in clusters around elevated walkways and exterior gathering areas that are accessed both directly off the walkway and from the water. The conjoined clusters are often connected back to land by a single walkway that is the primary access out of the community rather than into it. Most people access the buildings directly from boats in the water by climbing ladders attached to the walkways. This method of accessing their homes is generally not elaborate or celebratory, but simply a means of moving from the lower elevation of the water to the higher elevation of the structure. The simple plank construction is similarly ad hoc but differences are expressed through the varied use of cladding. The various materials that clad the simple stick-built frames indicate the relative prosperity of the family occupying the structure, with more "affluent" families employing more long-lasting and dense materials.

While our physiology dictates that we cannot survive water settings without mediation, communities can grow and develop skills enabling them to build, live, and work despite atypical conditions. The challenge for indigenous communities such as the Bajau is how their structures perform in severe weather and how fast they can return to normal life after experiencing extreme climate-change-related events. Bajau construction practices demonstrate that humans can adapt to living in extreme coastal environments. Architecturally their techniques are primitive but the community structure is resilient and robust.

In Venice, Italy, we find another example of a city grown out of the water. The Venetian architect Carlo Scarpa successfully and deliberately engaged with water for many of his most successful designs. Of particular note is the simple but elegant Memorial to the Partisan Woman which sits on the edge of the lagoon along the Viale dei Giardini Pubblici. In this design, which was always intended to be part of the water, Scarpa composed a series of Istrian stone blocks at differing heights as viewing platforms to the Augusto Murer sculpture of a woman lying in repose with her hands tied. The sculpture itself was originally placed on a floating platform that would allow it to rise and fall with the tide, never to be stranded above the water nor submerged below it, although it has since been modified to a fixed position due to the degradation of the hydraulic mechanism (Figure 22.3).

The sublime beauty of the design lies in its ever-changing configuration based on the elevation of the lagoon during tidal fluctuations throughout the day, as well as with the seasons. Scarpa's site is the water, and he creates an irregular mound of pedestals to sit within the dynamic of the tidal lagoon. While approaching the statue from land, she would be carefully framed by the water and pedestals; if viewed from the lagoon, she would be framed by the pedestals and the trees of the garden in the near distance. The placement and modulating heights of the stone pedestals creates a distinct tension for the human observer as they would traverse up, down, and around to view the sculpture, the city beyond, and upon turning to return to land would see the vegetated garden. Its relation to the rest of the city and the gardens is clearly meant to signify that this monument is not like others; that what it is portraying and what the inhabitants and visitors experience is of the water, not of the city. In this way, Scarpa is not engaging with water as a problem to be solved, but rather as matter to be explored and built within. The design explores the edge condition

FIGURE 22.3
Monument to the Partisan Woman by Carlo Scarpa of 1965. (Image by G. Dodds.)

as a place where ground is defined both by floating structures that move with fluctuating water levels and fixed elements that mediate the changing levels.

The pragmatism of contemporary Dutch culture has led to some novel explorations of construction techniques on water. The long history of Dutch houseboats has inspired architects to create several variations of this concept for recent projects that embrace living on the water. For example, the Dutch firm Waterstudio.NL has been working for over a decade on the idea of floating foundations. The project merges aspects of the Dutch houseboat, including the floating base or foundation, with industrialized techniques of dock construction including piles driven into the sea floor and attached to the floating foundation. The attachment is dynamic and allows for vertical movement of the foundation. The entire structure is then topped with a traditional house or other structures. Koen Olthuis, the founder of Waterstudio.NL, has proposed this technology for adaptation to various types of situations in cities around the world. In effect, his proposal is a tethered barge and leaves the question of how one interacts with the water and connects to the land unexamined. In the Ijburg project Waterstudio.NL has developed the communal aspects of neighborhood by creating a plinth or island on which houses are constructed and share the space of the artificial ground (Figure 22.4). These projects are largely speculative real estate developments and the communal aspects, as seen in the Bajau settlements, will take time to develop. One is left with the feeling that these floating Dutch projects do not reach the level of synthesis with the water nor achieve the sense of place that both the Bajau and Scarpa examples embody.

Architectural tactics for living on or above the water may seem to most fully embrace water as a site, yet it should not be assumed that just because a building is floating on or above the water, that it is of the water.

FIGURE 22.4
Ijburg housing development in Amsterdam. (Zoetnet, CC by 2.0, https://www.flickr.com/photos/zoetnet/.)

Architectural Tactic: Building on the Edge

The second temporal condition is perhaps the most vexing: the edge between land and water in which building sites are both land and water. Our collective instinct has been to reinforce and harden the edge with highly engineered infrastructures, often leading to a perceptual disconnect between people and their watery environment. In Venice and Amsterdam, this disconnect is impossible due to their unique urban sites.

The architects of Venice used the tactic of building land in order to create their structures and open spaces of the city. While the original settlement is said to be built on the marshy ground of the lagoon, whatever land there was has long since disappeared. In the renovation of the Querini Stampalia Palace, Scarpa used ideas about water to drive the design as he sought to bring the water into the "ground" floor through literal and symbolic means, creating a connection from the public canal to the private gardens within. In fact, the commission's basis was the anticipation of future flooding due to the intrusion of the sea through the first floor and into the garden behind. While Scarpa's design may seem to be very minimal, he dramatically altered the conditions he found in order to create a place where water would be known as a source of light and joy, rather than one of destruction.

Scarpa's most direct intervention was the bridge from the Campiello Querini across the Rio de Santa Maria Formosa. This intentional reordering of the entrance to the palace draws attention to the relationship of the building with the canal. Because the ground floor is slightly lower than the Campiello, the procession of the body across the bridge and down into the building heightens awareness of the water below. In addition to this, Scarpa installed a series of steps just inside new steel filigree water gates, allowing the waters of the canal to flow directly into the structure. He also created a series of steps leading into (or out of) the water. A technique Scarpa often deployed when seeking to draw attention to a particular moment in a sequence, the offset steps are more symbolic than practical as the gates are not intended as a passage for boats and humans, but rather to mark and articulate the changing water-level of the canal (Figure 22.5).

Moving through the first floor into the garden, water is incorporated through channels and fountains, but only after being significantly elevated through introduced fill. The newly mounded rear garden remains protected from intrusion by uncontrolled canal water, yet maintains the connection to the water of Venice. In these few elegant interventions, Scarpa allows the visitor to sensorily engage with the water, including the bold move of bringing the canal water directly into the building at the portego. Giuseppe Mazzariol reportedly told this story about Scarpa's inspiration for the project: "*One morning in 1961 at the Querini Stampalia, I asked him to keep water outside the palace … He looked at me and after a pause he said: 'Inside, inside! Water must be inside, like everywhere in the city. We just need to control and use it as a shining and reflecting substance. You will see the light reflections on the yellow and purple stuccos on the ceiling. That is so gorgeous!'*" (http://www.querinistampalia.org/eng/the_mark_of_carlo_scarpa.php)

Architectural Tactic: Building in the Floodplain

The third temporal condition found on the coast is "dry" land, where flooding has not habitually been anticipated or where water inundation happens infrequently: the

FIGURE 22.5
Floodgate at the Querini Stampalia. (Tommaso_men, CC by 2.0, https://www.flickr.com/photos/tommaso_men/8043014824.)

floodplain. This condition is where many coastal cities and buildings are found today, with the increased frequency of flooding or even permanent new bodies of water, but not full convergence with the sea. The impracticality of mounding land to safely elevate existing buildings or entire cities means that this expanding third condition will become the territory of new innovations in the field of architecture. Historically, this condition can be found in areas where land has ostensibly been reclaimed from the sea, as in the Netherlands, or in places like New Orleans where infrastructure has been built to protect land that the sea is trying to claim anew. As an attempt to escape the effects of flooding, humans would have traditionally retreated to higher ground, further away from rivers and other water bodies. The lack of naturally occurring high ground in a delta has seen people eventually building their own higher ground, which was convenient for protection from invaders, as well as from the natural elements. There are hundreds of examples of constructed mounds adjacent to low sloping river beds and coastal zones around the world that predate the written word. But in our post-industrial global environment, with so much building stock already in existence close to water and sea levels rising, previously dry land can be subject to inundation from seasonal and event-based flooding. Designers are going to be challenged by no longer being able to control and resist flows of water, but will instead have to devise ways to work within sites that accept water as a part of the ground we inhabit. Contemporary work in the in-between of the floodplain is just getting started; work in the Netherlands, vernacular buildings of Louisiana, and contemporary design competitions offer some approaches for how we might build in this expanding condition.

In *The Natural History*, Pliny the Elder described the land and peoples of the Netherlands thusly:

> Here a wretched race is found, inhabiting either the more elevated spots of land, or else eminences artificially constructed, and of a height to which they know by experience that the highest tides will never reach. Here they pitch their cabins; and when the waves

cover the surrounding country far and wide, like so many mariners on board ship are they: when, again, the tide recedes, their condition is that of so many shipwrecked men, and around their cottages they pursue the fishes as they make their escape with the receding tide.

Pliny the Elder

From their founding as a people, the Dutch have engineered their landscape, successfully holding back the reclaimed land from the sea and holding back the floods of rivers and tides. The common approach has been to assume that humans can control nature and engineer every situation. The traditional Dutch building type was predicated on the constructed mound. In cities, that would take the form of a site made dry by the control of groundwater and in more rural conditions, people would build their own walls and fill them in, in order to create a land-based site. In the industrialized age, the Netherlands is considered a world leader in coastal architecture because of their radical engineering which allows people to live below sea level but still remain competitive in commerce and culture. The ability to construct a ground, out of the sea, is a hallmark of the Dutch.

More current thinking in the Netherlands includes the Room for the River project. In this project, sea level rise due to climate change becomes partially accommodated by a restructuring of river edges that had previously been robustly leveed and channelized. While still heavily reliant on an engineered solution, the project does allow for more natural processes than any Dutch project in the nineteenth and twentieth centuries.

In the early days of what became the United States, settlers would typically seek relatively high ground adjacent to the coast on which to settle and build there. In low lying conditions, such as the Mississippi River Delta, settlers would often build on naturally occurring ridgelines created by ancient flooding and then work to protect surrounding lands from annual flooding by creating canals and levees. Still, even those measures could not always guarantee that flooding was not a risk, even though flooding might only occur every 10 years, rather than every year. In the event of flooding, the construction methods employed in the early settlements of the United States greatly reduced the damage caused by flooding and facilitated the return to normal life following a flood event.

One of the primary components of early coastal U.S. construction was locally harvested heavy timber, such as cypress in Louisiana. Native to deltaic swamps, cypress trees have evolved to become resistant to damage from water and pests commonly found in the environment. These types of trees have a cellular structure that includes a large amount of sap which hardens into an extremely durable, non-toxic, and water impermeable material. When harvested and dried, the resulting lumber hardens and becomes virtually impermeable to flood, fire, and pestilence. Given the water-resistant characteristics of the dried wood, it was a suitable material for use in foundational pilings and structural elements, as well as it's familiar use in boats of all types.

To infill heavy timber frames, early residents of Louisiana used a material known as bousillage, as seen in Figure 22.6. A combination of horse hair or Spanish moss and clay, the resulting material would be packed between the heavy timbers and allowed to dry before being coated with a protective lime-based plaster or stucco. The bousillage infill of heavy timbers also provided the benefit of thermal density, meaning that it did not transmit temperature differences between interior and exterior, leaving spaces cooler in the summer and warmer in the winter. The stucco mixture of lime and sand provided an additional layer of resistance to water damage and pests, but required regular maintenance to address its tendency to crack and spall over time. The labor-intensive process of creating and placing bousillage assemblies was replaced over time with more industrialized

FIGURE 22.6
Heavy timber and bousillage in a Louisiana structure. (Image by J. Erdman.)

construction processes but there is not an assembly process available today that is as non-toxic and resilient as bousillage. When properly covered in stucco, the bousillage structure is virtually impermeable to water and as an added advantage, any water that does penetrate is still able to dry because the traditional stucco material is breathable.

Plaster and lath construction is similar to the system used in bousillage construction although not as thermally inert. Because the lath replaced the bousillage infill in terms of structure, the cavity between timbers or other structural framing would simply be either air or a lightweight material. Plaster and lath was used across the United States until the 1950s when plasterboard (also known as drywall and gypsum board) was introduced as a speedier method of making interior walls and ceilings. Although slower to install, once in place, plaster has certain strengths that make it more resilient than plasterboard in flooding and humid environments. Chemically, plaster is similar to concrete in that, once hardened, it is irreducible to its original form. This makes it resilient in humid and wet climates because it will not deform when wet and is breathable, so it will not mildew as easily. It is also more resistant to fire. For these reasons, many older homes are more resilient to flooding than their recently built neighbors.

Found ground, constructed ground, material choices, and methods of assembly can greatly impact how well a building might withstand an occasional flood event. In this temporal condition, architectural tactics of mounding can elevate not only buildings, but our own sense of control and our ability to shape the flow of water to our will.

Again, we return to Carlo Scarpa who offers designers an example of how we might build in the temporal condition of land that floods infrequently. His Camping Fusina project of

1951 was designed as a landscape of permanent pavilions providing core needs to visitors along the edge of the Venetian lagoon. The project combines tactics of a slight mound with durable materials and detailing to resist effects of ponding during heavy rains. Although the pavilions are primarily single story, Scarpa also created several stair components allowing access to rooftops that are worth study.

Using the humble materials of brick, concrete, and gravel, Scarpa's design for the bathing areas reveal his appreciation for water. The detailing along the ground of the area is comprised of concrete stepping stones set into a gravel bed. The open-air sinks and other water vessels are formed concrete with simple but elegant shaping with some made of deep basins and others with shallow and ridged basins. The different heights and shaping of the basins provides accommodation for washing of bodies, clothes, and camping equipment with careful channeling of water both within the basins themselves as well as the ground surrounding. Paving patterns of the rough concrete paths allow rain and other ground water to percolate naturally and pedestrians to still move through the space without splashing in standing water. The slightly raised concrete entry to the brick shower pavilion slopes almost unnoticeably to the ground with a small step up. Again, this modest but distinct separation of the walking path from the surrounding ground reveals the nature of water to pool and stand without denying human connections with the environment (Figure 22.7).

Drawing on a vernacular detail found in many buildings, Scarpa employs a concrete pad from which the two visible staircases ascend. One of the stairs, a small metal spiral, elegantly springs from a concrete base which simultaneously lifts the corrosion-prone metal out of any flooding, as well as providing an anchor for the post around which the stairs rise. The second stair is located at a multi-purpose building set atop a concrete plinth and is further defined by a steel and timber pad from which the stairs ascend. This articulation of the stair, with the demarcation of threshold rendered in steel and concrete, both ties the stair to the ground while also elevating it beyond simple pragmatics of vertical ascension.

With the advent of modern, industrialized building, materials and methods shifted toward more globalized systems that created buildings that were cheaper and easier to construct, but lacked the resilience of earlier vernacular examples. Contemporary building practices differ significantly from their historical, pre-industrial methods because industrialization is predicated on "efficiency" which most often means standardization across the marketplace on a national scale. The International Building Code, which is in effect in all 50 of the United States, sets the standards of building construction. Contemporary residential construction is largely controlled by industry associations and manufacturers of home building products who have little regard for the nuances of place and climate, particularly that on the coasts and in floodplains. Their operational strategy has consistently sought to nationally (and globally) standardize a set of building materials that often make no sense for a given climate. Much of the extensive and expensive damage suffered by coastal homes during hurricanes, and even through annual seasons, is due to this standardization. Standardization has tended to focus on the construction needs of populations located off the coasts and north of the subtropical climates found in the American South.

The United Nations Environmental Programme issued a guidebook on climate change adaptation under the Technical Needs Assessment series titled "Technologies for Climate Change Adaptation—Coastal Erosion and Flooding." This book offers an overview of the global concerns relative to flooding. The authors argue that there are three approaches to flooding: protect, accommodate, and retreat. Protection measures include hard and soft infrastructures such as seawalls, dikes, beach, and marsh renourishment among the strategies possible. Accommodation measures include regulating building development,

FIGURE 22.7
Camping Fusina shower by Scarpa. (Seier+seier, CC by 2.0, https://www.flickr.com/photos/seier/8909822051/in/album-72157601329828130/.)

flood proofing, and hazard mapping. Retreat measures include pushing development inland and managing coastal realignment.

The Federal Emergency Management Agency is the leading source for information and research regarding floodplain design for architects and planners in the United States. The agency provides documents as well as software that enable better planning for everyday as well as emergency events. The "Technologies for Climate Change Adaptation" guidebook makes extensive use of the FEMA guidelines as they identify two methods of flood-proofing: wet and dry. They also point to the National Flood Insurance Program for further guidance on how to approach the technical details of building structures in a flood zone.

Spatially elevating structures using pilings, piers, or walls can provide opportunities for new programmatic elements that enhance individual use and community functioning when the elevation is high enough to allow for occupation below. For instance, the space can be used as a garage for vehicles and replaces the typical attached garages found in suburban homes. The space can also function in the same way that porches operate, as a mediating space between the public space of the street and the private space of the dwelling, business, or institution. However, culturally it may not be desirable for building owners to leave the space below their structures open, and building codes allow for both closed and open options. In Louisiana, the open space option has been embraced, and, driving along roads in the coastal zone, it is clear that residents are taking advantage of the shade and protection provided by their elevated homes to create the equivalent of a porch with attendant furniture, grills, and vehicles. However, in the more densely built neighborhoods of Norfolk, Virginia, residents most often choose to enclose the space beneath their houses once elevated, creating strange architectural hybrids and a reduced connection to the surrounding streets and community.

Raising a structure on pilings involves driving deep foundational shafts and placing all inhabited elements on the pilings. Pilings can be made of a variety of structural materials but are typically wooden or concrete. Wooden pilings are susceptible to degradation from pests and water exposures. Reinforced concrete pilings are more expensive and can be susceptible to structural degradation due to improper manufacturing and installation as well as salinity. The benefits of elevating a building with piers include increased spatial flexibility below the enclosed and protected space above, optimal cost, and significant structural efficiency. Pilings are also the preferred method of elevation in areas with fast moving floodwaters that carry the potential of debris.

Conclusion

Cultures around the globe share the story of a deluge that erases the collective sins of humankind along with any trace of civilization. After the cleansing waters recede, a heroic figure (someone designated by a god to prepare for the coming inundation) builds a new culture on a new site. In this trope, water is both destructive retribution and life affirming renewal. Indeed, as climate change and population growth continue unabated, the temporal nature of site and situation in the architect's task has grown increasingly imperative. Low-lying coastal and riverine areas remain particularly susceptible to flooding created by ill-considered siting and other clumsy artifice, particularly in low-lying rural areas. Architecture should play a key role in building and rebuilding in these dynamic conditions. As this essay has demonstrated, architecture has a variety of strategies and tactics within our reach that not only address the technical demands of a watery coastal landscape, but also respond to the human need for the primal joy that can be found in proximity to water.

As previously stated, the threat of flooding continues to rise because of both climate change and population growth, in other words, more people on the earth combined with increased storm activity means that more people than ever before will experience either the possibility or reality of flooding in their homes, workplaces, and communities. Climate change experts currently estimate that average global sea level rise will be at 900 millimeters by the year 2100 and could be as high as 1.8 meters. In addition to sea level rise, the earth's population is growing exponentially, and the majority of the growth is happening in coastal environments. For the more than 40% of the global population living within 100 km of the coast (UNEP) and 10% subject to coast flooding (McGranahan et al., 2007), the most cost-effective and expedient way to accommodate flooding is to build above it.

Yet, even in the face of increased risk of flooding, the dangers associated with a mass migration to higher ground are even more profound. "Climate change is the greatest security threat of the 21st century," said Major General Munir Muniruzzaman, chairman of the Global Military Advisory Council on climate change (Carrington, 2016). So, it is imperative that architects and other designers of the built environment get involved with the technological and cultural implications of building above the ground and with water. The majority of our human activities rely on some connection to the ground plain for transportation, navigation, commerce, and congregation, so what does it mean if we can no longer rely on the solidity of the ground?

In the 1970s, architects addressed the impacts of industrialization on the environment with ideas about environmental design and the Whole Earth Catalog. In more recent decades, architects have embraced the concept of sustainability, primarily as a concern with energy flows and calls to reduce energy use through programs such as the 2030 Challenge and LEED (Leadership in Energy and Environmental Design) certification. Now, climate change has progressed to the point where we are faced with the complex concepts grounding the idea of *resilience*. According to the noted planning scholar Simin Davoudi, sustainability implies a world that is predictable and static, a world in which one can aspire to reach equilibrium and maintain it. Resilience, on the other hand, is built on an understanding that the world is in flux and constantly changing … meaning that designers must approach resilience in a fundamentally different way without losing the core values that define our disciplines and profession.

Efforts in the realm of "sustainability" have achieved significant advances over the last two decades. The energy-reduction focused 2030 Challenge has been launched, and many large and small architectural practices have signed on. The U.S. Green Building Council's LEED certification program provides architects, owners, and occupants with a measuring stick for sustainability. But critics of LEED will call out the programs for their reduction of sustainability to a series of checklists and tradeoffs. Sustainability initiatives such as these are definitely admirable, but they do not recognize that our environment is more dynamic than ever and that buildings and communities must be designed to be adaptable if they are to survive. According to Simin Davoudi, "It appears that resilience is replacing sustainability in everyday discourses in much the same way as the environment has been subsumed in the hegemonic imperatives of climate change" (Davoudi et al., 2012).

Author Andrew Zolli, in his book *Resilience: Why Things Bounce Back*, addresses the conceptual underpinning of resilience as systemic, meaning resilience is not determined by one factor or behavior, but is rather about applying ecology-based systems thinking to the whole network of interrelated factors. What makes resilience different for most architects is thinking through the entire building including site and context, and anticipating the myriad of uses, conditions, and changes that will be necessary across a shortened timeframe. Systems thinking, as outlined by Zolli, offers some guidance about strategies and tactics for approaching resilient design that is holistic and better able to respond to the dynamic environment created by accelerated climate change.

Architects and architectural educators should approach resilience as both a challenge and an opportunity to elevate our discourse, to return to the primary questions of our discipline in how we define shelter and our connection to the natural environment. Architects have a unique ability to order and design within and for complexity. The first step is understanding that answers will not come easily.

References

Burns, C. and A. Kahn. 2005. *Site Matters: Design Concepts, Histories, and Strategies*. New York, NY: Routledge.

Carrington, D. 2016. "Climate change will stir 'unimaginable' refugee crisis, says military." *The Guardian*, December 1, 2016. https://www.theguardian.com/environment/2016/dec/01/climate-change-trigger-unimaginable-refugee-crisis-senior-military (accessed July 4, 2017)

Codello, R. and J. Dezio. 2009. "Carlo Scarpa's monument to the Partisan woman." *Future Anterior* 6(1), 38–48.

Crabbe, M.J.C. 2006. "Challenges for sustainability in cultures where regard for the future may not be present." *Sustainability: Science, Practice & Policy*, 2(2), 57–61.

Davoudi, S., K. Shaw, L. Jamila Haider, A.E. Quinlan, G.D. Peterson, C. Wilkinson, H. Fünfgeld, D. McEvoy, L. Porter, and S. Davoudi. 2012. "Resilience: A bridging concept or a dead end? 'Reframing' resilience: Challenges for planning theory and practice. Interacting traps: Resilience assessment of a pasture management system in Northern Afghanistan. Urban resilience: What does it mean in planning practice? Resilience as a useful concept for climate change adaptation? The politics of resilience for planning: A cautionary note." *Planning Theory & Practice*, 13(2), 299–333, doi: 10.1080/14649357.2012.677124

Dodds, G. 2004. "Directing vision in the landscapes and gardens of Carlo Scarpa." *Journal of Architectural Education* 57(3), 32–42. February.

Dodds, G. Landscape and Mirror: Landscape and Garden in the Work of Carlo Scarpa. Unpublished.

Fondazione Querini Stampalia. 2014. The Mark of Carlo Scarpa, June 5 – November 23, 2014 exhibition description. http://www.querinistampalia.org/eng/the_mark_of_carlo_scarpa.php (accessed July 4, 2017)

Frampton, K. 1990. "Rappel a L'Ordre, the case for the tectonic." *Architectural Design* 60(3–4), 19–25.

Frampton, K. and J. Cava. 2007. *Studies in Tectonic Culture: The Poetics of Construction in Nineteenth and Twentieth Century Architecture*. Chicago, IL: Graham Foundation for Advanced Studies in the Fine Arts.

Linham, M. and R.J. Nicholls 2010. *Technologies for Climate Change Adaptation: Coastal Erosion and Flooding*. TNA Guidebook Series. UNEP/GEF. http://www.tech-action.org/-/media/Sites/TNA_project/TNA%20Guidebooks/Technology%20English/TNA_Guidebook_AdaptationCoastalErosionFlooding.ashx?la=da (accessed July 4, 2017)

McGranahan, G., D. Balk, and B. Anderson. 2007. "The rising tide: Assessing the risks of climate change and human settlements in low elevation coastal zones." *Environment and Urbanization* 19(1), 17–37. April 1.

Pliny the Elder. *The Natural History*. Book XVI, Chapter 1. http://www.perseus.tufts.edu/hopper/text?doc=Perseus%3Atext%3A1999.02.0137%3Abook%3D16%3Achapter%3D1

World Health Organization. 1991. *Surface Water Drainage for Low-Income Communities*. Geneva: World Health Organization. http://www.who.int/iris/handle/10665/39775

Zolli, A. and A.M. Healy. 2013. *Resilience: Why Things Bounce Back*. New York, NY: Simon & Schuster Paperbacks.

23

The Hard Habitats of Coastal Armoring

Richard L. Hindle

CONTENTS

Hard (Adj.)
– solid, firm, and rigid
– requiring a great deal of endurance or effort

Ecology is not always soft, vulnerable, weak, and on the verge. Many marine species are inherently resilient, tough, and tenacious, often existing in the most turbulent, deepest, and hardest of places. Similarly, coastal infrastructure such as seawalls and breakwaters that employ solid, firm, and rigid materials to armor coastlines are sometimes amenable to the habitat needs of marine species. This is the realm of "hard habitats"—a reconciliation of urban armature and marine ecosystems through the design of advanced structures that protect against rising seas and storms while providing habitat and refuge for marine species. This essay explores the scale, scope, and future outlook for the design of "hard habitats" in urbanized marine environments and points toward the productive collaboration between marine scientists, materials research, and designers. For landscape architects and urbanists concerned with the ecology of cities, hard coastal infrastructure provides an exciting frontier through which to explore the built ecologies. For marine ecologists and other scientists, the fabrication of novel urban intertidal ecosystems provides new sites for experimentation and testing. This convergence of expertise and knowledge is occurring at exactly the moment when coastal infrastructure promises to multiply exponentially around the world, making the development of resilient and environmentally sensitive armoring especially salient. Many of the advances in hard habitat creation have focused on seawalls and breakwaters. Integration of marine habitat requirements with such hard coastal infrastructure promises to incrementally improve, or reconcile, the ecology of urban marine environments. The essay begins with an overview of the problem associated with extensive coastal armoring and scope of the potential solutions, provides a survey of relevant scientific literature, summarizes design principles for ecological seawalls and breakwaters, highlights built projects, and tracks technological innovation through patents.

Hard Coastal Armoring: Extent of the Problem and Scope of the Opportunity

Precedents for "hard habitats" are abundant in the urbanized intertidal zone. We need not look any further than the spontaneous species assemblages associated with boat wrecks or the robust colonization of piers and docks by mollusks for evidence of hard surfaces and extreme conditions harboring novel ecologies in the Anthropocene. Fortunately the habitat potential of hard surfaces is more than anecdotal. Pioneering research on the ecology of urban marine structures confirms that the proliferation of species on artificial structures can be altered through design, and new technologies have, and will be developed that combine sound ecological science with advanced marine construction. The question moving forward is how, not if, we can advance the *positive* interrelations of coastal infrastructure and marine habitats.

Urbanized waterfronts present a unique design challenge at the interface of marine and terrestrial systems. Few would disagree that the urban intertidal zone and vulnerable coastlines must sometimes be armored with rigid seawalls, breakwaters, and embankments, to protect our vital cultural and economic interests. Yet, it is widely accepted that conventional construction practices alter marine environments through the introduction of foreign materials and novel three-dimensional morphologies (i.e., form) into ecosystems that have evolved distinct abiotic and biotic characteristics that are often different from the constructed infrastructure that replaces them. The scale and extent of the design challenge presented by a hardened coastal zone is enormous (Chapman, 2003). Thousands of miles of armored edges and countless artificial marine structures currently exist globally, and much more is in the process of being built or planned, as sea level rise and storm surges increase and coastal development hastens. The artificial materials and structures associated with coastal armoring impact marine environments on micro-, meso-, and macroscales, yet currently there is no coordinated strategy or framework to integrate knowledge about these diverse sites or holistically evaluate their global impact. And, as the use of "soft" engineering approaches, such as constructed sand dunes, wetlands, mangroves, and other naturalistic coastal typologies are sometimes limited by the realities of urban sites, we should consider "hard habitats" among the coastal resilience tool kit.

The extent of global coastal armoring is difficult to accurately assess, but specific urban areas provide insights about generalized conditions around the world. The marine environment of Sydney Harbor, for example, is estimated to be 96% urbanized by walls, piers, wharves, jetties, docks, and other structures, with more than 50% of the shoreline composed of seawalls (Chapman et al., 2009a,b). In California, America's most populated state, an "astonishing" 177 kilometers (km) of the entire 1,770 km coastline is armored. In highly urbanized areas, such as the four southern counties of California (Ventura, Los Angeles, Orange, and San Diego), the issue is exacerbated, with more than 33% of the 360 km coastline armored with riprap and concrete walls (Griggs, 2009) (Figure 23.1a). This urban condition is of course not unique to Sydney or southern California. Most major harbors and urban waterfronts are heavily armored with variable configurations of steel sheet piles, cut stone, riprap, concrete, and timber, leading to extensive "ocean sprawl" that negatively impacts ecological connectivity and distribution of species (Bishop et al., 2017). Urban regions, including New York, Singapore, Jakarta, Hong Kong, San Francisco, and every other major coastal city, have stabilized and armored their waterfronts, radically altering the abiotic and biotic conditions of the intertidal zone (Figure 23.1b). Although significant distances often separate major cities, the aggregative effect of the hardened anthropogenic coastline is larger than a singular site or foreshore, making the urban intertidal and coastal zone "ground zero" for the destruction, and possible reconstruction, of urban marine ecology.

The pervasiveness of artificial materials and simplified built-form of the urban intertidal zone has shifted the ecology of marine systems. This change is often cataclysmic. Imagine if you will the effect of replacing vast, gently sloping, mucky planes of seagrass, marsh plants, and fine sediment with vertical steel sheet piles or monolithic concrete seawalls, or, alternately, the replacement of naturally eroding limestone and extensive mangrove swamps with immovable vertical concrete structures that obliterate sediment and groundwater exchange, and stabilize a once dynamic intertidal condition. Vertical structures truncate the intermediated boundary between water and land, altering the edge morphology in horizontal and vertical profile and diminishing habitat areas (Figure 23.2). Armored edges also restrict the movement and exchange of groundwater, alter salinity gradients and pH levels, limit sediment transport, replace vegetated slopes and planes, and simplify the heterogeneity of microclimates, ultimately altering intertidal habitats. Integration of habitat criteria into urban marine infrastructure may never reverse this damage, but it may help to bridge the ever-expanding schism between urbanization and marine ecology.

The extent and ubiquity of anthropogenic hard structures in the marine environment is emblematic of human geologic and ecologic "agency" in the Anthropocene. In southern

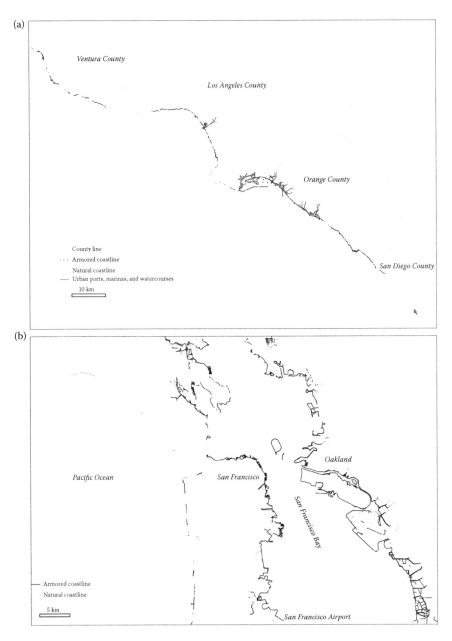

FIGURE 23.1

(a) Map of southern California armored coastline in the southern counties, showing the relationship of natural and anthropogenic structures. The image also highlights harbors and ports that contribute significantly to the urbanized intertidal zone. (Diagram by Kate Lenahan. Data source: California Coastal Commission. "Coastal Erosion Armoring." 2014. Web. 10 April 2017.) (b) Map of the San Francisco Bay area showing extent of artificial marine structures and flood control in the region. The image shows that approximately 50% of the Bay area's coastline is altered by seawalls, levee, berms, and other water control structures. (Diagram by Kate Lenahan. Data sources: San Francisco Estuary Institute (SFEI). "San Francisco Bay Shore Inventory: Mapping for Sea Level Rise Planning GIS Data." 2016. Web. 12 May 2017. California Coastal Commission. "Coastal Erosion Armoring." 2014. Web. 10 April 2017. National Oceanic and Atmospheric Administration. "NOAA Medium Resolution Shoreline." 2016. Web. 12 May 2017. Engineered levees, floodwalls, berms, shoreline protection structures, and water control structures.)

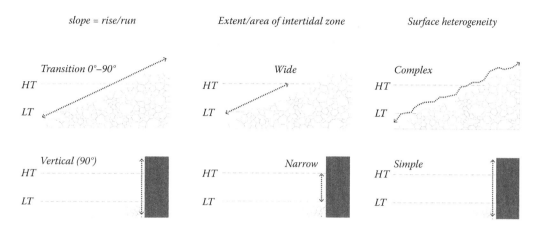

FIGURE 23.2
The shift from gradual shorelines to vertical walls shrinks and simplifies the intertidal zone. Vertical structures decrease the area of the intertidal and reduce the amount of habitat. Artificial seawall structures also simplify the form of the intertidal zone, reducing topographical heterogeneity. (Diagram by Lenahan, K. 2009. Based on similar drawings from McDonnell, M.J. Amy K. Hahs, and Jürgen H. Breuste (Eds.), *Ecology of Cities and Towns: A Comparative Approach.* Cambridge: Cambridge University Press, p. 159.)

California the extent of solid artificial armoring is commensurate with that of analogous naturally occurring geological features. Approximately 29% of southern California's coastline is classified as naturally rocky, and an estimated third is currently armored with manmade rip-rap in the form of jetties, breakwaters, and armored shorelines (Pister, 2009). Although species abundance and composition was found to be similar between riprap and naturally rocky coastlines in parts of California, the divergence in ecological structure is exacerbated in areas that were not formerly rocky. Material incongruence is therefore most commonly associated with typologies such as sandy beaches, soft bottoms, and alluvial cliffs composed of loosely consolidated sand and gravel which are often the most likely to be armored due to their structural instability. This problem is not unique to California. Studies conducted in the Mediterranean further illustrate that hard marine structures in areas with naturally soft bottoms invite invasive species and ultimately reduce biodiversity (Vaselli, Bulleri, and Benedetti-Cecchi, 2008). Likewise, in China's Yangtze river estuary, artificial hard surfaces were shown to facilitate the migration of rocky intertidal species throughout an estuary previously composed of marsh habitat—a problem that will only be exacerbated with climate change (Dong et al., 2016). Given the ability of species to migrate in association with hard coastal infrastructure, it may be concluded that the biogeography and migration of rocky intertidal species would also be radically expanded by the proliferation of coastal infrastructure in response to the threats of sea level rise and increased storm frequency. As we extrapolate to the future, the problem only appears to intensify as sea levels rise and storm surges increase.

Impacts of hard coastal infrastructure on the marine environment are truly multi-scalar as design details have the potential to impact larger scale ecological relationships. This is most succinctly illustrated in debates generated by European plans for offshore wind farms, which promise to supply sustainable energy but also require extensive marine infrastructure. As Europe plans for this massive investment, a critical dialogue has emerged around the foundation structures used to support the turbines. The reef effect (i.e., attractiveness for marine flora and fauna) of the footings are a major concern given the ubiquity of the proposed structural foundations (Petersen and Malm, 2006). Researchers concluded that at the microscale the chemical makeup and relief of the hard substratum play an important role in the composition of the epibenthic (i.e., top of the sea floor) community by impacting the formation of biofilms, anchorage of mollusks and seaweeds, and ultimately the recruitment of species that feed and forage on the structures. At the mesoscale, the size, slope, form, wave exposure, and depth relative to tidal fluctuations were identified as impacting the species associated with the footing of the turbines, and recommendations were made to use novel geometries, boulders, and artificial reefs to create habitat and reduce the negative impacts of scouring on the structure. At the macroscale, the composition of the species assemblages associated with each structure, distance between structures, size of the structures, in combination with local conditions were hypothesized to shift the ecology of the marine environment toward those commonly associated with rocky intertidal zones and reefs.

The multi-scalar and aggregative effect of artificial hard surfaces and substratum in the marine environment is hard to deny, from wind farm foundations to bulkhead walls of port lands. This results simply from the divergence of human construction practices and materials with those of unadulterated marine environments. Although the ubiquity of hard materials used in coastal armoring may be cause for alarm, effective design and technological innovations have been developed and implemented to ameliorate these effects. These hard habitats provide refuge, anchorage, and food sources for marine organisms, while helping to stabilize and protect human habitats and vital infrastructure. It is clear that armoring the intertidal zone is often essential and that the impacts are extensive and persistent. In this context, the aggregative effect of ecologically reconfiguring individual seawalls, breakwaters, and coastal infrastructure is significant on a global scale.

The Novel Ecology of Hard Materials, Vertical Surfaces, and Artificial Marine Structures

Soft engineering with plants and naturalistic landforms has its structural and conceptual limits, especially in highly urbanized areas where space is tight and waterfronts are often multifunctional. Of course, restoration of the water's edge using principles derived from local ecology is often preferable, but this is not always possible in dense urban areas. In cities, hard materials and their associated habitats play an important role in marine environments—like it or not.

Literature on the ecology of hard surfaces has advanced rapidly in the field of urban ecology. Terrestrial and marine ecologists alike have recognized the ecological potential of hard surfaces in urban systems, and many of their lessons and observations offer comparative insight. Terrestrial urban ecologists have focused on the spontaneous species diversity and ecosystems services associated with hard urban surfaces such as walls, stone surfaces, pavements, and rubble piles. In the urban landscape, vegetation commonly associated with rocky and disturbed natural habitats are found to thrive in urbanized hard habitats, including early successional genera of lichens, bryophytes, ferns, and other pioneer species for which these anthropogenic landscapes are habitat analogs (Lundholm, 2011). In this new urban ecological order, cracks in the sidewalk become habitat analogs for rock fissures, rubble piles are surrogates for natural areas of erosion, walls supplant cliff faces, and the broader anthropogenic urban landscape becomes a novel habitat with the potential to reconcile urbanization, industrialization, and ecology (Lundholm and Richardson, 2010).

Importantly, many of these novel habitats and spontaneous and volunteer species offer the same ecosystems services as highly engineered and costly designed systems (Tredici, 2010).

Within this burgeoning area of ecological research, particular attention has been given to walls as ubiquitous features of the urban landscape for their ability to serve as habitat for different species and support "non-standard cosmopolitan assemblages," as well as for their capacity to be redesigned to increase their ecological function (Francis, 2011). The ecology of walls is more than just a contemporary green fad. In Europe, the "novel" ecosystems of old stone walls are studied for their ecological resilience, artifice, longevity in the landscape, and diversity of species resulting from complexity of rock types, variations in form, and age (Collier, 2013). Around the world, related studies affirm the habitat value of vertical hard surfaces. In Hong Kong, research on the ecology of walls shows that traditional construction techniques with open stone joints allow for vibrant *Ficus* communities and thriving vertical ecologies, while modern walls of monolithic concrete limit growth (Jim, 1998; Jim and Chen, 2010). What emerges from a survey of this exciting literature is a new ecological sensibility relating to hard surfaces, vertical urban structures, and their associated novel species assemblages.

Advances in the ecology of terrestrial hard surfaces are mirrored in the study of marine environments, where research indicates that hard surfaces and artificial structures, such as seawalls, may become analog or surrogate habitats for marine species. The relations between marine organisms, artificial materials, and the morphology of artificial structures has been researched at least since the early twentieth century, when scientists attempted to identity factors that contribute to the fouling, or colonization, of marine structures and vessels by sessile (i.e., fixed) organisms. Researchers at the Woods Hole Oceanographic Institute conducted studies on the attachment of sedentary marine organisms to plastics, glass, woods, metals, linoleum, and other materials with additional tests on glass surface textures ranging from flat to ribbed and factrolite (Pomerat and Weiss, 1946). In these early experiments, variable materials and textured surfaces were arrayed in a harbor or bay to determine the rate and density at which organisms attached to the artificial surfaces. The research concluded that material composition played an important role in the surface fouling (i.e., colonization) dependent on the porosity of the material. Although the study was concerned with the primary factors contributing to the unintended colonization of marine structures, it paved the way for further research that confirmed the significance of differences in artificial materials on the abundance of marine species (McGuinness, 1989).

Many of the organisms observed in the first material studies are early colonizers and play an important role in the trophic webs of functional marine environments, providing a base for further development of the community, as well as evidence of artificial materials supporting novel ecological assemblages. Field studies conducted in the 1970s furthered this research, linking successional dynamics and community structure of marine organisms to ubiquitous urban infrastructure and establishing an ecological framework for artificial surfaces and objects. The studies documented the abundance and distribution of sessile epifaunal species on marine pilings and importantly related temporal and spatial scales to habitats developed in association with manmade marine structures (Karlson, 1978). Tim M. Glasby and Sean D. Connell published the first comprehensive survey on the subject of urban structures as marine habitat in 1999, which pointed toward the habitat potential of seawalls and coastal armoring.

Even though artificial reefs have been researched throughout the twentieth century and many of the findings are echoed in the early literature on artificial marine structures, Glasby and Connell's, (1999) paper indicated an important shift in focus on urban infrastructure as a progenitor of a novel marine ecology. The paper vividly illustrates the effects of piers, pontoons, walls, and other structures on marine species assemblages. The findings of this provocative early research are clearly summarized in the following quote:

> There is the potential, therefore, for these new habitats to influence or impact upon other marine species directly and indirectly. For example, fishes are known to aggregate around structures in the water column. The different assemblages of organisms growing on urban structures could influence the types of fishes that recruit to these areas. Organisms living in sediments can be influenced by species in surrounding habitats. Seagrasses are known to be sensitive to shading, sedimentation, and regimes of water flow, all of which may be altered by the addition of urban structures.

Glasby and Connell, 1999

Although tentative, Glasby and Connell's survey provided sound ecological context for ongoing research on artificial marine structures. A second survey, published in 2009 by M.G. Chapman, David Blockey, Julie People, and Brianna Clynick, offers a similar analysis with one significant difference—the recognition that design and engineering may facilitate the *creation* of ecologically sensitive coastal infrastructure. In the 10 years between the articles, researchers had advanced the study of artificial marine structures and concluded that species diversity increased with the availability of microhabitats such as those provided by deteriorated walls and therefore may be intentionally designed (Chapman et al., 2009a,b). The agency of design is therefore integral to the production of novel urban marine habitats, collapsing the boundaries between anthropogenic and native ecologies.

The novel marine habitats associated with urban structures are not without controversy. Researchers have identified common problems associated with the ubiquity of artificial marine habitats. Often the species associated with these structures do not exactly mirror the species of nearby natural habitats. Given the perennial comparison of artificial seawalls with unaltered marine environments, there is a tendency among scientists to "consider the impact of built structures as 'negative' if they are colonized by invasive species, but 'positive' if they attract native species (particularly fish) even if the native species would not live in the area were it not for the infrastructure" (Chapman and Underwood, 2011). In this context, the concept of reconciliation ecology, pioneered by terrestrial ecologists, is especially salient. Reconciliation ecology "discovers how to modify and diversify anthropogenic habitats so they harbor a wide variety of wild species. In essence it seeks techniques to give many species back their geographical ranges without taking away ours" (Rosenzweig, 2003). It has been used as an ecological framework through which to comprehend the ecological value of green roofs and vertical gardens in urban environments, which do not fall within strict definitions of restoration ecology (Francis and Lorimer, 2011). However imperfect from the perspective of purist ecology, the novel hard habitats associated with coastal armoring are here to stay and will multiply as sea levels rise, storms surge, and urbanization multiplies. The habitat potential of structures such as seawalls and breakwaters are at the center of this debate and are now the subject of rapid technological innovation.

Seawalls and Breakwaters as Marine Habitat

Seawalls and breakwaters are commonly used to protect urban waterfronts, harbors, and foreshores, and have radically altered the marine environment. These structures armor the shoreline against erosion, preventing inundation of low-lying areas, but they also alter the form of the coast and intertidal zone. Given their ubiquity in coastal armoring, seawalls and breakwaters also have the greatest potential to positively impact urban marine environments.

Seawalls refer to shore-parallel structures designed to stop erosion and retreat of the shoreline, limit inundation, and ameliorate wave action (Kraus, 1988) (Figure 23.3a). Seawalls are often vertical walls or steep revetments (embankments), primarily made of concrete, natural stone, riprap, steel, and even treated timbers. They are located at the intertidal zone

FIGURE 23.3

(a) Seawalls are artificial structures located at the interface between land and water. They often simplify the urban intertidal zone by straightening the horizontal (plan) and vertical (section) profiles of the water's edge. Alternation of the intertidal zone by conventional seawalls greatly reduces the area of marine habitat. The images shown here are of San Francisco Embarcadero seawall. (Photograph [left] by Richard Hindle, Author. Aerial [right] Source: Google Earth.) (b) Breakwaters are artificial structures located away from the shore, designed to reduce wave energy, protect vital infrastructure, and/or reduce erosion. Breakwaters are ubiquitous features of ports and harbors, often creating an entirely new intertidal zone and coastline profile. Given that they are located away from the shore, they have water on both sides of the structure. The images shown here are of a typical breakwater in Coffs Harbor, Australia. (Photograph [left] by Richard Hindle, Author. Aerial [right] Source: Google Earth.)

Microscale

Mesoscale

Macroscale

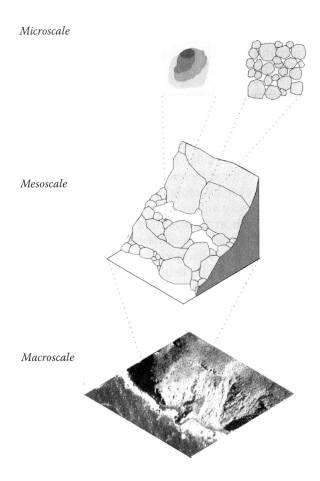

FIGURE 23.4
A diagram of topographical heterogeneity, illustrating the concept at the macroscale (entire cliff face), mesoscale (rocky tidal pool), microscale (rock pore/hole/surface texture). The design of artificial marine habitats often requires heterogeneity across scales to allow for overlapping spatial niches. Topographical heterogeneity also facilitates creation of diverse light/shade, salinity gradients, and temperature in a given area. (Diagrams by Kate Lenahan and Richard Hindle, Author.)

vegetation and substrate baskets to support emergent and aquatic vegetation (Dyson and Yocom, 2015).

The chemical composition of artificial materials is another important factor in creating artificial habitats. Modified concrete mixtures, for example, have been shown to promote colonization by marine species on infrastructure depending on pH, density, and the addition of fibrous matrices (Ido and Shimrit, 2015). Material choices for coastal armoring should consider the precedents and ongoing research generated by the U.S. National Artificial Reef Program established in 1984, which provides information on the function, compatibility, stability, and availability of materials to be used in artificial habitat creation. Diverse materials have been tested as part of the program for the development of artificial reefs including concrete, steel, wood, shell, rock, fiberglass, and waste products such as old cars, parts of gas platforms, tires, and byproducts from coal and oil combustion (Subcommittees, Lukens, and Selberg, 2004). Even though a considerable number have been tested, the effective use of artificial materials in habitat creation continues to evolve, and now includes a class of materials designed specifically to encourage the growth of marine organisms. The term, biomineralogy, is often used to help explain the interrelationship between biological systems and minerals at different hierarchical levels in an ecosystem. Studies have shown that biomineralogy plays a role in marine community development by impacting primary colonization as well as later stages of community development (Bavestrello et al., 2000). Biofilms, or accumulations of bacteria and other microorganisms, are not only essential in the early colonization of an artificial surface, but they also facilitate the settling and recruitment of macroscopic organisms. Because they are sensitive to the chemical composition of substrate materials, this chemical makeup impacts the utility of artificial materials as habitat.

Successful examples of artificial marine habitats are now abundant in scientific literature, making it possible to derive a set of generalized design principles: (1) The heterogeneity of a surface, and formal complexity of a structure increase the potential spatial niches that can be occupied by marine species; (2) Marine structures impact the environment across scales, making it important to consider the detailed rugosity and porosity in small patches in addition to the overall horizontal and vertical profile; (3) Design details that increase variations in moisture, salinity, light, texture, and temperature, are often beneficial to species diversity by creating varied abiotic conditions; (4) Materials play an important role in species establishment and colonization through the creation of biofilms, and processes of biominerology and variability in materials can positively impact species diversity; (5) Mimicking naturally and locally occurring forms, processes, and materials can improve the suitability of a structure. Although each new project will require knowledge of local ecology and coastal processes, these criteria establish a starting point for design development, and a lens through which to evaluate the ever-growing list of pilot projects and new technologies.

Hard Habitats of Coastal Armoring: Precedent Projects

Pilot projects have been constructed around the world to test the viability of ecologically engineered seawalls, breakwaters, and bulkheads. Some of the earliest, and most established prototypes are found in Sydney, Australia (Figure 23.5). The experimental walls constructed in New South Wales are summarized in a 2009 publication entitled "Environmentally Friendly Seawalls: A Guide to Improving the Environmental Value of Seawalls and Seawall-lined Foreshores in Estuaries," which offers an introduction to the methods and results of the program (Wiecek, 2009). The success of these early prototypes and consensus in scientific literature has lead to the construction of seawalls and breakwaters with integrated habitat globally. Not only do these projects help meet sustainability goals, but they also provide researchers with real world experiments to further evaluate the impact of ecologically engineering hard coastal infrastructure. The precedent projects included here offer an introduction to the diversity of project types currently in process or recently completed. Although the project sites are geographically disparate, they are linked by the integration of marine habitat into the hard surfaces employed in coastal armoring and urbanized waterfronts. Collectively, they represent the first wave of hard habitats merged with coastal armoring.

FIGURE 23.5
Ecologically designed seawall at McMahons Point, Sydney Australia. Tide pools are integrated into the block assembly, providing habitat for intertidal marine species. (Photograph by Richard Hindle, Author.)

Habitat Panels: Seattle, Washington

Seattle's waterfront is in the process of being redesigned, with a new master plan developed by James Corner Field Operations. The need to repair large sections of dilapidated seawall, which threatened important infrastructure, served as a major catalyst for the project. The redesigned seawall will enhance habitats along the urban waterfront and improve public amenities. In preparation for the new seawall design, researchers compared three types of panel relief, including flat panel, sloped steps, and a "fin" pattern, in addition to two surface textures and an untreated wall. Early experiments confirmed that the textured and stepped panels supported more diverse communities than the existing seawall, with densities of species like mussels on the flat panels resembling those in pre-existing habitats. As a result, Seattle will incorporate habitat panels into a large expanse of seawall. The artistic team Haddad|Drugan developed a design for the seawall panels (shown in Figure 23.6) based on results from the experimental test panels and in consultation with a range of experts. The project, led by the City of Seattle Department of Transportation, involved a diverse set of consultants, including Parsons engineering as the prime consultant, and Magnusson Klemencic Associates (civil engineers) who led the overall seawall/public realm design. The University of Washington also conducted experiments and research that led to the design parameters for texture size, depth, and shelf configuration. Haddad|Drugan took the parameters and applied a conceptual and aesthetic interpretation to them that resulted in a 3-D computer model for the actual texture. The result is a visually compelling seawall

FIGURE 23.6
Example of a seawall design in Seattle, Washington, that uses texture and shelves to promote the growth of marine life through the increase of surface area and incorporation of crevices. (Photograph Courtesy of Haddad|Drugan LLC. Photo by Laura Haddad.)

that meets habitat requirements. Importantly, the city plans to monitor the seawall panels for several years after construction, generating data needed to design future ecologically beneficial seawalls in the region and around the world (Goff, 2010).

Mussel Beach: East River, New York City

Ken Smith Landscape Architects designed the Mussel Beach ecological habitat demonstration project as part of the East River Waterfront in New York City (see Figure 23.7). The project is composed of a folded terrain on Pier 35 that spans from the pier's deck level through the intertidal zone. The folded slope is composed of specially textured, precast concrete panels embedded with rocks to serve as habitat for the river's native mussel population. An upland riparian planting flanks the intertidal constructed habitat, and a footbridge crosses the area for public access and viewing. The geometry of the overall habitat increases intertidal area, and the diversity of textured surfaces provides refuge and anchorage for marine species. Significant features of the habitat are its proximity and integration of sewer outlets and the use of precast modules in construction. The precast concrete panels allowed for the development of a diversity of textures and rock patterns that increase to topographical heterogeneity of the surface as well as provide a diversity of materials for marine organisms. The Project team included the City of New York (client),

FIGURE 23.7
Mussel Beach at Manhattan's East River Waterfront Esplanade. (Photograph courtesy of Ken Smith Landscape Architects. Copyright Peter Mauss/Esto.)

Ron Aleveras (ecologist), ARUP (engineering), HDR (engineering), SHoP (architects), and Tillotson (lighting design).

"Green" Breakwater Habitat: Cleveland, Ohio

The Green Breakwater at Cleveland Harbor is a pilot project to test the viability of ecological breakwater blocks to function as marine and intertidal habitat. The project was developed as part of the United States Army Corps of Engineers "Engineering with Nature" program, which aims to develop environmentally sensitive solutions to engineering problems. The Cleveland Breakwater is 7.5 km long, providing for safe navigation and protection of the harbor. The "green" breakwater project modifies the design of standard concrete toe blocks to create habitat opportunities. The redesigned concrete blocks feature grooved surface textures, dimpled surface textures, and indented surfaces for refuge and spawning. Sample blocks were installed in 2012 and 2014, and results from monitoring suggest they stimulated an increase in algae species (*Cladophora* spp.) as well as a variety of aquatic invertebrates (Dreissenid mussels, oligochaetes, amphipods, etc.). The blocks are currently being maintained and evaluated for use in other breakwaters around the Great Lakes (Figure 23.8) (Fredette et al., 2014).

FIGURE 23.8
Precast concrete blocks for use in seawall construction. (Photograph courtesy of the United States Army Corps of Engineers Buffalo District.)

Living Breakwaters: Staten Island, New York

SCAPE Landscape Architecture DPC developed the Living Breakwaters project and program as part of their winning entry for HUD's Rebuild by Design Competition. The Living Breakwaters proposal envisions a series of breakwaters flanking the Staten Island shoreline. Community outreach and education are integrated with the design through programs that educate and cultivate the coastal defense structures. The breakwaters are configured to provide habitat at the macro- and microscale through a series of "reef streets" that offer habitat complexity and heterogeneity to host finfish, shellfish, and other marine species. The SCAPE team developed the Living Breakwaters concept for the U.S. Department of Housing and Urban Development's Rebuild by Design (RBD) initiative. SCAPE's approach is unique for its integration of resilient coastal infrastructure with habitat enhancement techniques and environmental stewardship. A pilot project is being developed for the South Shore of Staten Island using special ECOncrete habitat blocks. The concrete mixture for the blocks is specially formulated to encourage marine organisms, and the surface of the block is textured to enhance surface heterogeneity. The blocks are arranged to provide spatial complexity and variability (Figure 23.9).

FIGURE 23.9
Building ecological resiliency. (Photograph Courtesy of SCAPE Landscape Architecture DPC.)

Fish Habitat Enhancement Devices (FishHEDs): Rhinebeck, New York

The Hudson River Sustainable Shorelines program has developed a series of demonstration projects to make science-based information available about best shoreline management techniques. As steel sheet piles were identified as exhibiting almost no habitat value, prototypes for Fish Habitat Enhancement Devices (fishHEDs) were installed at Rhinecliff Landing in Rhinebeck, New York. The pilot project aims to improve fish and invertebrate habitat within the corrugations of the steel bulkhead wall while maintaining human use and function of the dock. The team designed, built, and installed the fishHEDs in 2015 to increase habitat complexity and diversity on the site (Figure 23.10) (https://www.hrnerr.org/fishheds.html).

Artificial Tide Pools: Brooklyn, New York

Brooklyn Bridge Park, designed by Michael Van Valkenburgh Associates, integrates precast tide pools into the riprap slopes at Pier 4. The ECOncrete tide pools are intended to increase intertidal biodiversity by improving habitats for species that that are not typically associated with riprap, specifically by providing refuge during low tide. The patented concrete mixture composite closely resembles the texture and makeup of rock and coral. It facilitates the growth of algae, seaweed, oysters, and other marine life, which offers habitat, breeding grounds, and food sources for fish, crabs, and other organisms. Since the precast tide pools are specifically designed for use with riprap slopes, they can be easily integrated with existing construction practices (see Figure 23.11).

FIGURE 23.10
FishHEDs mounted on bulkhead. (Photograph courtesy of Hudson River Estuary Program and New York State Department of Environmental Conservation.)

Patent Innovation and the Development of New Habitats

Technology plays an important role in the development of artificial habitats, as it does in all ecologically designed and engineered systems. In this sector of "ecotechnology," patents are an effective method of tracking innovation while offering insights about what the manmade habitats of the future may look like. Since hard habitats are entirely fabricated, they are liable to undergo many phases of design iteration, research, and development before new best practices and preferred systems are adopted. Irrespective of this sector's infancy, patent innovation parallels, outpaces, and even predicts scientific discoveries. Artificial reefs, for example, have been the subject of scientific inquiry and patent documentation since at least the 1960s. Many of the advances made in artificial reef technology, such as techniques to enrich surface textures, provide three-dimensional refuge, and attract fish, may now be utilized for the creation of habitats on seawalls and breakwaters. Technological innovation can sometimes outpace scientific discovery. For example, marine ecologists first altered seawalls to integrate habitats such as rock-pools in 2005, yet a similar system for integrating habitat in seawalls was patented in 1992, leaving a 13-year lag between invention and materialization of the idea in the built environment.

FIGURE 23.11
ECOncrete tide pool. (Photograph Courtesy of ECOncrete.)

Technology and ecology continue to converge; yet debates persist about the role and viability of innovative materials and construction systems in marine ecosystems. For instance, a survey of "alternative" shoreline devices published in 2012 questioned the environmental impact and efficacy of non-conventional technologies such as artificial seaweeds or net groins (Pilkey and Cooper, 2012). Yet, similar systems are now being studied for their capacities to create habitat for species like seahorses, which introduce filamentous textures to the marine environment that facilitate anchorage of marine species (Hellyer, Harasti, and Poore, 2011). Notwithstanding scientific opinion, this sector of technology and industry continues to grow. Companies such as ECOncrete are developing new concrete compositions and marine habitat modules, with pilot projects around the world (http://www.econcretetech.com). To date, there are hundreds of patented artificial habitat and shore protection systems that address complex environmental problems, from the dissipation of wave energy, to the construction of large ports in open water. The question moving forward is how to best implement, test, and evaluate novel construction systems and new materials.

Patent innovation in hard habitats can be reduced to two main categories: (1) systems and complex assemblages that create novel three-dimensional form to serve as habitat and/ or stabilize coastal infrastructure by mitigating wave energy or reducing scouring, and (2) material compositions or production processes that modify the chemistry or physical properties of substrates to catalyze growth of marine organisms. The patents listed below provide a sampling of the techniques and technologies related to the creation of hard habitats (see Figure 23.12a–g).

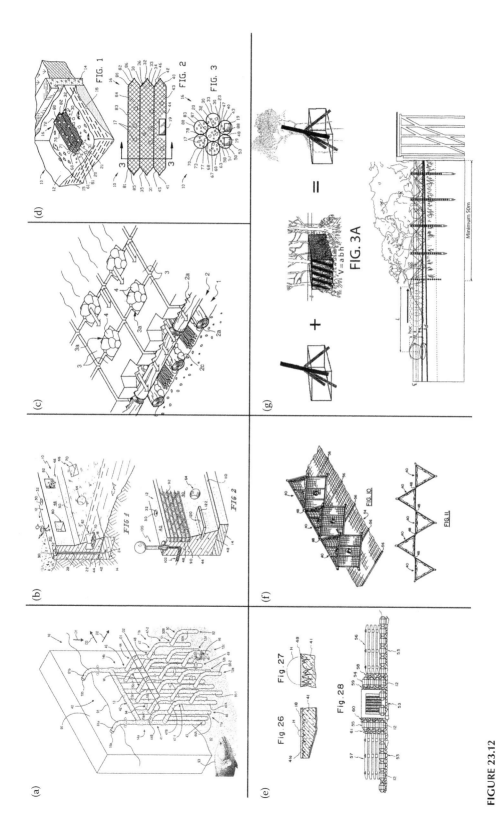

FIGURE 23.12
(a–g) Patent images for coastal armoring integrated with marine habitats. (http://patft.uspto.gov.)

Three-Dimensional Structures and Complex Assemblages

U.S. 8,635,973: Artificial Mangrove Assembly (2014)

The artificial mangrove assembly provides habitat and refuge that replicates the three-dimensional structure of mangrove roots. The system provides habitat complexity throughout the water column directly in front of a seawall and can be affixed to the surface of existing walls to improve intertidal habitat. Each artificial root element is secured to the mounting assembly, extending downward though the water column, and eventually contacting the bottom through a respective distal end. In nature, the roots of the mangrove provide habitat for marine flora and fauna such as fish, crustaceans, and birds. Moreover, they trap aquatic nutrients for marine life and act as a substrate for colonization. Naturally occurring mangroves are rapidly disappearing for a number of reasons, including climate change, barnacle infestation, weeds, pollution, logging, oil exploration and extraction, shrimp aquaculture, tourism, and urban development. The artificial mangrove assembly proposes a formal substitute that can foster habitat complexity comparable to that of this threatened system.

U.S. 5,125,765: Seawall Construction (1992)

The patent for "Seawall Construction" is an early design concept for a rigid wall with integrated habitat for marine species. The seawall is formed to have a plurality of recesses in the lower front panel, providing nesting places and refuge for marine life and other associated wildlife. A diverse array of recessed geometries, shelves, and textures enhance habitat complexity for marine species. The design also permits the movement of organisms through the wall so they may build habitat in the soil and stone behind it. This type of connectivity would simultaneously permit the exchange of groundwater with the water body. It is important to note that this seawall patent was developed more than 10 years before its first prototypes were tested in Sydney.

U.S. 6,746,177: Block and a Riparian Improvement Structure Inhabitable for Aquatic Life (2004)

The riparian improvement structure is designed to stabilize shores, allow for the exchange of water and oxygen between soils and water, and provide habitat for marine organisms. It consists of a series of blocks and wooden materials that accommodate the exchange of oxygen and water through the system, facilitate the movement and anchorage of marine organisms, and stabilize the shore. The structure creates an environment suitable for aquatic life such as fish and crustaceans by incorporating void spaces, diverse textures, and organic materials into the assembly. The blocks are each provided with a water passage through a series of grooves. This configuration admits flow through the riparian improvement structure to accelerate the exchange of water containing a sufficient amount of dissolved oxygen for aquatic life, thereby creating an environment suitable for its growth.

U.S. 5,007,377: Apparatus and Method for Marine Habitat Development (1991)

This patent discloses a method for the development of a marine habitat through the growth of mollusks adjacent to urban waterfronts. The system and strategy comprise a plurality of retaining members made of a mesh wall that may be filled and colonized by mollusks and other marine organisms. This early design for "oystertecture" aggregates permeable bags of adult mollusks to seed a reef in a desired location. The permeability of the bags permits the growth of mollusks inside, through, and outside the retaining structure. The patent also anticipated that other marine organisms might pass through and occupy the reef bags, cultivating a functional reef habitat. Importantly, the reef modules can be arranged as necessary along an urban shoreline and may be incorporated into seawalls.

U.S. 4,508,057: Algal Culturing Reef Unit, Artificial Reef Unit, and Artificial Culturing and Fishing Field Unit (1985)

The artificial algal culturing reef unit aims to raise fish or culture algae, fishes, and shellfishes and is typically submerged in a shallow sea zone. Because these units must resist high waves and fast oceanic currents usually encountered in this environment, they are made of concrete. Although concrete materials are suitable because of their strength and cost, an alkali is necessarily emitted from their surfaces that is extremely harmful to diatoms and algae, as well as fishes and shellfishes. The algal culturing reef unit uses a layer of iron sulfate or acid and iron oxide powders to penetrate the surface of the concrete blocks, whereby the alkali from the concrete is oxidized to create a surface chemistry preferred by algae and shellfish and thereby facilitates rapid colonization.

U.S. 5,269,254: Method and Apparatus for a Growing Oyster Reef (1993)

The patent proposes a method for oyster reef formation by setting seed oysters on cultch material, such as recycled oyster shells. The placement of seeded cultch material in permeable panels forms a vertical wall through which water may flow. The modules are made of welded wire metal frames and mesh material and arranged in a stable triangular form that may be arrayed along the shoreline. The structures can be configured in a manner that effectively accretes sediment, and in conditions favorable for oyster growth, become living walls that grow oyster reef. The thin panels of cultch material and metal allow water passage, but also accumulate sediment.

U.S. 8,511,936: Method and Apparatus for Coastline Remediation, Energy Generation, and Vegetation Support (2013)

The reinforced mangrove infrastructure integrates living mangrove plants into a biomechanical structure for coastline remediation. A structural framework supports mangrove plants such that they may take root, forming a biomechanical skeleton that may catalyze the development of a mangrove forest and habitat for oysters and other marine organisms. The system may be incorporated into the existing coastline morphology for restoration of degraded mangrove habitat within the native range of the species. It can also be adapted for use in combination with existing coastal infrastructure, such as seawalls, to dissipate wave energy from storm surges and sea level rise.

Chemical Modification of Materials

U.S. 8,312,843: Artificial Material Conducive to Attract and Grow Oysters, Mollusks, or Other Productive and/or Stabilizing Organisms (2012)

The artificial construction material facilitates the setting and growth of oysters, other mollusks, and other organisms for the purposes of food production and creation of marine infrastructure such as artificial reefs/breakwaters. The material is a composite concrete that acts as both an attractant and nutrient source for mollusks and other aquatic organisms. The binder for the composite consists of both cement and an organic material with several admixtures that modify the physical and chemical properties of the cement. The organic component is selected to attract and feed aquatic organisms, and may include cottonseed, peanuts in the shell, animal byproducts, slow release fertilizers, and other materials containing the desired levels of nitrogen, fat, and sugar. Because the material releases nutrients in a form usable by microorganisms in its vicinity, quality grades of organic material that may not otherwise be suitable for consumption or normal animal feed may be used in the artificial material. Previously useless material, such as cottonseed high in free fatty acid and peanuts below acceptable grade, can be incorporated into the composite material to manage waste products, as well as beneficially grow mollusks and other organisms.

CA 2,901,149: Methods and Matrices for Promoting Fauna and Flora Growth (2014)

The invention proposes a marine infrastructure comprising a concrete matrix having a pH of less than 12, which is beneficial to the generation of biofilms, and supplemental admixtures that promote the growth of fauna and flora in the aquatic environment. This is reported to include endolitic and epilitic, anaerobic and aerobic flora and fauna (lichens, fungi, mosses, and blue-green algae). In addition to its pH requirement, the concrete matrix is modified to have a roughness grade beneficial to the anchorage of marine organisms and a compressive strength that meets construction standards.

DE 69806616: Stone Material for Submerging into Water, Method of Production Thereof, and Method of Forming Submarine Forest (2003)

This patent discloses a method of producing stone material for submerged conditions to cultivate a submarine forest of seaweeds and algae. The procedure to prepare the stone mixture integrates a granular slag generated in the steel production process, applies carbonized treatment, and uses the resulting carbonates as a binder. In prior processes calcium contained in slag is eluted (washed out) in the sea, raising the pH of seawater in the periphery. Compared with concrete products, the agglomerated slag obtained in iron- and steel-making processes is more suitable as a block for a seaweed bed or the like due to its surface properties. However, it has the same degree of function (adhesion and viability of marine algae) as natural stone, and it does not by design promote the growth of marine algae.

U.S. 20150230434: Application of Green Technology Techniques to Construct a Biodegradable Artificial Reef (2015)

This invention provides a cellulose-based surface that is coated in nutrients to promote the rapid growth of marine microbes at the base of the marine food chain. The reef material intends to catalyze rapid growth by providing mineral based substrate attached to the cellulose-based material which is denser than water to allow the entire structure to sink; both the cellulose- and mineral-based materials will degrade rapidly, leaving behind nucleation sites for microbes, corals, invertebrate collections, and more. The reef is constructed entirely from biodegradable materials, and production costs are economical. To achieve the goals, a cellulose-based material is soaked in nutrients, combined with a mineral-based biodegradable substrate, sunk in an aquatic or marine environment, and used to function as a nucleation and nutrition site for a variety of organisms. As opposed to other approaches for constructing artificial reefs, this method utilizes green technologies' principles to stimulate the rapid colonization of the structure by the microbial community, the bottom of the food chain, shortly after being submerged. Tests have been conducted in Florida as part of a pilot project.

(Note: *Additional information on each patent, or expanded searches, are available on the United States Patent and Trademark Office website* www.uspto.gov, *or the European Patent Office website* https://worldwide.espacenet.com/.)

Conclusion

The integration of habitat within the armored intertidal zone attempts to reconcile the schism between marine ecosystems and the need for urban infrastructure, creating fertile ground for research, experimentation, and technological innovation. The triple threats of sea level rise, increased storm surges, and ongoing development pressure will hasten the demand for coastal armoring and expand the hardened anthropogenic intertidal zone, ultimately shifting the ecology of urban marine environments. Although fraught with risk, proliferation of coastal armoring also provides opportunities to create novel habitats in highly altered environments and offers alternatives to soft engineering in areas where space is limited or structural requirements are stringent. It is also worth noting that these new habitat types occur at the convergence of capitol investment, infrastructure, and environment cost typically associated with urban waterfronts. Given these economic and environmental factors, it is a safe assumption that innovation will continue in this relatively new sector of technology.

Coastal infrastructure integrated with habitat necessitates collaborative research, design experimentation, and creates business opportunities. Patent submissions indicate that new materials and structural systems are being invented as scientific observations continue to verify the novel habitat benefits of coastal infrastructure. Simultaneously, creative precedent projects are being developed that provide real world experiments for urban marine scientists to verify. The intersection of expertise, capitol, and technology, evident at the urban intertidal zone, reveals a unique "innovation model" for physical urban infrastructure in which new technologies beget novel ecologies that improve the city and incentivize investment. This convergence of factors has recently been coined as a "blue" framework for the eco-engineering of urban marine environment in which technological progress, investment, primary research, and ultimately new forms of multifunctional infrastructure are coordinated to improve urban intertidal ecology (Mayer-Pinto et al., 2017).

Of course, hard habitats are not merely a scientific or technical proposition, they are in fact cultural in nature, requiring a shift in consciousness and socio-technical and socio-ecological evolution. Culturally, our recent consideration of the ecologies of hard surfaces and coastal armoring represents a significant change in attitude from a strict division between marine and urban environments toward a kind of mutualism. Whether or not this technological form of reconciliation ecology can drastically improve the ecology of cities is yet to be determined at a significant scale, but in this early phase of innovation it is best to dream big as the next phase of coastal infrastructure will undoubtedly be mightier, harder, and more extensive than the last. And, as awareness about the threats to urbanized marine environments is foregrounded by shifts in our changing climate, *hard habitats* will provide a valuable framework for sustainability and potentially a robust catalyst for future innovation. We can already observe positive trends in patent submissions, built precedent projects, and experimental fabrication and scientific research projects that are pushing the boundaries of ecology and technology in global cities. And, the next generation promises to be even move exciting and foreword looking. For example, experiments in 3-D printing of artificial reefs conducted by Fabien Cousteau, grandson of Jacques Cousteau, suggest that synthetic reefs composed of calcium carbonate may be created to augment natural reefs using additive manufacturing. And, in the San Francisco Bay Area, students at UC Berkeley have developed fabrication processes for complex seawall modules using dissolving and biodegradable formwork which allows for the creation of heterogeneous habitat surfaces and voids not currently achievable using conventional construction practices. These projects are only the beginning of a concerted effort by researchers, designers, and city makers, to rebuild the ecology of a rapidly urbanizing planet and materialize the "hard habitats" of the future.

References

Bavestrello, G. et al. 2000. "Bio-mineralogy as a structuring factor for marine epibenthic communities." *Marine Ecology Progress Series* 193:241–249.

Bishop, M.J. et al. 2017. "Effects of ocean sprawl on ecological connectivity: Impacts and solutions." *Journal of Experimental Marine Biology and Ecology* 492:7–30. http://dx.doi.org/10.1016/j.jembe.2017.01.021 (accessed June 12, 2017).

Bozek, C.M., and D.M. Burdick. 2005. "Impacts of Seawalls on Saltmarsh Plant Communities in the Great Bay Estuary, New Hampshire USA." *Wetlands Ecology and Management* 13(5):553–568.

Chapman, M.G. 2003. "Paucity of mobile species on constructed seawalls: Effects of urbanization on biodiversity." *Marine Ecology Progress Series* 264:21–29.

Chapman, M.G., and A.J. Underwood. 2011. "Evaluation of ecological engineering of "armoured" shorelines to improve their value as habitat." *Journal of Experimental Marine Biology and Ecology* 400(1–2):302–313.

Chapman, M.G., and D.J. Blockley. 2009. "Engineering novel habitats on urban infrastructure to increase intertidal biodiversity." *Oecologia* 161(3):625–635.

Chapman, M.G., and F. Bulleri. 2003. "Intertidal seawalls—New features of landscape in intertidal environments." *Landscape and Urban Planning* 62(3):159–172.

Chapman, M.G. et al. 2009a. "Comparative effects of urbanisation in marine and terrestrial habitats." *Ecology of cities and towns: A comparative approach*, pp. 51–70.

Chapman, M.G. et al. 2009b. "Effect of urban structures on diversity of marine species." *Ecology of Cities and Towns: A Comparative Approach. Cambridge University Press, Cambridge*, pp. 156–176.

Collier, M.J. 2013. "Field boundary stone walls as exemplars of "novel" ecosystems." *Landscape Research* 38(1):141–150.

Coombes, M.A. et al. 2015. "Getting into the groove: Opportunities to enhance the ecological value of hard coastal infrastructure using fine-scale surface textures." *Ecological Engineering* 77:314–323.

Dong, Y.-W. et al. 2016. "The marine "great Wall" of China: Local- and broad-scale ecological impacts of coastal infrastructure on intertidal macrobenthic communities." *Diversity and Distributions* 22(7):731–744.

Dyson, K., and K. Yocom. 2015. "Ecological design for urban waterfronts." *Urban Ecosystems* 18(1):189–208.

Evans, A.J. et al. 2017. "Stakeholder priorities for multi-functional coastal defence developments and steps to effective implementation." *Marine Policy* 75:143–155. January 2017.

Firth, L.B. et al. 2013. "The importance of water-retaining features for biodiversity on artificial intertidal coastal defence structures." Ed. Omar Defeo. *Diversity and Distributions* 19(10):1275–1283. Web (accessed June 12, 2017)

Francis, R.A. 2011. "Wall ecology: A frontier for urban biodiversity and ecological engineering." *Progress in Physical Geography* 35(1):43–63.

Francis, R.A., and J. Lorimer. 2011. "Urban reconciliation ecology: The potential of living roofs and walls." *Journal of Environmental Management* 92(6):1429–1437.

Fredette, T.J. et al. 2014. *Epifaunal Community Development on Great Lakes Breakwaters: An Engineering with Nature Demonstration Project.* DTIC Document. http://www.usace.army.mil/Media/News-Archive/Story-Article-View/Article/583427/usace-buffalo-district-a-proving-ground-for-engineering-with-nature/

Glasby, T., and S. Connell. 1999. "Urban structures as marine habitats." *Ambio* 28(7):595–598.

Goff, M. 2010. Evaluating Habitat Enhancements of an Urban Intertidal Seawall: Ecological Responses and Management Implications. Diss. University of Washington. http://nsgl.gso.uri.edu/washu/washuy10003.pdf.

Griggs, G.B. 2009. The effects of armoring shorelines—The California experience. *Puget Sound Shorelines and the Impacts of Armoring—Proceedings of a State of the Science Workshop.* N.p. 77–84.

Hellyer, C.B., D. Harasti, and A.G.B. Poore. 2011. "Manipulating artificial habitats to benefit seahorses in Sydney Harbour, Australia." *Aquatic Conservation: Marine and Freshwater Ecosystems* 21(6):582–589.

Humphries, A.T. et al. 2011. "Testing the effect of habitat structure and complexity on Nekton assemblages using experimental oyster reefs." *Journal of Experimental Marine Biology and Ecology* 409(1):172–179.

Ido, S., and P.-F. Shimrit. 2015. "Blue is the new green—Ecological enhancement of concrete based coastal and marine infrastructure." *Ecological Engineering* 84:260–272.

Jim, C.Y. 1998. "Old stone walls as an ecological habitat for urban trees in Hong Kong." *Landscape and Urban Planning* 42(1):29–43. Print.

Jim, C.Y., and W.Y. Chen. 2010. "Habitat effect on vegetation ecology and occurrence on urban masonry walls." *Urban Forestry & Urban Greening* 9(3):169–178.

Karlson, R. 1978. "Predation and space utilization patterns in a marine epifaunal community." *Journal of Experimental Marine Biology and Ecology* 31(3):225–239.

Kozlovsky, R., and Y.J. Grobman. 2017. "The blue garden: Coastal infrastructure as ecologically enhanced wave-scapes." *Landscape Research* 42(5):439–454. Web (accessed June 12, 2017).

Kraus, N.C. 1988. "The effects of seawalls on the beach: An extended literature review (AUTUMN 1988)." *Journal of Coastal Research* (4):1–28.

Larkin, D., G. Vivian-Smith, and J.B. Zedler. 2006. "Topographic heterogeneity theory and ecological restoration." *Foundations of Restoration Ecology* 142–164.

Lavender, J.T. et al. 2017. "Small-scale habitat complexity of artificial turf influences the development of associated invertebrate assemblages." *Journal of Experimental Marine Biology and Ecology.* Web (accessed June 12, 2017).

Loke, L.H.L. et al. 2014. "Complexity for artificial substrates (CASU): Software for creating and visualising habitat complexity." *PloS One* 9(2):e87990.

Loke, L.H.L. et al. 2015. "Creating complex habitats for restoration and reconciliation." *Ecological Engineering* 77:307–313.

Lundholm, J. 2011. "Vegetation of urban hard surfaces." *Urban Ecology: Patterns, Processes, and Applications* 93–102.

Lundholm, J.T., and P.J. Richardson. 2010. "Mini-review: Habitat analogues for reconciliation ecology in urban and industrial environments." *Journal of Applied Ecology* 47(5):966–975.

Mayer-Pinto, M. et al. 2017. "Building 'blue': An eco-engineering framework for foreshore developments." *Journal of Environmental Management* 189:109–114.

McGuinness, K.A. 1989. "Effects of some natural and artificial substrata on sessile marine organisms at Galeta Reef, Panama. Marine ecology progress series." *Oldendorf* 52(2):201–208.

Nichols, C.R., and R.G. Williams. 2009. *Encyclopedia of Marine Science*. New York, United States: Infobase Publishing.

Petersen, J.K., and T. Malm. 2006. "Offshore windmill farms: Threats to or possibilities for the marine environment." *AMBIO: A Journal of the Human Environment* 35(2):75–80.

Pilkey, O.H., and J.A.G. Cooper. 2012. "'Alternative' shoreline erosion control devices: A review." In: Cooper J., Pilkey O. (Eds.), *Pitfalls of Shoreline Stabilization*. Coastal Research Library, vol 3. Dordrecht: Springer, pp. 187–214.

Pister, B. 2009. "Urban marine ecology in southern California: The ability of riprap structures to serve as rocky intertidal habitat." *Marine Biology* 156(5):861–873.

Pomerat, C.M., and C.M. Weiss. 1946. "The influence of texture and composition of surface on the attachment of sedentary marine organisms." *The Biological Bulletin* 91(1):57–65.

Rosenzweig, M.L. 2003. "Reconciliation ecology and the future of species diversity." *Oryx* 37(2):194–205.

Subcommittees, Artificial Reef, R.R. Lukens, and C. Selberg. 2004. Guidelines for Marine Artificial Reef Materials. http://www.okaloosafl.com/sites/default/files/doc/dept/public_works/reefs/reef_guidelines.pdf.

Tredici, P.D. 2010. "Spontaneous urban vegetation: Reflections of change in a globalized world." *Nature and Culture* 5(3):299–315.

Vaselli, S., F. Bulleri, and L. Benedetti-Cecchi. 2008. "Hard coastal-defence structures as habitats for native and exotic rocky-bottom species." *Marine Environmental Research* 66(4):395–403.

Wiecek, D. 2009. *Environmentally Friendly Seawalls A Guide to Improving the Environmental Value of Seawalls and Seawall-Lined Foreshores in Estuaries*. Sydney: Department of Environment and Climate Change NSW on Behalf of Sydney Metropolitan Catchment Management Authority. . http://www.environment.nsw.gov.au/resources/estuaries/pubs/090328-Seawall-Guide-2012-Reprint.pdf

24

Armatures for Coastal Resilience

Kristina Hill

CONTENTS

> The edge of the sea is a strange and beautiful place.
>
> **Rachel Carson**

> Geography is destiny.
>
> **Napoleon Bonaparte**

Introduction

Cities built on the edges of tidal estuaries are the most vulnerable to the processes of coastal change. Understanding these estuary cities involves a study of opposites—brittle structures made of concrete and steel trying to claim stable territory, while oceans increasingly claim the same locations with saltwater and dynamic wave energy. Cities that try to simply block out the ocean with concrete and steel will eventually lose; as Herman Melville wrote of the sea, no power but its own controls it. Resilience and adaptation will be a long game, requiring many strategies better than fragile structures.

In North America, estuary cities originated as outposts of trade and resource control during colonization by European nations. The first components of their ocean edges were often built more quickly than well. For example, San Francisco's waterfront was built using a fast-and-cheap berm of piled-up rocks as the basis for its seawall (Figure 24.1) (GHD-GTC Joint Venture, 2016). Near downtown, this rock berm is interspersed with the wooden hulls of ships abandoned by gold miners who were far more interested in claiming resources elsewhere. The ship hulls and rock berm were both covered with sandy fill scraped off the

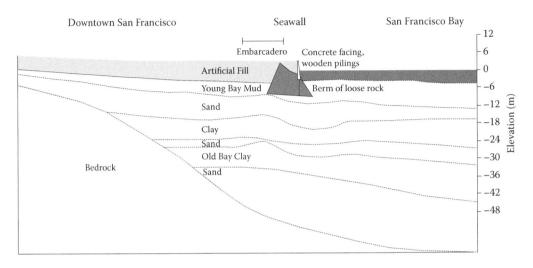

FIGURE 24.1
The existing San Francisco seawall was hastily constructed in the nineteenth century, during the gold rush era. It is literally a berm of loose rocks, with wooden and concrete piles driven into it. Concrete panels are used as a seaward facing material. The wall holds back the sandy fill on the landward side and provides a quay for boats to dock. It needs to be replaced, at an estimated cost of US$4–5 billion. The question is, how high should it be built? (Image courtesy of the author.)

city's pre-urban dune fields, and wooden wharves were constructed toward the ocean from the fill. It was not the kind of foundation designed to last in a seismic region, and its first full-scale replacement is underway—estimated at a cost of US$4–5 billion (Watts, 2016). In 150 years, geologic resilience has replaced geologic exploitation as the goal for infrastructure, and the "time problem" has been reframed as one of longevity rather than speed.

Reframing a coastal city's needs as longevity raises several new questions in a rapidly changing global environment. How many generations will a new seawall serve? How long do we want to be indebted to pay for it? What if our children have to pay for it longer than it is effective? What, if anything, do present generations owe the future? A century and a half after San Francisco was founded, the tangle of questions around resilience is genuinely cultural, not simply technical.

American coastal cities expanded from bare colonial outposts using housing and street design standards that have nothing to do with the dynamics of the sea. In the 1850s, roads and wharves were built on fill made of sand (Booker, 2013). For the next 100 years, the strategy was to fill the shallow waters of San Francisco Bay and other estuaries with municipal waste and building rubble, placing profitable industries, public works, and private housing on top. The typology of buildings and infrastructure was not specific to an estuary environment, except for the presence of working docks and wharves. American estuary cities don't reveal the dynamics of the landscape they are built in, except through occasional failures—such as the collapse of buildings constructed on filled land in an earthquake event or highway lanes flooded by high tides. Our streets and structures are mute, inexpressive of the real world around them except through failure. These structures blunt or even remove our awareness of the processes associated with our estuary environment, rather than heighten our sense of the strange beauty Rachel Carson wrote about at the edge of the sea (Carson, 1955).

Building cities that reduce our awareness of the environment was never a good idea and is now a practice that reduces our cultural capacity for resilience. In an era of rapid environmental change, humans will need to use all animal senses, all emotional, cultural, and intellectual resourcefulness, to focus and prepare. In 2017, we watched Houston, Miami, and San Juan, Puerto Rico, flooded by extreme storms and saw Mexico City heavily damaged by earthquakes, along with the rhythm of monthly high tides blocking roadways in Virginia and California. The aesthetic experience of this may best be captured by the original description of the sublime—an experience of magnitudes beyond comprehension and of processes beyond our control (Doran, 2015). The desire for greater resilience is a clear response to the failure of the modern project to control "nature." If floods can't be controlled by technology, cities need to emphasize a new set of strategies.

As a result of the changing global climate, all cities—not only estuary cities—can be said to be moving to a new planet. Storm forecasters struggle for words to describe weather events that are more intense than they have ever seen. Permanent changes in sea levels and precipitation patterns alone (and there are many more) will affect high tide levels, groundwater, food supply chains, transportation networks, water supply, and waste treatment systems. The breadth and complexity of these extreme events and permanent changes literally prevent us from knowing what specific changes will happen in a particular location. Without that ability to know how these dynamics will become location-specific, our urban design and planning epistemology shifts away from a deterministic frame and toward a heuristic frame.

My own writing and projects have tried to explore these five questions, as the beginning of a heuristic frame for reasoning about resilience: (1) What is stable, and what is likely to change quickly? (2) What kinds of spatial armatures can we use to establish relatively fixed territories? (3) How can we organize ourselves to implement adaptation strategies? (4) What kinds of physical designs and ways of occupying space can help us learn faster? (5) Do we have a broadly shared understanding of what resilience means?

In this chapter, I begin with the last question, then use the example of Christchurch, New Zealand, as a kind of rhyzomatic node linked to other places and ideas to help reason through some possibilities for design. This chapter closes with a discussion of the implications of sharing an understanding of resilience more broadly, with a wider public that focuses on justice arguments rather than design or technology.

What Is Resilience?

The concept of *resilience* as a property of a human bodily organ dates back to the seventeenth century, when it was used by Francis Bacon and others to describe the action of rebounding, recoiling, or returning to an original position (Oxford English Dictionary, 2014). At that time, it was used in physiology to refer to a quality of human lung tissue. By the nineteenth century, scientists used the term in a formalized definition of material elasticity. The term was later used in ecology, where *resilience* refers specifically to the capacity of an ecosystem to return to a previous state after a disturbance (Allaby, 2010). *Resilience* has also referred to the ability of people to recover from or resist a shock, since the nineteenth century, when it was synonymous with robustness and adaptability (Oxford English Dictionary, 2014).

Among North American planners and designers, "resilience" came into common usage after 2005, when major storms first caused extensive damage in New Orleans. Usage expanded dramatically after Hurricane Sandy hit the New York region in 2012. In Europe, scientific awareness of climate change and biodiversity losses created an impetus for the

Stockholm Resilience Centre, which was founded in 2007 with the mission to study the complex interdependencies of humans and their environment. The Stockholm Centre uses "resilience" in a broader sense that includes biodiversity, while most North American urban planners do not.

It is important to note that the concept of "strategic resilience" dates to the era immediately before Hurricanes Katrina and Sandy. According to the authors of an influential article in the *Harvard Business Review* in 2003, large industrial corporations need resilience. They claimed that the success of a private company, even a very large one, relies on its ability to dynamically reinvent business models and strategies as circumstances change (Hamel and Valikangas, 2003). They wrote, "strategic resilience is not about responding to a one-time crisis. It's not about rebounding from a setback. It's about continuously anticipating and adjusting to deep, secular trends that can permanently impair the earning power of a core business. It's about having the capacity to change before the case for change becomes desperately obvious." The authors proposed that a corporation could adopt the philosophy and tactics of strategic resilience to pursue the goal of "zero trauma" to the corporation.

What would it look like if cities pursued this same goal, of implementing resilience to achieve "zero trauma?" Cities would have to be actively managed, but perhaps take on some of the recommendations for big corporations—for example, allowing smaller, nimble units to self-organize and pursue shared goals in unique ways, experimenting to find success. Making room for lots of small pilot projects, some of which might fail—instead of trying to implement new standardized changes on the whole city, all at once. Clearly, cities would face different challenges in a climate of experimentation than corporations would, such as attending to equity in the use of public funds.

Maybe the right scale and aesthetic of experimentation for resilience is different for cities. In the next section, I'll argue that it requires an understanding of the landscape armatures of a coastal city, and that it should tinker with the barriers to shared prosperity—including the impacts of repeated flooding, or earthquake risks, on housing markets.

Physical Geography: The Ultimate Armature

Geography provides a context for the evolution of resilience strategies. Napoleon may have been the first to say that geography is destiny, precisely because he and many other territorial strategists have learned that technology and strategy operate within geologic conditions. For example, the contemporary Irish word for topography is "dinseanchas," but this is a word with an older set of more complex meanings that could be described as something more like "strategic landscape knowledge." It was the kind of knowledge prized by military chiefs and bards alike, because it allowed them to both plan actions and tell the story of those actions according to the terrain. Knowing where a river was narrow enough to cross, and in which months, or where wetlands could bog down troops, or where a tide might cut off or create access to a coastal headland was critical strategic knowledge. Contemporary people need to recover the capacity to synthesize and tell stories like these, but about the dynamics and structures in our own urban landscapes that will allow us to adapt successfully to a new climate. Estuary cities need to know their own stories and tell them.

Estuaries lie inside what are sometimes referred to as geologic basins. These basins are lower than the surrounding rock, and often are in the process of subsiding. Like shallow bowls that tilt toward the ocean, basins provide pathways both for the water draining out of the higher ground around them and for saltwater tides to enter. These basins may be shaped by fault zones, as in the San Francisco Bay, or by a long, shallow tilting of the earth's crust, as in the Boston area. Cities have been built in basins because they're relatively flat, with access to both freshwater and saltwater harbors.

To visualize the structures and dynamics that matter to coastal adaptation, it is helpful to think about geological patterns as armatures cutting through a three-dimensional volume of material. Geology produces a connected regional anatomy analogous to bones and muscles, circulatory systems, and skins. These can be deep or shallow and are often expressed at the surface if you know what to look for. A drastic contrast in materials, such as a contact between old volcanic rocks and recent sandy gravels deposited by rivers, can form an important structural boundary. Faults cut through all materials, redirecting the flow of surface water to create small lakes, and altering underground flows of groundwater and seismic energy. When seismic energy is generated, materials change their behavior—sand, clay, and fill materials can behave like a liquid as an energy wave passes through them.

Reading Landscape Armatures

In 2011, the coastal New Zealand city of Christchurch experienced a major earthquake that killed 185 people and left thousands of structures in ruins, destroying 80% of the city's underground infrastructure (New Zealand History, 2011). As the city plans for its future, its elected officials and designers also must consider sea level rise. Christchurch offers an example that cities of the American West Coast can and should learn from, because their major earthquakes are coming. By reviewing the conditions that shape future options for Christchurch, we can see how synthesizing the geology of an urban estuary's basin creates a context for adaptation to sea level rise.

Christchurch was built in a basin that drains to the Pacific Ocean on the east coast of New Zealand (Figure 24.2) (Brown et al., 1995). The basin is bounded by sandstone mountains to the west, accreted "terranes" that arrived when the Pacific plate collided with the Australian plate, and a peninsula formed from an old volcanic cone to the south. A braided river, the Waimakariri, forms the northern boundary of the urbanized plain, which is itself made of successive river deposits of outwash gravels and sands originating in mountain glaciers that have mostly melted away. Stable ancient dunes fan out at the point where the riverbed begins to slant more steeply toward the western mountains and run southward from the Waimakariri's mouth. Two smaller rivers, the Avon and the Heathcote, drain the city and empty into its estuary. A sand spit supplied by the sediment load carried by the Waimakariri closes off most of the estuary mouth, blocking out the ocean's waves (Figure 24.3).

While the western mountains are known to contain many faults, the earthquake of 2011 occurred on a "blind" fault, one that was previously unknown (Bradley and Cubrinovski, 2011). It stretched down toward the ocean from the sandstone mountains to the west, along the volcanic Port Hills on the southern boundary of the city's basin. When the fault slipped, it created a rippling energy wave with a magnitude of 6.7, which does not sound catastrophic—but it was. Geologists think that part of the energy wave reflected off the volcanic rocks to the south, cracking large boulders off the bluffs, and returned toward the city—leading to the fastest vertical acceleration ever recorded in an earthquake event and a

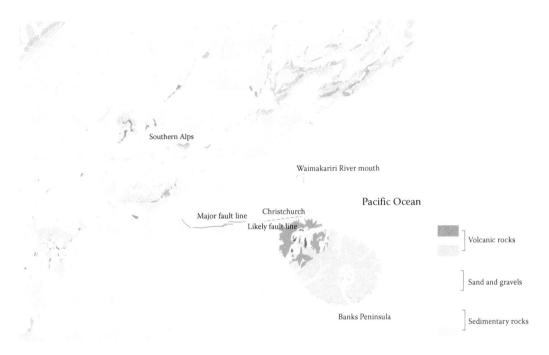

FIGURE 24.2
The Christchurch geologic region contains a set of armatures at a larger geographic scale that mark structures produced by a longer geologic timeframe of processes. The steep slopes of the sedimentary Southern Alps produced huge outwash plains that provide useful sands and gravel to Christchurch as rivers carry them to the coast. (Image courtesy of the author.)

very high lateral acceleration. And because the city was built on the sands and gravels of an old glacial outwash plain with underground water (groundwater) very close to the surface, the ground movement produced extensive liquefaction (Arnold, 2016). Cars were literally sucked down into the streets, which resolidified around them when the shaking was done. Most of the people who were killed were in buildings that collapsed, and a few died when boulders crashed through homes. More than 5,000 damaged homes were purchased and removed to prevent future loss of life, particularly in the so-called "Red Zone" along the Avon River where liquefaction was dramatic.

The question of coastal resilience is significant in relation to seismic events like the Christchurch earthquakes because many active tectonic margins occur along coastal areas with large cities. From Anchorage to Santiago and Seoul to Christchurch, the active tectonic margins of the Pacific plate create vulnerability to tsunamis and earthquakes that can not only cause solid earth to behave like a liquid but can permanently change the tilt direction of flat plains. In Christchurch, the land near to the south near the quake's epicenter tilted up by half a meter; in the area around the Avon River, closer to the city center, the land subsided by half a meter. What was flat became tilted, dropping the basin a little further, and creating a larger area that could be flooded by high tides as sea levels rise.

Christchurch has some geologic advantages as well. For one, it is well south of the main belt of cyclone activity in the South Pacific, which are more likely to hit northern Australia than New Zealand's South Island. Not having to design coastal adaptation to accommodate or block cyclone waves is a major advantage, since everything they build can be smaller to start and be raised incrementally. Second, the steep sandstone Southern Alps to the west are

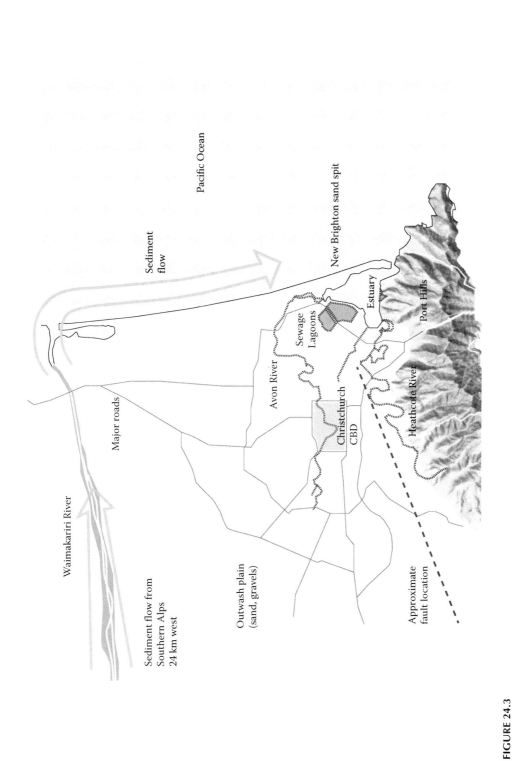

FIGURE 24.3

Mapping the armatures of the Christchurch landscape reveals a fishnet of roadways, one big fault line, the accreting sand spit, a radiating set of smaller rivers, and the basin boundaries created by the Waimakariri River to the north and the Port Hills to the south. (Image courtesy of the author.)

FIGURE 24.4
Sand is already accreting onto the tip of the New Brighton sand spit. This diagram illustrates the process by which sand could close the estuary mouth, as more sand is delivered by waves over time. Eventually, rising seas could overtake this process, however, and erode the sand gate. (Image courtesy of the author.)

soft and eroding. They provide a constant supply of sand and gravel to the coast—carried by the "conveyor belt" of the Waimakariri River. In its current position to the north, this big, braided river supplies sand to the spit that protects the Christchurch estuary from waves. The beach along this spit is growing wider naturally, as are its dunes (Figure 24.4). This sand and gravel supply could be used to build protective landforms along the inside edge of the estuary to prevent flooding from rising seas. The Japanese have used river gravels to build superdikes that are extra wide and seismically stable for safe building construction (Figure 24.5). The Dutch use sand to build protective dunes and widen their

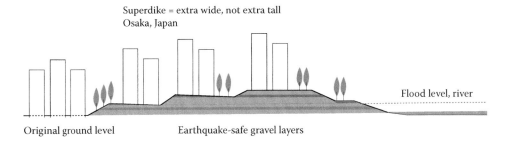

FIGURE 24.5
The Japanese superdike strategy uses layered gravels to achieve seismic resilience, using locally available gravels that are carried down from the mountains by steep rivers. (Image courtesy of the author.)

beaches artificially, as their main defense against sea level rise. Those are the kinds of advantages that make Christchurch a place where landforms can be used for long-term adaptation.

Superdikes and Sand Gates

Suppose, for instance, that the city's estuary was ringed with wetlands instead of the small walls typical of waterfront residential landscapes. Wetlands that are only 40 meters wide are likely to reduce incoming wave heights by as much as 70% during the year's highest tides (BCDC, 2013; Bay Institute, 2013). The shallow substrate "trips" the waves and the vegetation slows them by friction. Proposals have been made arguing that levees would only have to be half as high, if they were fronted by gently-sloping marshes (Bay Institute, 2013). If those "habitat wedges" of marsh can be designed to grow as sea levels increase, it is not hard to add to the levee on the landward side as well—adding up to a biologically active version of the Japanese superdike that does multiple kinds of work at once. These wedges of habitat could protect inland areas from storms, provide habitat, help to filter water quality, sequester carbon from the atmosphere, and provide recreational landscapes for people. Building a wider "wetland superdike" may cost about half of what a conventional levee would cost, per kilometer (Bay Institute, 2013).

On the ocean coast of cities like Christchurch, sand can be used to build a similar kind of habitat wedge. The Dutch refer to the width and height of sandy beaches and dunes as a "prism," representing and tracking changes in its sectional dimensions using the three sides of a long triangle. The coast is treated as a volume that changes over time through the addition of sand, not a two-dimensional line that must be preserved because it existed at a specific moment in time. Geomorphological theories of beach dynamics put this prism into the context of a dynamic equilibrium, in which accretion and erosion are like two directions the same system can travel in—depending on the direction and magnitude of waves, as well as the mean sea level. In places where sand is accreting, a rising sea level could push back against today's balance of forces and tip the dynamics toward erosion (Figure 24.6). But because those theories were developed in a sectional representation of the beach, using only two dimensions as a way of simplifying complex processes, we do not really know that accreting beaches will become erosional. Public agencies need to engage in reversible pilot projects on sandy coasts in order to learn and be ready for surprises as the global climate shifts.

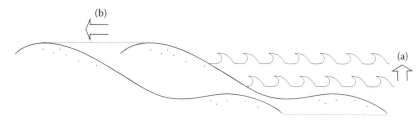

FIGURE 24.6
Bruun's Rule predicts that as sea level rises 1 unit of height (a meter, for example), the sandy profile of the beach will migrate inland 80–100 times that distance—unless it is blocked by a wall, in which case it would likely erode. (Image courtesy of the author.)

In some landscapes, like Christchurch, sand carried by rivers and waves is available as an asset that grows the beach and extends sand spits. But that is an unusual asset. On most sandy urban coasts, the beach is eroding because coastal structures like jetties at boat marinas redirect the flow of sand outward, away from the shore, or because dams higher up in river systems trap sand before it reaches the coast. Even without these structures, many sandy shores are eroding simply because there are no more Ice Age glaciers pushing sand into the system. To use sand on an eroding coast is to take on a role like that of Sisyphus in the Greek myth of a human being forced to roll a rock up a hill only to watch it roll back down, for all of eternity. This may sound like a bad thing at first. But one of the reasons for the longevity of this myth is that it represents something fundamental about being human—that mortality means all the projects we "roll uphill" will eventually roll down again, to be taken up (or not) by our children. Adaptation is a multi-generational process, a process of relearning the world in each successive generation.

Relearning a contemporary version of a landform-based approach to coastal resilience relies on our ability to recognize that landscapes have armatures—spatial elements that strengthen and support the processes we rely on. A ring of wetlands around an estuary can be seen as an armature. Expanding the width of these wetlands and buttressing them with earthen levees is a way of hybridizing a natural system with an urban system, particularly when buildings and roads are built on top. These hybrids make sense. People get to see the dynamics of their environment, instead of hiding behind walls that keep them unaware of the changes happening around them. We become something more unabashedly like gardeners of our larger environment, learning from the past forms of a landscape rather than trying to "freeze" the processes of change. When the world begins to change more rapidly, we can engage with it to a greater degree—adding to and incrementally reshaping our own habitat. Understanding the strength and value of landscape armatures, of landforms in a geological and ecological context, is the key to our success.

One of the most difficult coastal adaptation questions is what to do where a large body of water enters the open ocean (as at an estuary or river mouth), or where a tributary river or stream enters an estuary. Some major port cities have built iconic storm surge barriers that open and close mechanically at the mouths of estuaries and tributaries (Hill, 2015). These are big, expensive pieces of concrete and steel infrastructure with complex moving parts, like the Rotterdam Barrier. They were designed for a world that did not have rapid sea level rise. The footings of the structures that support the moving parts, like the footings of any wall, are sized to the height of the structure. The mechanical movement itself is effective within a range of sea levels but cannot be extended above that range. Unless these storm surge barriers are built for 100 years of sea level rise (which may be as much as 3 meters above today's storm surges), future generations will have to replace them. The Thames Barrier already has to close much more often than it once did and will eventually have to be closed all the time unless the shores of London's metropolitan areas are themselves raised (Reeder and Ranger, 2011). As the sea rises, mechanical barriers will have to remain closed because the land behind them is not able to handle the higher water levels when they open (Walsh and Miskewitz, 2013).

In addition, tide gates and surge barriers alter the ecology and geomorphology of the water bodies and ecosystems on the landward side (Giannico and Souder, 2005). Even if a tide gate is only closed at high tide, the closure starves inland marshes of the sediment deposits that would come primarily on the incoming tide, and which they need to both sustain their biological diversity and grow upward to keep pace with sea level rise. These inland waters also often decrease in water quality, because they are no longer "flushed" as effectively by high tides. Invasive species that can tolerate lower water quality often expand

FIGURE 24.7
The New Brighton sand spit in Christchurch, New Zealand. (Image courtesy of the author.)

in the area they occupy. In other words, a gate that opens and closes is not an effective long-term solution to adaptation at a tributary or estuary mouth. What alternatives exist?

The tip of the sand spit that protects Christchurch's estuary is also growing, like the width of the beach and the height of the dunes (Figures 24.7 and 24.8). It is a normal thing for many coastal lagoon openings to be closed by sand spits seasonally, during months when the volume of sediment transported along the shore increases. Some of those sand closures on lagoons open naturally when high river flows force the sand out of their way. Others are literally bulldozed open by people, when freshwater entering the lagoon from the landward side threatens to cause flooding. What if we thought of that as an alternative to today's tide gates, a feature we might call a "sand gate?"

Sand mounds could be placed artificially at the mouth of a small tributary, allowing small amounts of water to flow through the sand at a regular rate, while keeping out the erosional energy and height of waves. It is a common natural landform in dry climates, where tributaries may only flow seasonally and rarely have the force to push the sand out of their way. During low flow months, the freshwater disappears into the sandy delta at the mouth, slowly percolating through to the estuary or open ocean. In a rare heavy rain event, the water can push the sand open—or bulldozers can. A sand gate would mimic natural processes, and have multiple benefits in addition to blocking waves, such as providing some filtration of the freshwater as it passed through, providing habitat for sand-nesting birds, and giving both humans and animals a way to cross the tributary mouth without a bridge.

FIGURE 24.8
View of the New Brighton sand spit from the south, at Shag Rock. (Image courtesy of the author.)

The use of "sand gates" as flexible barriers at the tidal mouths of tributary rivers is a new proposal, a landform for which I have become the first tentative advocate. Like tide gates, they would change the flow of sediments and fish. They might not work and should be tested using reversible pilot projects. But sand gates are an example of a flexible landform approach that, like wetlands, dunes, and beaches, can be raised incrementally.

The work we do to build resilience by artificially enlarging coastal landforms would be a legacy for future generations, a foundation they can build on or expand—rather than a liability that they will still be paying for after it has ceased to function. In the worst maladaptive case, fixed structures that are built too small for future conditions might represent an additional liability because not only are they insufficient, but they also have to be removed and replaced with a newer and more expensive structure. The concrete-and-steel foundations of walls and tide gates could easily be undersized and limit the capacity to raise their height.

In contrast, experimental landforms are part of our human legacy; we have learned to make them from thousands of years of living on the coast (Hill, 2011). Dunes and wetlands can be created with many laborers and lots of shovels, or few laborers and large mechanical equipment. What we need to rediscover is that these landform experiments belong in urban districts adjacent to coastal structures as well. When we pair an urban district strategy with an expandable coastal edge strategy, we will form a hybrid that (like many hybrids in plant ecology) is more robust than either of the original types (Hill, 2015).

Ponds and Canals

There are two driving processes that require us to rethink the goal of achieving dry ground behind new coastal structures.

FIGURE 24.9
Flooding during an extra-high tide on Plover Street, Christchurch. Seawater from the estuary is flooding into the street and being conveyed from there into the neighborhood. (Image courtesy of the author.)

One is the storm surge (salty) or rain-driven (fresh) flooding that occurs at the mouths of rivers and small tributaries (Figures 24.9 and 24.10). Some of that flooding is permanent, as the mean sea level rises and low tides are higher. Some of it is temporary, as surge- or rain-driven floodwaters recede when the storm is over. If we cannot manage this type of coastal flooding completely with a sand gate or mechanical gate, we are left with two choices: build long dikes along the edges of the river mouth, which can fail catastrophically if an event exceeds the designed height and strength of those dikes, or redesign the adjacent urban districts to be resilient to temporary flooding.

Under normal circumstances (i.e., without sea level rise), designing for temporary flooding would be enough—surge- or rain-driven floods would indeed be temporary. But as sea levels rise, some saltwater flooding will become permanent. Temporary fresh- and saltwater flooding will occur farther inland. These kinds of flooding are familiar but will show up in surprising places. What is unfamiliar is that the freshwater "lens" we call the water table will rise on top of rising seawater. Seawater is denser than freshwater. The lighter-weight freshwater that is stored in the ground after many rainfalls will be forced upward. This rising water table will be the second major driver of flooding in coastal areas (Rotzoll and Fletcher, 2012).

A rising water table reveals the extent to which the design of conventional urban districts relies on having a predictable "dry zone" immediately below the ground surface. Pipes that convey sewage and rainwater away from buildings and roadways can be infiltrated by

FIGURE 24.10
A woman tries to clear a storm drain on Tern Street in Christchurch, New Zealand during the high tide event. As groundwater rises, drain inlets like this one will be filled and will not drain unless special flaps are installed on the ocean outlet of drainpipes, and unless pipes are replaced when they crack. (Image courtesy of the author.)

groundwater if it rises to surround them. If the pipes are not regularly replaced or repaired, they will have cracks that make infiltration by groundwater very common. Seismically active regions often have cracked pipes. Countries like the United States, where local communities do not regularly maintain or replace their underground infrastructure, have cracked pipes. Cities where major flooding occurs, like New Orleans, can also have cracked pipes from the weight of water sitting on top. In short, many and possibly most cities have cracked underground pipes that are meant to carry waste and excess water away. These pipes will not function for either purpose if they are filled with groundwater most of the time. Basements and toilets can back up with raw sewage; rainfall can cause more serious flooding because it has nowhere to go. If we lose the shallow dry zone where most cities place critical urban infrastructure, we can also lose the function of the infrastructure.

Urban areas where land has sunk or "subsided" offer us a preview of that future. Using pumps to turn wetlands dry in the Netherlands and the U.S. Gulf Coast led to very significant land subsidence, because the dry soils have less volume. Pumping up large amounts of groundwater for drinking water, irrigation, and industry—as well as drilling for oil and natural gas—have also caused major land subsidence in coastal areas. In the California Delta, areas of land called "islands" that once sat above the level of the river waters were diked and dried for agriculture; now they sit several meters below the water level of adjacent rivers. In all of these areas, gravity can no longer do the work of removing wastewater or floodwater. The Dutch maintain a sophisticated system of pumps that allows them to legislate the depth of the water table, using a lot of energy and relying on a culture of skilled operators and frequent maintenance. Most other countries lack this deep-rooted culture, which includes political support for the costs of expensive mechanical infrastructure.

Before the development of highly efficient mechanical groundwater pumps, low-lying cities were designed with ponds and canals to collect and convey both stormwater and shallow groundwater. In cities with high water tables, like Amsterdam and Venice, canals were used to drain groundwater away from the interior of a buildable urban block. The same canals served as transportation infrastructure, bird habitat, and as locations for low-cost floating homes. In Chinese cities like Suzhou, a complex system of ponds was integrated with urban blocks and used to manage stormwater.

Ponds and canals are spatial strategies for coping with a shallow water table. Ponds are also often used in sequence, to manage the quality of water coming off the land and into the nearshore environment by providing filtration and sequestration of nutrients like nitrogen and phosphorus. Together, ponds and canals can perform as transportation and recreational resources and create stable sites for buildings. They are interesting as strategies in and of themselves, but what is really useful about ponds and canals is evident from a simplistic characterization of the way they are constructed.

The oldest idea in landscape architecture, and perhaps in human manipulation of the land, is to dig a hole and use the material from the hole to build a mound. Medieval castles were built on mounds, which were themselves built with the material that came from a trench surrounding the site—the defensive moat. With rising sea levels and water tables accompanied by more frequent river flooding, coastal cities should embrace the beneficial reuse of sediment from dredging to build superdikes with wetland habitat wedges on the estuary side. But we can do more than that. We can avoid the creation of a complex system of pumps whose failure would be a disaster by digging ponds and canals behind the superdikes and using that sediment as well to build the coastal landforms higher over time. With all that we have learned in recent years about the use of pontoons in heavy construction of bridges and wind turbines, it has become clear that ponds can support entire urban districts—districts that are genuinely adapted to flooding, because they simply rise and fall on the smooth water surface of a pond that has no waves.

Ponds surrounded by superdikes on the estuary side can provide safe waters for floating pontoons that support whole urban blocks, with 3- to 5-story buildings (Figure 24.11). By placing the ponds behind the superdikes, a water-based urban district can accommodate both a high water table and flooding from rainwater. In fact, the ponds can be used to remove flooding problems in adjacent neighborhoods by receiving the stormwater from

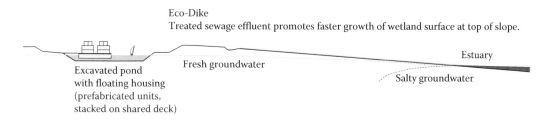

Eco-Dike
Treated sewage effluent promotes faster growth of wetland surface at top of slope.

Excavated pond
with floating housing
(prefabricated units,
stacked on shared deck)

Fresh groundwater

Estuary

Salty groundwater

FIGURE 24.11
An "eco-dike" is a superdike (an extra-wide dike) with a wetland on its seaward face. (Image courtesy of the author.)

existing streets. The ponds would serve as a district-scale infrastructure for managing stormwater, while expressing the level of the groundwater and providing a recreational amenity. Perhaps best of all, for places like Christchurch and San Francisco, floating urban blocks would provide protection from seismic energy waves using the "cushion" of the water. Instead of putting buildings on deep, expensive foundations in urban fill areas, those buildings can be floated on water displacement foundations—pontoons, with flexible infrastructure lines linking them to fixed infrastructure on land. This means that instead of engineering soils to prevent liquefaction, housing can be built that anticipates liquefaction—by floating on a liquid surface. When lateral spreading occurs, it would only be the edges of the pond and the land under the adjacent roadways that slump—not the ground under the buildings (Figure 24.12).

From an ecological perspective, the opportunity to collect stormwater from a larger area translates into an opportunity to filter it before it enters an estuary. The canal and pond systems can allow sediment that has pollution attached to it, like phosphorus or metals, to drop to the bottom in designated locations, and periodically remove that polluted sediment with a hydraulic dredge truck. Green infrastructure systems can be extended out from the ponds on surface streets and parking areas to increase the filtration performance of the whole landscape, letting the ponds collect only the overflow sediment from extreme rainfall events.

Now imagine that these ponds, which are providing safe space for new housing and recreation while helping to clean urban runoff water, are "built upon" in a modular way—with pontoons and prefabricated housing units that can be craned in and stacked, and then craned out if the location is overwhelmed by a faster sea level rise. The prefabricated housing units can be restacked elsewhere with no loss in value, no abandoned pile foundations, no underground pipes left behind. New sets of ponds can be dug farther up slope, where conventional housing cannot be sustained because of rising water tables and tributary flooding. And the cut material can be used for expanded superdikes, continuing to protect the built environment while building a higher foundation for wetland habitat. All this adaptation using landforms can be accomplished incrementally, so that each generation can add layers of material without having to pull out and replace the last generation's investments. We would genuinely be building a foundation for the future, something that is a legacy instead of a liability.

Extending the Strategies

Almost all coastal urban areas need to rethink both their shorelines and their adjacent urban districts. Taking a landform-based approach offers the advantages of being incremental, creating multi-benefit edges that support habitat and recreation, and being translatable to economies where extensive grading work is done with shovels instead of bulldozers.

The alternatives have serious drawbacks. Building walls in an environment with wave energy often leads to loss of habitat on both the ocean and land sides of the walls. In general, walls of all kinds are brittle—they can fail catastrophically and will require replacement instead of placing new layers of material to increase their height over time. Landforms can fail too, but the superdike model offers an example of how going "big" (really, going wide) can reduce or eliminate the risk of sudden failures. Landform materials offer multiple benefits, as in the way that sand allows slow percolation of water, with filtration benefits,

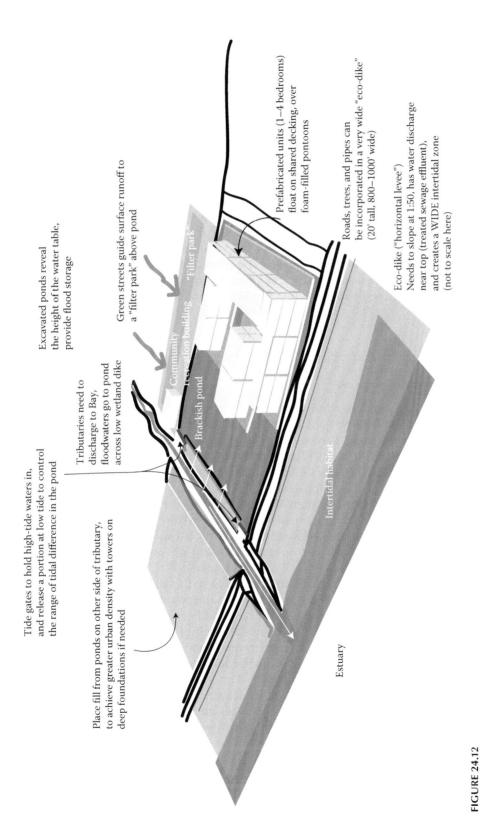

Tide gates to hold high-tide waters in, and release a portion at low tide to control the range of tidal difference in the pond

Excavated ponds reveal the height of the water table, provide flood storage

Green streets guide surface runoff to a "filter park" above pond

Prefabricated units (1–4 bedrooms) float on shared decking, over foam-filled pontoons

Place fill from ponds on other side of tributary, to achieve greater urban density with towers on deep foundations if needed

Tributaries need to discharge to Bay, floodwaters go to pond across low wetland dike

Roads, trees, and pipes can be incorporated in a very wide "eco-dike" (20' tall, 800–1000' wide)

Eco-dike ("horizontal levee") Needs to slope at 1:50, has water discharge near top (treated sewage effluent), and creates a WIDE intertidal zone (not to scale here)

Community recreation building

"Filter park"

Brackish pond

Intertidal habitat

Estuary

FIGURE 24.12

Artificial armatures of topography can be used to create protected ponds for floating housing. Creek water would be directed away from the pond, while stormwater would be directed into it for temporary storage and to flush the water of the ponds. Tide water could be captured and held to mute the vertical difference of the water level as tides rise and fall. Groundwater would be expressed by the excavation of the pond and become an asset rather than a flooding and liquefaction hazard. (Image courtesy of the author.)

while blocking out waves. Removal of existing urban districts may work in some areas and may be the best solution. But in many cases, it will be expensive to remove foundations, underground pipes, and contaminated soils, and relocation may be too difficult a process from a political standpoint to complete in a timely way. Allowing people the option of buying new houses in an infrastructure of ponds has proven popular in the Netherlands and can be in many other areas as well.

In constrained coastal urban areas where land is at a premium, such as Singapore, concrete caissons are being filled with sediment and used to build what is effectively a hybrid wall/landform combination. A caisson is essentially a concrete container, open at the top and bottom, that sits with its base buried in the sand and serves to hold fill material. Like a honeycomb made of concrete and fill, an interlocking series of these containers is used to create a stable base for buildings. The purpose of these structures is to act as a stable base for building and road construction. In deeper water, this may be the only real option. But a series of caissons can also serve as an outer ring of defense against waves and storm surge, with ponds on the inland side. Ponds can be constructed behind the outer ring of caissons that provide shallow water habitat and a wide range of recreational benefits, along with the potential for adding floating blocks of buildings and floating roadways.

Being Resilient and Adaptive

Our climate is changing, and as a result, everyone and everything exists in a changing environment. Strategic resilience is a concept that was proposed specifically to address this need to be nimble and adapt. Ecological resilience is a different concept, which refers to the ability of an ecosystem to recover from an event. But recovering from an event is no longer enough. It is necessary, but not sufficient.

Because climate change is driving global sea levels higher, coastal resilience is a condition that can no longer exist without adaptation. We must be adaptive to long term, permanent changes like higher seas as the foundation for our capacity to be resilient to events. I would argue that it is useful to distinguish between these terms, because the strategies we use to adapt may both be distinct from and influence the strategies we use to recover from temporary disaster events such as hurricanes and earthquakes.

One example of why it is important to distinguish between these two goals is the question of how high coastal cities should build their defenses, and whether those are landforms or concrete-and-steel walls. The answer depends on whether those cities' urban districts are vulnerable to major damage and loss of life when flooding occurs, or whether they are designed adaptively so that they can function even when flooding occurs. Without deciding on a strategy for urban districts inside the coastal barriers, there is no logical way to decide how high the coastal defenses should be.

We studied the cost of raising coastal barriers all around San Francisco Bay, so that everyone could see the enormous cost of using levees and walls versus other strategies (Hirschfeld and Hill, 2017). The numbers are big: approximately US$57 billion for adapting to 1 meter of sea level rise, along 1,300 km of shoreline where flooding would otherwise occur. We found that one of the largest single influences on cost for conventional flood control structures will be the requirement to build seawalls and levees high enough to prevent water from flowing over the top in extreme events, a phenomenon known as "overtopping." The cost of building walls and levees to prevent flooding in events with a

1% probability of occurrence would be almost twice the cost of building the same structures to protect against typical tide heights with a 1-m higher sea level. The strategic implication of these findings is that it makes sense to consider allowing temporary, extreme events to overtop coastal structures and cause some temporary flooding in adjacent urban areas. If those urban areas themselves are required to adapt to this condition, so that people will be safe and urban functions can continue during a flood event, coastal structures can be smaller and more multi-functional—providing habitat and recreational value as well as flood reduction.

Building "floodable" urban districts will allow people to maintain an awareness of the coastal dynamics where they live and may result in helping urban populations become more truly resilient over time. If typologies such as floating urban blocks and floating roadways or secondary mobility systems allow people and property to remain safe during temporary flood events, and also protect functional urban infrastructure during the event, our ability to live near the coast during this time of rising global sea levels and rising groundwater will be enhanced.

Floodable urban districts can use pond excavation to expose shallow groundwater, revealing its seasonal and year-to-year fluctuations, as well as to provide room to store temporary floodwaters due to overtopping from coastal surges. Floating urban blocks in these permanent ponds would provide safety from seismic events as well. Canals can extend from the ponds into existing dense urban areas, helping to drain away their high water table as well as excess rainfall. A filtration zone of wetland ponds can be used uphill of the "housing ponds" to capture contaminants in stormwater as it flows downhill, keeping the ponds relatively clean. Ponds at the shoreline can also store enough rainwater to remove upland neighborhoods from flood zones that are defined by rainfall accumulation, protecting people in those neighborhoods and allowing them to avoid expensive insurance costs.

The key is to achieve coastal resilience in a way that kick-starts a long-term process of adaptation. Any proposal for resilience that does not consider this long-term need for change over the next two centuries is at best a delaying tactic, and may be actually maladaptive— like the construction of brittle concrete-and-steel walls that can fail catastrophically or will need to be replaced by future generations at great cost. Otherwise, future generations could conceivably have to replace these massive structures while they are still paying off the bonds used to finance them, if sea level rises faster than the estimates being used to design those structures.

The Justice Rule for Resilience

In order to hold a broader public conversation about resilience it may be critical for designers and planners to consider justice arguments. These include arguments about who pays and who benefits, as well as arguments about whether change needs to happen now, affecting today's occupants of seashore homes, or whether it can happen later instead.

The decisions urban regions make about coastal resilience and adaptation in the next 10 or 20 years should be grounded not only in geography, but in concepts of inter-generational and international justice. From a practical perspective, decisions that do not consider a justice context may well be unstable, and result in a waste of resources when they are eventually abandoned. And from a justice perspective, the generations and nations that have used the most fossil fuel and contributed the most CO_2 to the atmosphere owe a debt to other nations and to future generations. Today's children in coastal areas all over the world will inherit radically more difficult environments because of a legacy of consumption patterns in the developed world, even if people in developed countries changed to entirely renewable energy sources immediately. Seen from that perspective, coastal resilience can only be achieved if there is a transfer of wealth from today's generations to tomorrow's—by building a new coastal edge that is genuinely a legacy, a foundation for their future efforts, not a liability. Similarly, developed nations like the United States that have generated by far the most greenhouse gases must adapt in ways that are inexpensive enough to allow them to continue to pay for adaptation in other countries. Without this transfer of wealth between nations, crises caused by waves of refugees, resource shortages, and militarized territorial disputes will make adaptation much harder, if not impossible.

As we consider what coastal resilience means, both the armatures contained in landscapes and the underlying need for justice should be central elements of that definition.

References

Allaby, M. 2010. *A Dictionary of Ecology*. 4th ed. Oxford, UK: Oxford University Press. http://www.oxfordreference.com/view/10.1093/acref/9780199567669.001.0001/acref-9780199567669-e-6261?rskey=WT8PJ9&result=5123

Arnold, M. 2016. *Liquefaction: A Wider-Spread Problem Than Might Be Appreciated*. RMS. http://www.rms.com/blog/tag/new-zealand-earthquake/

Bay Conservation and Development Commission (BCDC). 2013. *Innovative Wetland Adaptation Techniques in Lower Corte Madera Creek Watershed*. Bay Conservation and Development Commission, San Francisco. http://www.bcdc.ca.gov/climate_change/WetlandAdapt.html

Bay Institute. 2013. "Analysis of the costs and benefits of using tidal marsh restoration as a sea level rise adaptation strategy in San Francisco Bay," Report issued February 2, 2013. http://bayecotarium.org/wp-content/uploads/cost-and-benefits-of-marshes-.pdf

Booker, M. 2013. *Down by the Bay: San Francisco's History between the Tides*. Berkeley, CA: University of California Press.

Bradley, B.A., and M. Cubrinovski. 2011. "Near-source strong ground motions observed in the 22 February 2011 Christchurch earthquake." *Seismological Research Letters* 82 (6):853–65.

Brown, L.J., R.D. Beetham, B.R. Paterson, and J.H. Weeber. 1995. "Geology of Christchurch, New Zealand." *Environmental & Engineering Geoscience* 23 (3). http://eeg.geoscienceworld.org/content/I/4/427

Carson, R. 1955. *The Edge of the Sea*. New York, N.Y: Houghton-Mifflin.

Doran, R. 2015. *The Theory of the Sublime from Longinus to Kant*. Cambridge, UK: Cambridge University Press.

GHD-GTC Joint Venture. 2016. *Earthquake Vulnerability Study for the Seawall Vulnerability Study of the Northern Seawall San Francisco, California*. Port of San Francisco.

Giannico, G., and J. Souder. 2005. Tide Gates in the Pacific Northwest: Operation, Types and Environmental Effects. Oregon Sea Grant Research Paper ORESU-T-05-001.

Hamel, G., and L. Valikangas. 2003. "The quest for resilience." *Harvard Business Review*, 9 (September):22.

Hill, K. 2011. "Climate-resilient urban waterfronts." In: Aerts, J., Botzen, W., Bowman, M., Ward, P. & Dircke, P. (Eds) *Climate Adaptation and Flood Risk in Coastal Cities*. London, New York: Earthscan Climate.

Hill, K. 2015. "Coastal infrastructure: A typology for the next century of adaptation to sea-level rise." *Frontiers in Ecology and the Environment* 13 (9):468–76. doi: 10.1890/150088.

Hirschfeld, D., and Hill, K.E. 2017. "Choosing a future shoreline for the San Francisco Bay: Strategic Coastal adaptation insights from cost estimation." *J. Mar. Sci. Eng.* 42 (5).

New Zealand History. 2011. Christchurch Earthquake Kills 185. https://nzhistory.govt.nz/page/christchurch-earthquake-kills-185

Oxford English Dictionary, 2014. "Resilience." *Oxford English Dictionary.* Oxford, UK: Oxford University Press.

Reeder, T., and N. Ranger. 2011. *How Do You Adapt in an Uncertain World? Lessons from the Thames Estuary 2100 Project.* World Resources Report Uncertainty Series. Washington, DC: World Resources Institute. www.worldresourcesreport.org

Rotzoll, K., and C.H. Fletcher. 2012. "Assessment of groundwater inundation as a consequence of sea-level rise." *Nature Climate Change* 3 (5):477–81. doi: 10.1038/nclimate1725

Walsh, S., and R. Miskewitz. 2013. "Impact of sea level rise on tide gate function." *Journal of Environmental Science and Health, Part A* 48 (4):453–63. doi: 10.1080/10934529.2013.729924

Watts, J. 2016. "San Francisco seawall upgrades could cost $5 billion." *The Bond Buyer*, April 13. https://www.bondbuyer.com/news/san-francisco-seawall-upgrades-could-cost-5-billion

Index

Note: Page numbers followed by *"fn"* indicate footnotes.

Printed and bound by CPI Group (UK) Ltd, Croydon, CR0 4YY

24/10/2024

01778288-0012